New Processes: Working towards a Sustainable Society

New Processes: Working towards a Sustainable Society

Editors

Pau Loke Show
Kit Wayne Chew
Aydin Berenjian

MDPI • Basel • Beijing • Wuhan • Barcelona • Belgrade • Manchester • Tokyo • Cluj • Tianjin

Editors

Pau Loke Show
Chemical Engineering
Khalifa University
Abu Dhabi
United Arab Emirates

Kit Wayne Chew
Chemistry, Chemical
Engineering and
Biotechnology
Nanyang Technological
University
Singapore
Singapore

Aydin Berenjian
School of Engineering
The University of Waikato
Hamilton
New Zealand

Editorial Office
MDPI
St. Alban-Anlage 66
4052 Basel, Switzerland

This is a reprint of articles from the Special Issue published online in the open access journal *Processes* (ISSN 2227-9717) (available at: www.mdpi.com/journal/processes/special_issues/sustainable_society).

For citation purposes, cite each article independently as indicated on the article page online and as indicated below:

LastName, A.A.; LastName, B.B.; LastName, C.C. Article Title. *Journal Name* **Year**, *Volume Number*, Page Range.

ISBN 978-3-0365-7681-7 (Hbk)
ISBN 978-3-0365-7680-0 (PDF)

Contents

About the Editors

Pau Loke Show

Professor Ir. Ts. Dr. Pau-Loke SHOW is the Full Professor of Biochemical Engineering at Khalifa University, Abu Dhabi, United Arab Emirates. He is also the former President and Founder of the International Bioprocessing Society based in Malaysia. He was the Director of the Sustainable Food Processing Research Centre and Co-Director of Future Food Malaysia, Beacon of Excellence, at the University of Nottingham Malaysia. Prof Ir. Ts. Dr. Show successfully obtained his PhD in two years after obtaining his bachelor's degree from Universiti Putra Malaysia. In the year of 2022, he was elected as a Fellow of the Institution of Chemical Engineers IChemE (FIChemE). He is currently a Professional Engineer (PEng) registered with the Board of Engineers Malaysia (BEM); a Chartered Engineer of the Engineering Council UK (CEng); a Corporate Member of The Institution of Engineers, Malaysia (MIEM); and a Professional Technologist (PTech) registered with the Malaysia Board of Technologists (MBOT). Prof Ir. Ts. Dr. Show obtained a Post-Graduate Certificate of Higher Education (PGCHE) in 2014 and is now a Fellow of the Higher Education Academy (FHEA) UK.

He has received numerous prestigious academic awards, including the recent Highly Cited Researcher Award; World's Rising Stars of Science; Ten Outstanding Young Malaysian Award; the Tan Sri Emeritus Professor Augustine S H Ong International Special Award on Innovations and Inventions in Palm Oil; the APEC Science Prize for Innovation, Research, and Education ("ASPIRE") Malaysia Award; the Malaysia Young Scientist Award; the DaSilva Award; JSPS Fellowship; Top 100 Asian Scientists; Asia's Rising Scientists Award; and the Young Researcher Award at the IChemE Awards. At present, he has published more than 850 journal papers in less than 10 years of his career. His current h-index is 75, placing him among the top leaders of his chosen field (Microalgae Technology).

Kit Wayne Chew

Dr Kit Wayne Chew is an Assistant Professor at the School of Chemistry, Chemical Engineering and Biotechnology, Nanyang Technological University, Singapore. He obtained his PhD from the University of Nottingham. His research and technological development interests include sustainable bioprocess engineering for food and pharmaceutical products, focusing on the utilization of algae biotechnology to develop cost-effective and environmentally friendly methods for the synthesis of functional ingredients. He also works on renewable energy, sustainability, waste management, and automation integrated bioprocesses. He has received several prestigious awards, including the Green Talents Award 2022, the Global IChemE Award for Young Researcher 2020, the Young Research Award at the IChemE Malaysia Awards 2020, the Postgraduate Prize Award 2019, and the Vice Chancellor's Medal 2019; was a Finalist for YSN-ASM Rising Star Award 2019; and ranked among the top 2% world top scientists in The Stanford List 2022/2021. At presnt, he has published more than 180 kournal articles and 15 book chapters.

Aydin Berenjian

Aydin Berenjian is a researcher in the field of biochemical engineering. Dr. Berenjian's research interests include bioprocess engineering, fermentation technology, and biosensors. He has published numerous papers in peer-reviewed journals and has also written book chapters on topics related to his field. In addition to his academic work, Dr. Berenjian has also worked on several industrial projects, including the development of novel fermentation processes for the production of vitamin MK7.

Editorial

Special Issue on "New Processes: Working towards a Sustainable Society"

Kit Wayne Chew [1,2], Aydin Berenjian [3,4] and Pau Loke Show [5,*]

1   School of Energy and Chemical Engineering, Xiamen University Malaysia, Jalan Sunsuria, Bandar Sunsuria, Sepang 43900, Selangor Darul Ehsan, Malaysia; kitwayne.chew@gmail.com
2   College of Chemistry and Chemical Engineering, Xiamen University, Xiamen 361005, China
3   School of Engineering, Faculty of Science and Engineering, The University of Waikato, Hamilton 3240, New Zealand; aydin.berenjian@waikato.ac.nz
4   Department of Agricultural and Biological Engineering, Pennsylvania State University, 221 Agricultural Engineering Building, University Park, PA 16802, USA
5   Department of Chemical and Environmental Engineering, Faculty of Science and Engineering, University of Nottingham Malaysia, Jalan Broga, Semenyih 43500, Selangor, Malaysia
*   Correspondence: pauloke.show@nottingham.edu.my

Citation: Chew, K.W.; Berenjian, A.; Show, P.L. Special Issue on "New Processes: Working towards a Sustainable Society". *Processes* 2022, *10*, 869. https://doi.org/10.3390/pr10050869

Received: 7 April 2022
Accepted: 26 April 2022
Published: 28 April 2022

Publisher's Note: MDPI stays neutral with regard to jurisdictional claims in published maps and institutional affiliations.

The idea of a sustainable society comprises a consumer society that considers the welfare of the planet for future generations. A sustainable society would include a clean environment with clean air and water; human appreciation of the natural environment; no dependence on fossil fuels; materials being fully recycled for the creation of products; and an interconnection between the ecosystem and the digital network. A society that can maximize the utilization of resources and sustain a zero-waste bioeconomy will enable growth in both per capita consumption and the population. Many societies have yet to develop a vision of what a truly sustainable society could look like. Important aspects such as basic material usage, new energy sources, production factory organization, developments in communication, transportation, and machinery, and the transition from science to industry could majorly contribute towards improving the technological advancements within a society. An Industrial Revolution in these areas would lead to changes to the domestic system of industrial production and increase overall wealth and production capacity. The creation of a sustainable society is essential to create a future for future generations in which the sustainable development goals (SDGs) will be achieved and the natural ecosystem of the Earth will be preserved. This Special Issue, entitled "New Processes: Working towards a Sustainable Society", showcases the advancements in sustainable processes and technologies that can provide an environmentally friendly management system. The Special Issue is available online at: https://www.mdpi.com/si/processes/sustainable_society.

1. Nanotechnology in Materials and Bioprocessing

Nanomaterials have recently emerged as a class of modern materials that have unique characteristics such as thermal, electrical, optical, and magnetic properties, which are highly desirable in applications to advanced materials and bioprocessing. The integration of nanomaterials into existing structures can significantly boost the capability of these structures, which is a big advancement in science and technology. Rezaei et al. [1] employed the incorporation of nano iron oxide into PCL nanofibers to create a composite scaffold with enhanced performance. This study showed that incorporating magnetite nanoparticles into PCL fibers greatly improves the cell attachment and promotes the cell growth rate, which will provide better biological performance, especially in liver tissue engineering. In the review by Taghizadeh et al. [2], the potential of iron-based nanostructures was analyzed and reviewed comprehensively. The structural properties, methods of synthesis, and various applications of iron-based nanostructures were discussed with regard to their utilization in medicinal and technological development. Transforming iron compounds so that they have nano iron structures yields better and useful properties that mean they can be employed in biomedical, industrial, environmental, agriculture, and engineering fields.

2. Sustainable Adsorption and Wastewater Treatment

Water resources are severely impacted by eutrophication and pollution caused by human activities and industrial processing. These harmful compounds can be discharged improperly into lakes, rivers, and oceans, where they have devastating effects on both human and aquatic health. Three papers in this Special Issue present adsorption techniques for use in wastewater treatment and in the slow release of nutrients. Alhogbi et al. [3] presented a green removal process for dye molecules from wastewater using activated carbon from waste. Palm tree fiber waste was used to synthesize activated carbon, which was found to be highly efficient in the removal of Congo red anionic dye and Rhodamine B cationic dye. The synthesized activated carbon also showed good reusability for up to five cycles and could be easily removed from the wastewater. Gheju and Balcu [4] also explored the use of an inexpensive walnut shell waste material as an adsorbent for the removal of iron and chromium heavy metals in water treatment. Fresh walnut shell powder was shown to be very promising in the removal of the heavy metals. The results indicated that it was suitable for application in water treatment processes as a sustainable solution can that reduce the amount of waste generated and assist as an efficient disposal method. Zhang et al. [5] focused on the use of ion exchange resin to control the adsorption and release of ammonia nitrogen in an attempt to produce a slow-release insecticide. Various parameters were altered so the response of the ion exchange resin could be observed, and it was found that the resin had a good adsorption effect on the ammonia nitrogen. This study provided insights into alternative raw materials for use in slow-release insecticides that can improve the utilization rate of biogas slurry and encourage the cleaner processing of insecticides.

3. Process Management and Analysis for Sustainable Biomass Utilization

Many technologies have been used to determine the sustainability of biomass utilization, either through economic, environmental, or social sustainability means. In different technologies, assessment methods and multi-criteria decision-making methods have been utilized to help researchers prioritize sustainable biomass utilization. Moy et al. [6] studied the life cycle assessment of bioplastic and paper straws in Malaysia. This study was carried out with the aim to environmentally assess the usage of plastic straws within the region, with emphasis on the global warming potential, acidification potential, and eutrophication potential. The outcome of the study indicated that bioplastic straws were favored as a material with less environmental impact than paper straws, which indicates that the materials used to make straws should be changed in Malaysia. Abad-Segura et al. [7] also introduced industrial process management for a sustainable society. This study analyzed many bibliographies regarding industrial process management and its sustainable effects on society; the work can serve as a platform for researchers to understand the growing trends in achieving sustainable societies and goals. Huy and Phuc [8] also discussed a structural model regarding the impact of corporate social responsibility (CSR) activities and how they encourage better organizational performance. Based on the data gathered from respondents in a public sector organization in South Vietnam, it was found that the integration of CSR activities into public sector management does impact the CSR disclosure and organizational performance, and implications for management were also outlined.

4. Optimization and Modeling for Enhanced Performance

Industrial processes in various fields and applications are complex, and to enhance them to enable better outputs, large amounts of resources and time are needed. Modeling and optimization tools have been used to evaluate the complexity of systems, develop strategies to optimize systems, select appropriate mathematical models that can represent systems, and to progress towards the commercialization of the process based on inputs from modeling. Usman et al. [9] demonstrated the use of a response surface methodology (RSM) optimization technique to evaluate the performance, exhaust emissions, and acoustics of a compression ignition engine when applied with hydroxyl-gas-enriched diesel. Several parameters were altered to determine the level of their influence on the engine. The RSM model was able to predict the suitable range of conditions based on the studied factors (brake thermal efficiency, brake specific fuel consumption, hydrocarbon content, carbon monoxide content, noise, and smoke). Hamzah et al. [10] also highlighted the use of varying alcohol concentrations on the precipitation of lignin from a bamboo-like stem crop. Different concentrations of ethanol was used to observe the purity of precipitated lignin from *Miscanthus x giganteus*. The study showed that the overall size of lignin aggregates decreased with lower ethanol concentrations. The study provided findings regarding the ethanol concentration range suitable for obtaining better-quality lignin aggregates.

This Special Issue covers a broad range of topics that all aim to contribute to the movement towards a sustainable society. For this to be achieved, the highest standard of wellbeing must be provided for the environment, humans, and the economy, through actions that are energetically, resource, environmentally, and socially sustainable.

We thank the Editor-in-Chief, Giancarlo Cravotto, and all of the contributors for their enthusiastic support of the Special Issue, as well as the editorial staff of *Processes* for their effort and assistance.

Funding: This work was supported by the Fundamental Research Grant Scheme, Malaysia [FRGS/1/2019/STG05/UNIM/02/2], MyPAIR-PHC Hibiscus Grant [MyPAIR/1/2020/STG05/UNIM/1], and Xiamen University Malaysia Research Fund [Grant number: XMUMRF/2021-C7/IENG/0033].

Conflicts of Interest: The authors declare no conflict of interest.

References

1. Rezaei, V.; Mirzaei, E.; Taghizadeh, S.-M.; Berenjian, A.; Ebrahiminezhad, A. Nano Iron Oxide-PCL Composite as an Improved Soft Tissue Scaffold. *Processes* **2021**, *9*, 1559. [CrossRef]
2. Taghizadeh, S.-M.; Berenjian, A.; Zare, M.; Ebrahiminezhad, A. New Perspectives on Iron-Based Nanostructures. *Processes* **2020**, *8*, 1128. [CrossRef]
3. Alhogbi, B.G.; Altayeb, S.; Bahaidarah, E.A.; Zawrah, M.F. Removal of Anionic and Cationic Dyes from Wastewater Using Activated Carbon from Palm Tree Fiber Waste. *Processes* **2021**, *9*, 416. [CrossRef]
4. Gheju, M.; Balcu, I. Sequential Abatement of FeII and CrVI Water Pollution by Use of Walnut Shell-Based Adsorbents. *Processes* **2021**, *9*, 218. [CrossRef]
5. Zhang, Q.; Liu, Z.; Petracchini, F.; Lu, C.; Li, Y.; Zhang, Z.; Paolini, V.; Zhang, H. Preparation of Slow-Release Insecticides from Biogas Slurry: Effectiveness of Ion Exchange Resin in the Adsorption and Release of Ammonia Nitrogen. *Processes* **2021**, *9*, 1461. [CrossRef]
6. Moy, C.-H.; Tan, L.-S.; Shoparwe, N.F.; Shariff, A.M.; Tan, J. Comparative Study of a Life Cycle Assessment for Bio-Plastic Straws and Paper Straws: Malaysia's Perspective. *Processes* **2021**, *9*, 1007. [CrossRef]
7. Abad-Segura, E.; Morales, M.E.; Cortés-García, F.J.; Belmonte-Ureña, L.J. Industrial Processes Management for a Sustainable Society: Global Research Analysis. *Processes* **2020**, *8*, 631. [CrossRef]
8. Huy, P.Q.; Phuc, V.K. Does Strategic Corporate Social Responsibility Drive Better Organizational Performance through Integration with a Public Sector Scorecard? Empirical Evidence in a Developing Country. *Processes* **2020**, *8*, 596. [CrossRef]
9. Usman, M.; Nomanbhay, S.; Ong, M.Y.; Saleem, M.W.; Irshad, M.; Hassan, Z.U.; Riaz, F.; Shah, M.H.; Qyyum, M.A.; Lee, M.; et al. Response Surface Methodology Routed Optimization of Performance of Hydroxy Gas Enriched Diesel Fuel in Compression Ignition Engines. *Processes* **2021**, *9*, 1355. [CrossRef]
10. Hamzah, M.H.; Bowra, S.; Cox, P. Effects of Ethanol Concentration on Organosolv Lignin Precipitation and Aggregation from *Miscanthus × giganteus*. *Processes* **2020**, *8*, 845. [CrossRef]

processes

MDPI

Article

Nano Iron Oxide-PCL Composite as an Improved Soft Tissue Scaffold

Vahid Rezaei [1], Esmaeil Mirzaei [1,*], Seyedeh-Masoumeh Taghizadeh [2], Aydin Berenjian [3,4] and Alireza Ebrahiminezhad [2,*]

1 Department of Medical Nanotechnology, School of Advanced Medical Sciences and Technologies, Shiraz University of Medical Sciences, Shiraz P.O. Box 71348-14336, Iran; v_rezaei@sums.ac.ir
2 Biotechnology Research Center, Shiraz University of Medical Sciences, Shiraz P.O. Box 71348-14336, Iran; taghizm@sums.ac.ir
3 School of Engineering, Faculty of Science and Engineering, University of Waikato, Hamilton 3216, New Zealand; aydin.berenjian@waikato.ac.nz
4 Department of Agricultural and Biological Engineering, 221 Agricultural Engineering Building, Pennsylvania State University, State College, PA 16802, USA
* Correspondence: e_mirzaei@sums.ac.ir (E.M.); a_ebrahimi@sums.ac.ir (A.E.)

Abstract: Iron oxide nanoparticles were employed to fabricate a soft tissue scaffold with enhanced physicochemical and biological characteristics. Growth promotion effect of L-lysine coated magnetite (Lys@Fe$_3$O$_4$) nanoparticles on the liver cell lines was proved previously. So, in the current experiment these nanoparticles were employed to fabricate a soft tissue scaffold with growth promoting effect on the liver cells. Lys@Fe$_3$O$_4$ nanoparticles were synthesized via co-precipitation reaction. Resulted particles were ~7 nm in diameter and various concentrations (3, 5, and 10 wt%) of these nanoparticles were used to fabricate nanocomposite PCL fibers. Electrospinning technique was employed and physicochemical characteristics of the resulted nanofibers were evaluated. Electron micrographs and EDX-mapping analysis showed that nanoparticles were well dispersed in the PCL fibers and no bead structure were formed. As expected, incorporation of Lys@Fe$_3$O$_4$ to the PCL nanofibers resulted in a reduction in hydrophobicity of the scaffold. Nanocomposite scaffolds were shown increased tensile strength with increasing concentration of employed nanoparticles. In contrast to PCL scaffold, nearly 150% increase in the cell viability was observed after 3-days exposure to the nanocomposite scaffolds. This study indicates that incorporation of magnetite nanoparticles in the PCL fibers make them more prone to cell attachment. However, incorporated nanoparticles can provide the attached cells with valuable iron element and consequently promote the cells growth rate. Based on the results, magnetite enriched PCL nanofibers could be introduced as a scaffold to enhance the biological performance for liver tissue engineering purposes.

Keywords: electrospinning; iron-enriched scaffold; magnetite nanoparticles; nanofiber; PCL scaffold; liver tissue engineering

Citation: Rezaei, V.; Mirzaei, E.; Taghizadeh, S.-M.; Berenjian, A.; Ebrahiminezhad, A. Nano Iron Oxide-PCL Composite as an Improved Soft Tissue Scaffold. *Processes* **2021**, *9*, 1559. https://doi.org/10.3390/pr9091559

Academic Editor: Maurizio Ventre

Received: 19 April 2021
Accepted: 5 August 2021
Published: 1 September 2021

1. Introduction

In recent years, tissue engineering (TE) has been recognized as a new approach in the biomedical sciences that focuses on repair, regeneration, and replacement of tissues [1]. TE consists of three fundamental portions including scaffolds, growth factors, and cells. By mimicking the extracellular matrix (ECM), scaffolds play the most important role in the TE techniques. Scaffolds with the help of growth factors provide a suitable site for cell attachment, growth, proliferation, and differentiation [2–7]. TE technologies have to overcome many limitations around the used scaffolds such as ineffective cell growth, inadequate production of effective growth factors, and inability to control cellular functions. Some scaffolds also suffer from poor biological, mechanical, and electrochemical properties. In addition, further studies and developments are required to achieve scaffolds with appropriate physiological structure [5].

Electrospinning is a well-known and simple technique for the construction of three-dimensional biomimetic nano-scaffolds with hierarchical fibrous architecture [3,8–10]. Due to the structural similarity of electrospun fibrous to the ECM, electrospinning has become one of the most common techniques for the fabrication of TE scaffolds [11]. This technique provides scaffolds with a large surface area, good stability, high porosity, and pore interconnections [8–10,12].

Up to now, many natural and synthetic polymers have been used in electrospinning [7,13–15]. Natural polymers (e.g., alginate, gelatin, and collagen) have not been widely used alone due to weak mechanical strength and inappropriate degradation rate. Nonetheless, these compounds have been frequently used in composite scaffolds. Synthetic polymers with better mechanical properties than the natural ones are a good substitute to increase the biodegradability and biomechanical properties of scaffolds. Poly-lactic acid (PLA), poly (glycolic acid) (PGA), poly (lactic-co-glycolic acid) (PLGA), polyurethane (PU), and poly(ε-caprolactone) (PCL) are the most prevalent synthetic polymers used in electrospinning techniques [7]. PCL, as FDA approved polymer, has received much attention in this regard due to its tunable biodegradability and biocompatibility [2,7,8,12,16,17]. As with most other synthetic polymers, low hydrophilicity is one of the major disadvantages of PCL polymer and reduces the cell affinity to the resulting scaffolds [7,11]. The addition of hydrophilic compounds such as gelatin, collagen, and polyethylene glycol (PEG) is a common approach to increase hydrophilicity of PCL scaffolds [18–21]. In addition, metallic and non-metallic nanoparticles such as silver, chitosan, titanium oxide, silica, and magnetite nanoparticles were also incorporated in PCL scaffolds to improve physicochemical properties [6,22–26]. Magnetite nanoparticles are one of the well-studied nanoparticles to improve physicochemical and biological properties of PCL scaffolds. In all previous investigations however, magnetite nanoparticles loaded PCL scaffolds were employed just for efficient mesenchymal stem cell proliferation, bone regeneration, and drug delivery [6,8,12,22,27]. Meanwhile, in our previous investigation, we found that magnetite nanoparticles promote the growth of Hep-G2 cell line as well as human hepatic cell line. These nanoparticles can act as nano iron sources to increase hepatic cell growth and proliferation [28]. So, particularly in the case of liver scaffolds, in addition to beneficial effects on the physicochemical properties, it seems fortification with magnetite nanoparticles can provide a scaffold with growth-promoting effects. This hypothesis was investigated in the current study by developing composite nanofibers that are incorporated with magnetite nanoparticles (Scheme 1). Besides physicochemical evaluations, in vitro investigations were also performed on the Hep-G2 cells as a liver cell line.

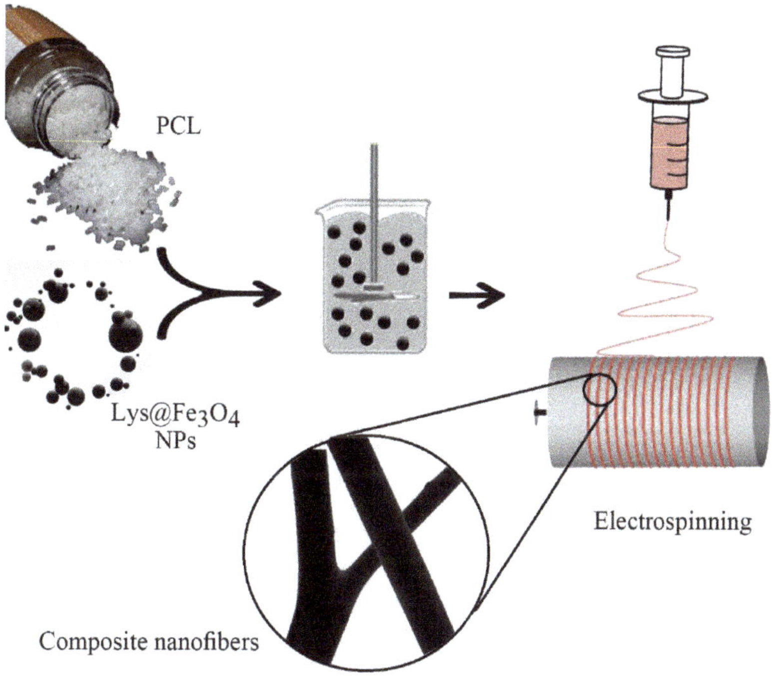

Scheme 1. Procedure for the fabrication of PCL nanofibers that were enriched with Lys@Fe$_3$O$_4$ nanoparticles.

2. Results

2.1. Characterization of Nanoparticles

As shown in Figure 1, the XRD pattern of the synthesized lysine coated magnetite (Lys@Fe$_3$O$_4$) nanoparticles represented characteristic peaks of magnetite at 30, 35.5, 43, 57, and 63 degrees in 2-Theta scale. The phase was identified by using PANalytical X'Pert HighScore Plus software version 3.0e (3.0.5) (PANalytical B.V., Almelo, The Netherlands) and COD databases bank. Analysis of the XRD graph was performed by using the Execute Search and Match tool under the Analysis menu and Magnetite 96-900-5839 reference code was the first phase candidate by 34 scores and all peaks were matched. Particle size analysis was done using the TEM micrographs and revealed that the Lys@Fe$_3$O$_4$ nanoparticles are in the range of 4–10 nm with an average size of 7 nm (Figure 2). FTIR evaluations were done in the region of 3800 to 400 cm^{-1} to investigate the chemical features of the nanoparticles. As shown in Figure 3, the strong band at 631 cm^{-1} is a characteristic peak for Fe-O bonds in the magnetic nanoparticles. Peaks of O-H groups appeared at 1621 cm^{-1} (deforming) and 3390 cm^{-1} (stretching). Peaks from C=O and N–H functional groups in the Lys coating were expected to appear at about 1630 cm^{-1} and 3000 cm^{-1}, respectively. However, these peaks were overlapped with the peaks from O-H deforming and stretching vibrations [29,30]. Previous studies showed that Lys and also other amino acids interact with the magnetite nanoparticles via their carbocycle groups and the side changes are exposed out of the nanoparticles [29–34].

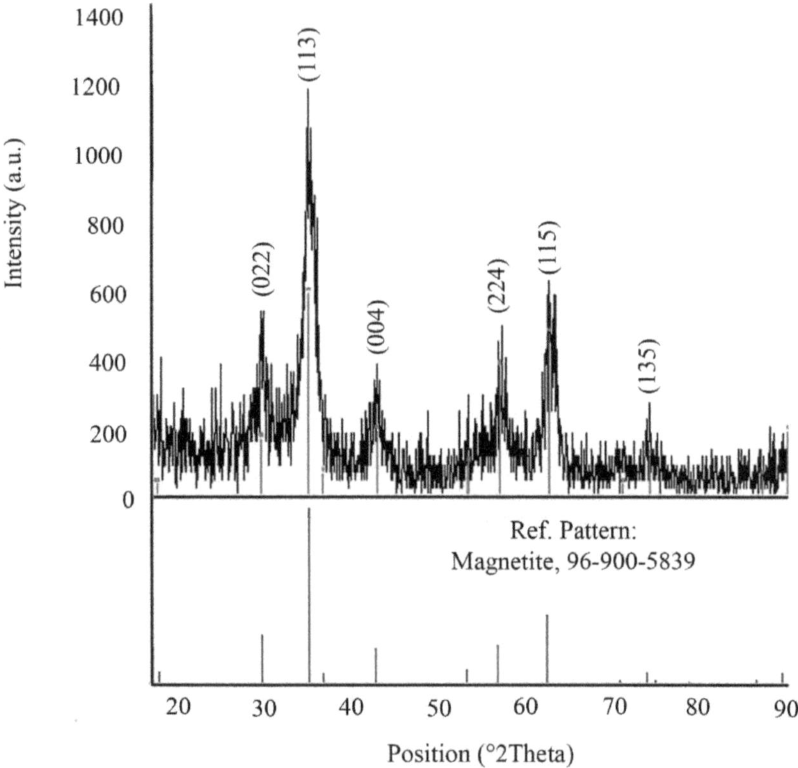

Figure 1. X-ray spectra of Lys@Fe$_3$O$_4$ nanoparticles.

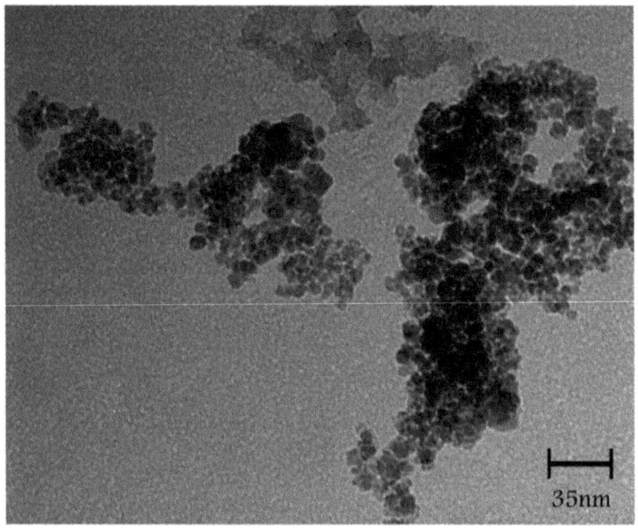

Figure 2. TEM image of the Lys@Fe$_3$O$_4$ nanoparticles.

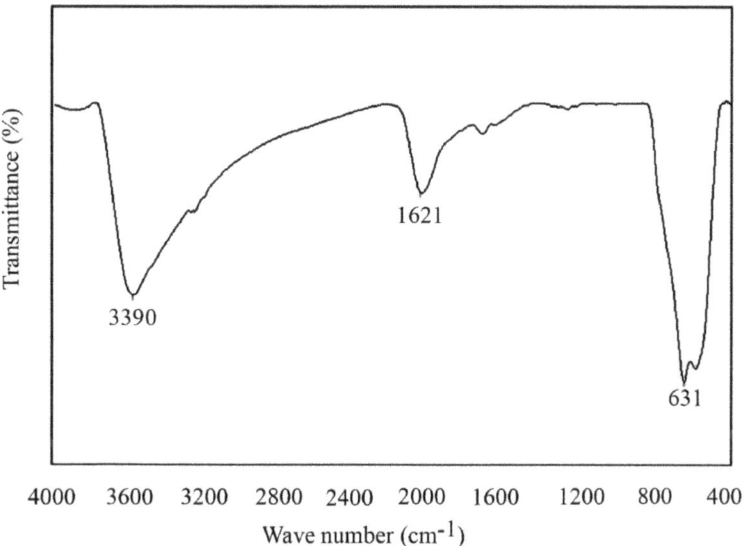

Figure 3. FTIR spectra of Lys@Fe$_3$O$_4$ nanoparticles.

2.2. Characterization of Scaffolds

Appearance of the PCL nanocomposite fibrous mats with various concentrations (0, 3, 5, and 10%) of Lys@Fe$_3$O$_4$ nanoparticles is illustrated in Figure 4. An increase in the nanoparticles concentration resulted in the darker brown membrane of fibers. SEM micrographs of the prepared fibers are provided in Figure 5 and illustrated a typical smooth and bead-free morphology. The diameters of the fibers were measured as 0.04–1.39 μm, 0.19–1.78 μm, 0.50–1.54 μm, and 0.65–1.59 μm, for PCL, 3MNP, 5MNP, and 10MNP, respectively. The average fiber diameters (AFD) were calculated to be 0.719 μm, 0.985 μm, 1.021 μm, and 1.122 μm, for PCL, 3MNP, 5MNP, and 10MNP, respectively. There are controversial data about the impacts of incorporated nanoparticles on the nanofiber diameter. Some investigations indicated that incorporation of nanoparticles (up to 15%) in the PCL nanofibers resulted in a decrease in the fibers' diameter but an addition of 20% nanoparticles resulted in the fibers with increased diameter [6,8]. Effects of additives on the diameter of electrospun fibers can be related to the alterations in viscosity of the PCL solution. It is confirmed that increasing the viscosity of the electrospinning solution would subsequently result in larger nanofibers [35–37].

The elemental analysis of specimens was performed by SEM-EDX as shown in Figure 6. EDX mapping analysis collected evidence to confirm the existence and dispersion of magnetic nanoparticles in the fiber specimens. Three elements (iron, oxygen, and carbon) were evaluated in specimens and results are provided in Table 1. As expected, by increment of magnetic nanoparticles concentrations, an increase in the fibrous iron content was recorded.

ATR-FTIR spectra of the fiber specimens are shown in Figure 7. The peaks that are due to functional groups in the PCL polymer appeared in 2943, 2865, 1722, 1240, and 1165 cm^{-1}. The peaks at 2943 and 2865 cm^{-1} were due to asymmetric and symmetric stretching vibrations of C-H bonds. The peak that is due to carbonyl groups stretching vibration appeared at 1722 cm^{-1}. The peaks at 1240 cm^{-1} and 1165 cm^{-1} could be defined for C-C and C-O bonds, respectively. Moreover, the characteristic peak of iron oxide nanoparticles that is due to Fe-O stretching vibration was recorded at 583 cm^{-1} in the fibers that contain magnetite nanoparticles. Recorded data from the ATR-FTIR analysis were very similar to the previously reported spectra [8,38,39].

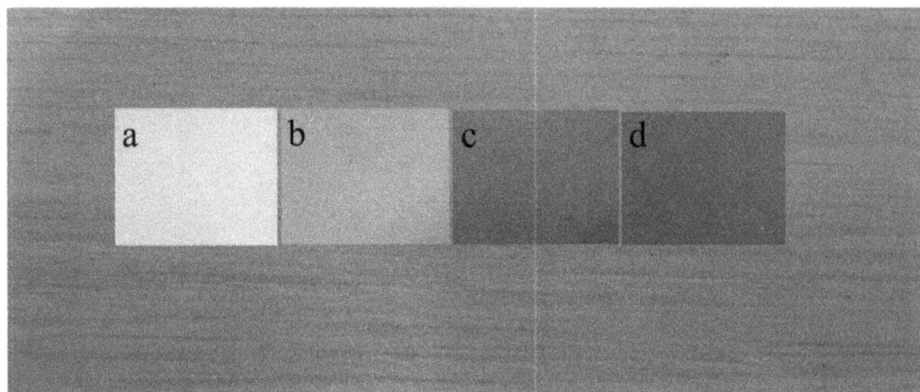

Figure 4. PCL and nanocomposite PCL fibrous scaffold membranes, (**a**): PCL, (**b**): 3MNP, (**c**): 5MNP, and (**d**): 10MNP.

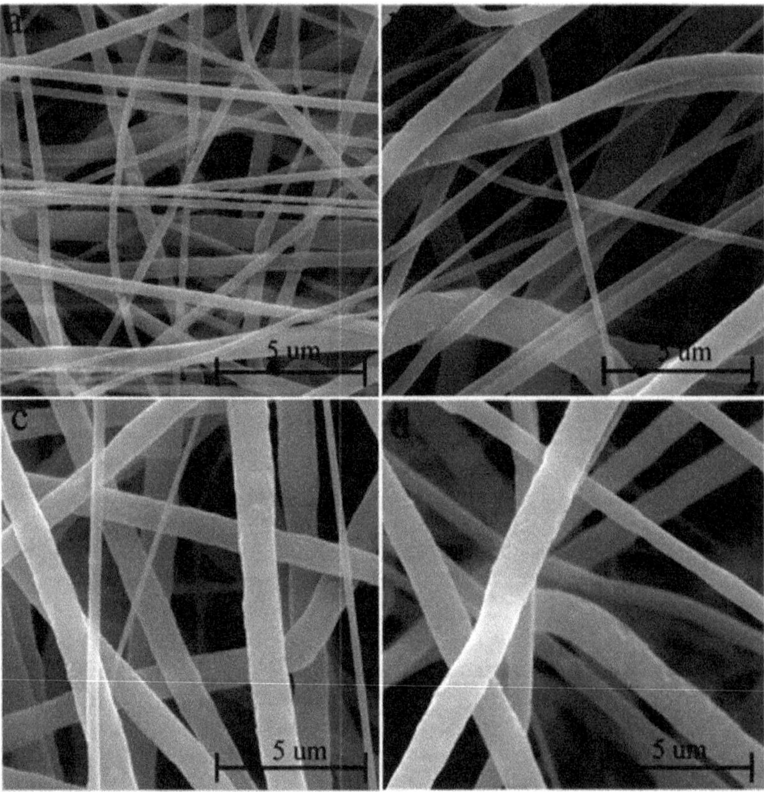

Figure 5. SEM micrographs of the fibrous scaffolds, (**a**): PCL, (**b**): 3MNP, (**c**): 5MNP, and (**d**): 10MNP.

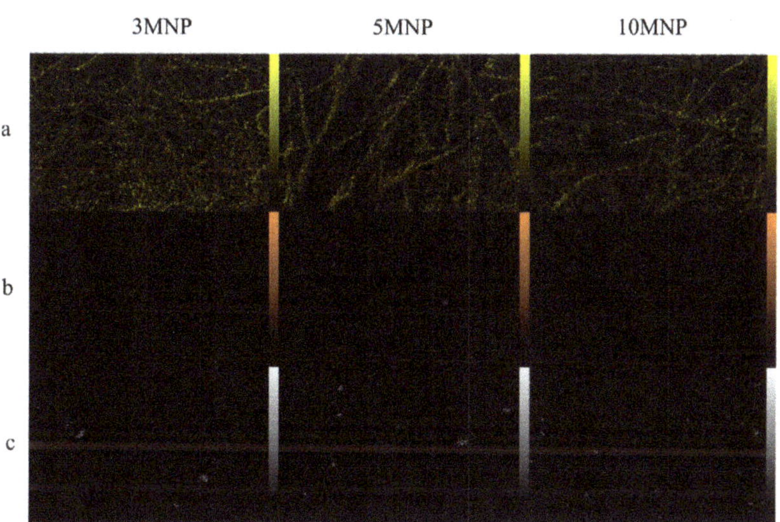

Figure 6. EDX-mapping analysis of nanoparticles incorporated scaffolds, (**a**): carbon, (**b**): oxygen, and (**c**): iron.

Table 1. Elemental content of the magnetite enriched PCL scaffolds resulted from EDX analysis.

Elements Composite	3MNP		5MNP		10MNP	
	A%	**W%**	**A%**	**W%**	**A%**	**W%**
Carbon	73.19	66.93	72.34	65.53	72.88	65.77
Oxygen	26.66	32.47	27.29	32.93	26.58	31.95
Iron	0.15	0.6	0.36	1.53	0.54	2.27

Abbreviations: A%: Atomic percent; W%: Weight percent.

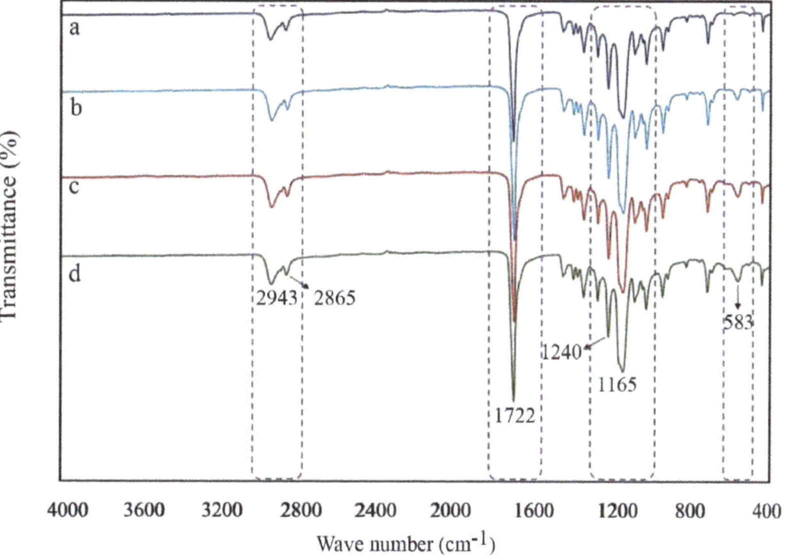

Figure 7. ATR-FTIR spectra of the fibrous scaffolds, (**a**): PCL, (**b**): 3MNP, (**c**): 5MNP, and (**d**): 10MNP.

The wettability of fibrous scaffolds was explored by contact angle testing (Figure 8). Incorporation of nanoparticles in the PCL polymer resulted in a concentration-dependent increase in the wettability of nanofibers. The contact angle of PCL fibers was measured to be 138.93°, while the angles for nanoparticles incorporated fibers were 132.21°, 127.64°, and 121.5° for 3MNP, 5MNP, and 10MNP, respectively. Previous investigations also showed magnetite nanoparticles have a concentration-dependent impact on the reduction of hydrophobicity of PCL nanofibers [6,8].

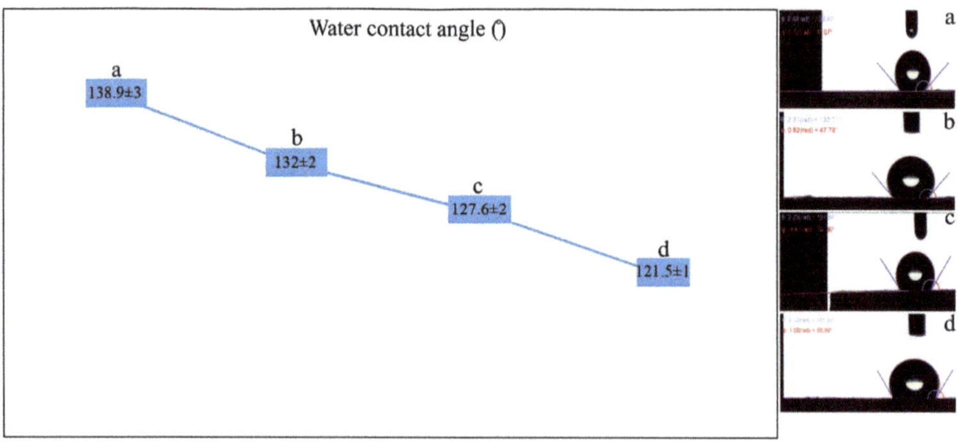

Figure 8. Contact angles of the fibrous scaffolds, (**a**): PCL, (**b**): 3MNP, (**c**): 5MNP, and (**d**): 10MNP, data were presented as mean± standard deviation (SD), *n* = 5.

Results from tensile testing were presented as an overlaid diagram and presented in Figure 9. Other mechanical properties such as elastic module, elongation in peak point and elongation in break point are provided in Table 2. These data indicated that incorporation of magnetite nanoparticles in the PCL nanofibers resulted in a significant increase in the scaffold tensile strength. Increase in the scaffold strength was directly dependent on the concentration of employed nanoparticles. In the current experiment, magnetite nanoparticles were added to the PCL scaffolds up to 10 wt%. Based on the results, it is obvious that high concentrations of incorporated nanoparticles can disturb the integrity of polymer fibers and hence reduce the tensile strength of the resulted scaffold [6]. Uniform dispersion of nanoparticles is the other critical point. Presence of beads in the fibers could possibly be due to nanoparticles agglomeration during the electrospinning process which will weaken the mechanical strength [40].

Table 2. Mechanical properties of fibrous scaffolds.

Samples	F (MPa)	E_{mod} (MPa)	Elongation$_{pp}$ (%)	Elongation$_{bp}$ (%)
PCL	1.98 ± 0.32	2.95 ± 0.66	49.41 ± 14	57.64 ± 16
3MNP	2.21 ± 0.37	2.38 ± 0.75	72.96 ± 17	117.63 ± 27
5MNP	3.38 ± 0.53	4.42 ± 0.9	59.85 ± 12	93.87 ± 23
10MNP	4.9 ± 0.62	6.41 ± 1.2	62.71 ± 13	81.65 ± 20

Abbreviations: F: tensile strength; E: elastic module; pp: peak point; bp: break point, data were presented as mean ± standard deviation (SD), *n* = 3.

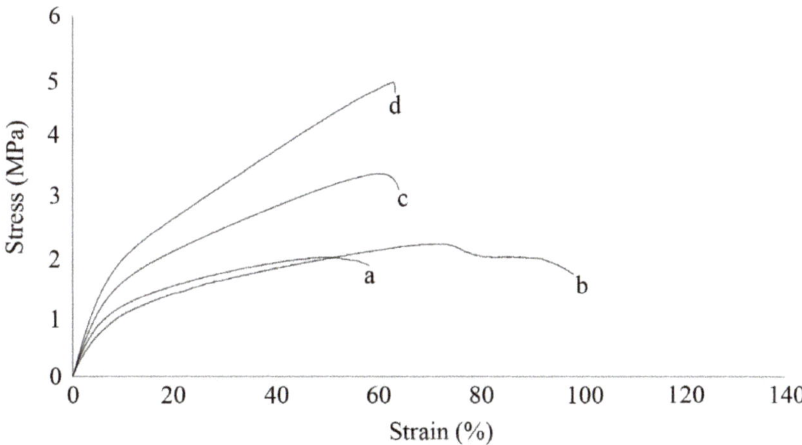

Figure 9. Tensile strength of the fibrous scaffolds, (**a**): PCL, (**b**): 3MNP, (**c**): 5MNP, and (**d**): 10MNP.

Evaluations about the crystal structure of prepared nanofibers were performed by XRD analysis as shown in Figure 10. Due to the semi-crystalline nature of the PCL polymer, two sharp diffraction peaks were recorded from the PCL and nanoparticles incorporated PCL scaffolds. In the case of nanocomposite scaffold, the diffraction peaks from magnetite nanoparticles were considerably eliminated and just the sharpest peak of magnetite crystals at 35.5° of 2-theta value was identifiable. These results indicated magnetite nanoparticles were completely embedded in the PCL fibers and there are no nanoparticles on the surface of fibers. Previous investigations about crystalline nanoparticles have shown similar observations. Coverage of crystalline nanoparticles with another crystalline material resulted in the elimination of diffraction peaks from the core structure [41]. Additionally, it is worth mentioning addition of magnetite nanoparticles to the PCL polymers reduced the sharpness of the PCL characteristic peaks. This finding indicated presence of nanoparticles in the fibers to some extent can disturb the crystal structure of the PCL polymer [42].

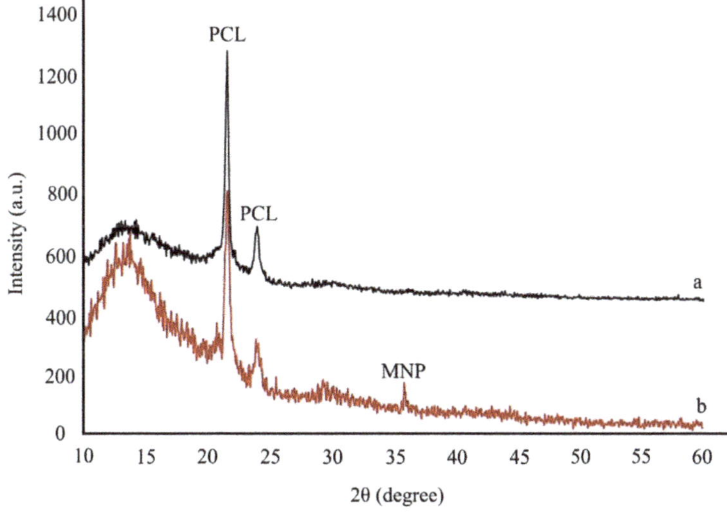

Figure 10. XRD patterns of PCL (**a**) and 5MNP (**b**) scaffolds.

Transmission electron micrographs of the prepared nanofibers were depicted in Figure 11. No aggregation of magnetite nanoparticles was seen within the nanofibers; therefore, it could be assumed the mixture of magnetic nanoparticles and PCL polymer that was used in electrospinning process was a homogeneous suspension. While magnetic nanoparticles are not colloidally stable in the organic solvents, presence of PCL polymer can be considered as a key point to increase the stability of nanoparticles in the employed solvent system (chloroform/methanol). Similar observation was also reported in the previous study on the fabrication of magnetic nanofiber scaffolds [6].

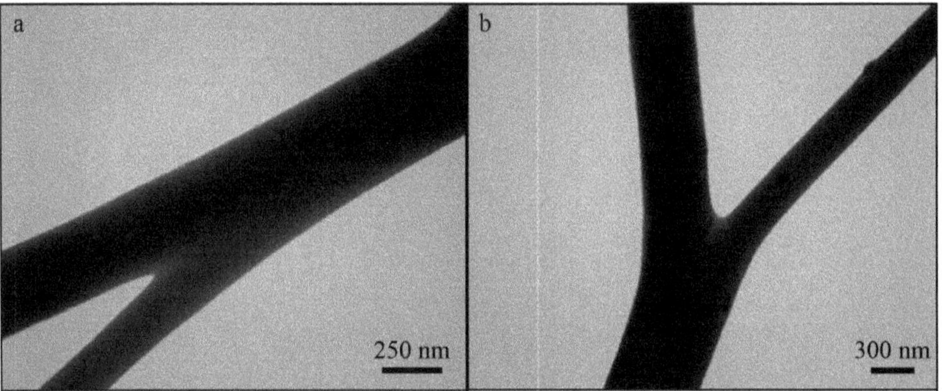

Figure 11. TEM images of PCL (**a**) and 5MNP (**b**) nanofibers.

TGA analysis was performed to evaluate thermal behavior of PCL and nanoparticles incorporated PCL nanofibers. As illustrated in Figure 12 major decomposition of PCL nanofibers was started at 340 °C and only 5.7% of the fibers remained at 440 °C. Reduction in the weight continued at a very low rate up to 600 °C and all the fibers were decomposed. Decomposition of nanocomposite fibers was started at 320 °C and just 4.13% of the fiber remained at ~360 °C. By increasing the temperature more than 360°, decomposition continued at a low rate and 1% remnant was recorded at 600 °C. This remnant is claimed to be the remnant from magnetite nanoparticles. However, similar to previous reports the recorded value is not well-matched with the employed concentration of nanoparticles [6]. The effect of magnetite nanoparticles on the nanofibers' thermal stability was not found to be concentration-dependent. In fact, nanoparticles incorporated fibers were reported to be more thermally sensitive than PCL nanofibers and no obvious difference was reported for nanocomposites with various concentrations of nanoparticles [6].

2.3. In Vitro Studies

Metabolic activity and viability of Hep-G2 cells on the prepared nanofiber scaffolds were investigated over 1, 3, and 5 days by MTT assay. Metabolic activity on the PCL scaffold was considered as a control (100% viability) and measured activity on the nanoparticles incorporated scaffolds was reported as viability percent in contrast to control (Figure 13). After one day of exposure of Hep-G2 cells to nanofiber scaffolds no significant difference was seen among PCL and nanoparticles enriched PCL scaffolds. However, after three and five days of exposure to the scaffolds, more viability was recorded in the well that contained nanoparticles enriched nanofibers.

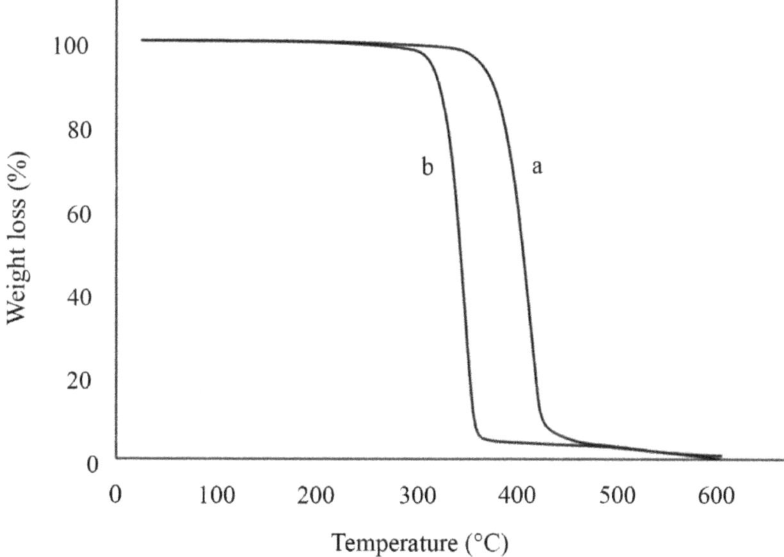

Figure 12. TGA analysis of PCL (**a**) and 5MNP (**b**) nanofibers.

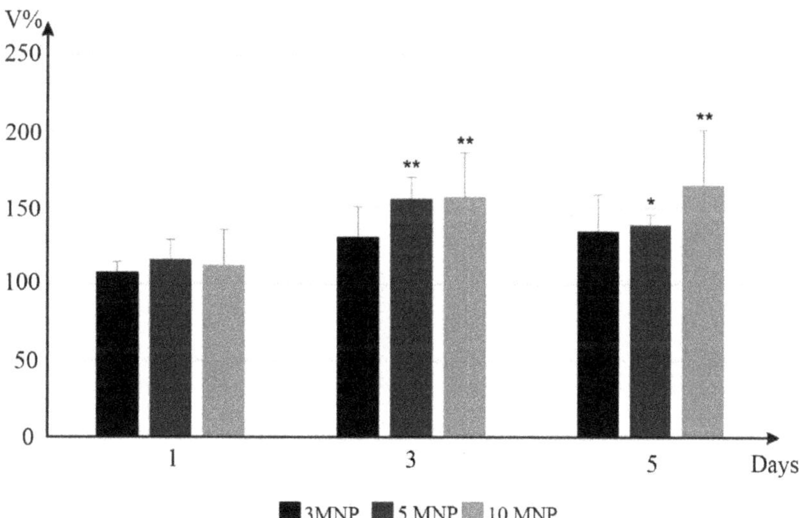

Figure 13. Viability of Hep-G2 cells over one, three, and five-days exposure to the magnetite enriched scaffolds, measured by MTT assay, data are represented as mean $\pm$ SD (n = 3), asterisks indicate significantly different as compared with the control group (* $p < 0.05$ and ** $p < 0.01$).

SEM was used to visualize the attachment of Hep-G2 cells on the prepared nanofibers. The investigation was performed after one and three-days exposure of cells to the nanofibers and resulting micrographs were provided in Figure 14. As expected, on the first day of exposure, cells retained their circular shape and did not tend to adhere to the scaffold. However, on the third day, the tendency of the cells to adhere on both PCL and nanoparticles enriched PCL scaffolds was increased and elongated flat cells were obviously identified. Due to the more hydrophilic nature of the nanoparticles enriched scaffolds, it can be concluded these scaffolds are more prone to cell colonization than PCL scaffolds. Additionally, it was observed the rate of elongation, expansion and even cell infiltration in the enriched scaffold was higher than the PCL scaffold.

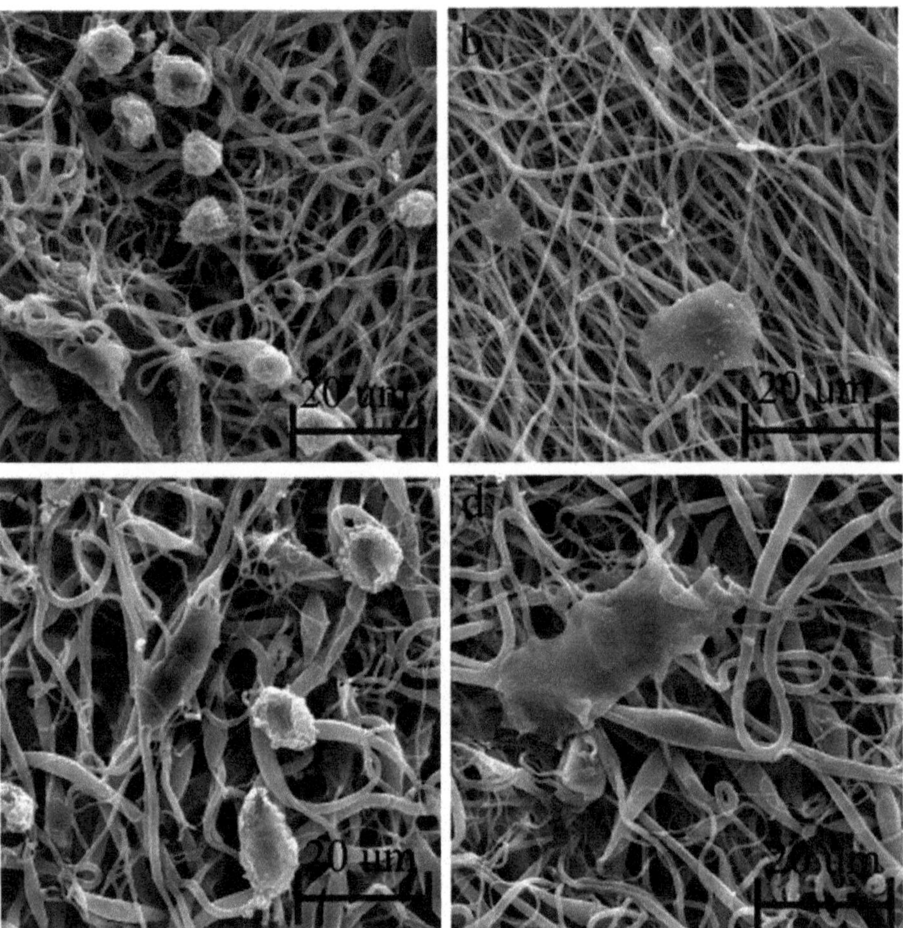

Figure 14. Cell attachment evaluations by SEM analysis, PCL fibers after one day (**a**) and three days (**b**) of cells seeding, 5MNP fibers after one day (**c**) and three days (**d**) of cells seeding.

3. Discussion

Previous investigations indicated that Lys@Fe$_3$O$_4$ nanoparticles can promote cell growth and proliferation in the liver-originated cell lines [28]. Magnetite nanoparticles were also employed for the fortification of hard tissue scaffolds to achieve improved physical and chemical characteristics. In the current experiment, magnetite nanoparticles were used to fabricate a soft tissue scaffold with improved physicochemical characteristics and also enhanced biological functionalities. Magnetite nanoparticles embedded in the PCL nanofibers can act as iron nano-sources and provide the hepatic cells with the adequate iron for enhanced growth and metabolism. It is interesting that an increase in the biologic performance of the scaffolds can be up to 150%. The scaffolds that were enriched with 5 or 10 wt% of nanoparticles provided significantly higher growth rate after three to five days of culture. In addition to iron supplementation, magnetite nanoparticles enhanced the hydrophilicity of PCL scaffolds and made the scaffolds more permeable to water dissolved nutrients. Additionally, more hydrophilic scaffold provides a better attachment site for the cells and results in an enhanced viability percent. These results are consistent with previous studies for application of composite PCL scaffolds for bone and dental tissue engineering (6, 12, 22, 34), but this investigation is the first effort for the fabrication of iron-enriched PCL nanofibers as a fortified soft tissue scaffold with enhanced biologic functionalities.

In addition to biological properties, incorporation of magnetite nanoparticles improves mechanical strength of the scaffolds. In this experiment, magnetite nanoparticles were added to the scaffold up to 10 wt% and it was seen that magnetite nanoparticles enhance tensile strength of the nanofibers. It has been shown higher concentrations (20%) of nanoparticles can disturb the integrity of the fibers and reduce the tensile strength [6]. These data indicated there is an optimal concentration for the incorporated magnetite nanoparticles to maintain polymer accretion and it was revealed that addition of magnetite nanoparticles to the PCL nanofibers resulted in a reduction in the nanofibers' thermal stability. It seems magnetite nanoparticles acted as a nano-catalyst and enhanced the rate of polymer decomposition. On the other hand, reduction in the crystallinity can be the other reason to reduce thermal stability of nanocomposite fibers [43]. Based on the XRD data, incorporation of magnetite nanoparticles in the PCL nanofibers disturbed crystallinity of the PCL polymer and less sharp diffraction peaks were recorded. These results indicated that nanoparticles can reduce the crystallinity of PCL nanofibers and therefore reduce fibers' thermal resistance.

In addition to nanomaterials, there are several reports for the fortification of PCL nanofibers with phytochemicals, biomolecules, and even viruses. For instance, in an experiment curcumin-loaded PCL fibers were fabricated by melt and solution electrospinning methods. It was shown that the approach for the fibers' preparation has an immense impact on the pattern for the curcumin release. The fibers that were fabricated by melt electrospun had a low rate for curcumin release. Meanwhile, solution electrospinning provided fibers with a burst release profile. The difference in the drug release pattern was discussed to be due to the difference in the crystalline feature of the fibers [44]. Prolong drug, doxorubicin, release pattern was also reported by the PCL fibers with high crystallinity. The structure was designed to be a star polymer made up of a poly (amido-amine) (PAMAM) core and PCL branches. The structure exhibited effective controlled toxicity over A43, HeLa, and MCF-7 cells lines [45]. Drug-loaded PCL fibers were also employed as an efficient delivery system for vascular implant application. In this regard, PCL fibers were loaded with cilostazol to achieve a system for drug dissolution and diffusion in combination with polymer relaxation [46]. PCL nanofibers were also employed for the delivery of biomolecules such as microRNA. The delivery system was developed as a nanofiber-mediated microRNA delivery system to control cells differentiation through a combination of fiber topography and gene silencing [47]. Fortification of PCL nanofibers was not just performed by the chemicals or biomolecules and there are some experiments that used viral particles in this way. Localized transduction was reported to be performed by using a virus-encapsulated electrospun PCL fibrous scaffold. Recombinant adeno virus that encodes green fluorescent protein (GFP) was embedded in PCL fibers via co-axial electrospinning. Subsequently, the viral particles were released through a porogen-mediated process [48]. Bacteriophages were another viral particle that were employed for the fortification of PCL scaffolds. Bacteriophage capsids were covalently immobilized on the PCL nanofibers to maximize the phages' tail exposure and increase their antibacterial activity. Authors introduced the system as a promising substitute for antibiotics in applications toward skin infections [49]. These data from previous reports and the current experiment indicated PCL nanofibers can be employed as biocompatible and flexible fibers for selective fortification and fabrication of novel high throughput scaffolds.

4. Materials and Methods

4.1. Materials

PCL (MW = 80,000 Da) and 3-(4,5-Dimethylthiazol-2-yl)-2,5-diphenyltetrazolium bromide (MMT) were obtained from Sigma-Aldrich (Darmstadt, Germany). Ferric chloride hexahydrate ($FeCl_3 \cdot 6H_2O$), Ferrous sulfate heptahydrate ($FeSO_4 \cdot 7H_2O$), ammonium hydroxide (NH_4OH) and Lys were purchased from Merck (Darmstadt, Germany). Chloroform ($CHCl_3$) and methanol (CH_3OH) were provided by Dr. Mojallali Chemical Complex (Tehran, Iran). Hep-G2 cell line was kindly provided by the Pharmaceutical Science Research Center, Shiraz University of Medical Sciences (Shiraz, Iran). Cell culture medium (RPMI-1640), trypsin, and fetal bovine serum (FBS) were also purchased from Gibco (Gibco Laboratories, Grand Island, NY, USA).

4.2. Synthesis of Lys Coated Magnetite Nanoparticles

Magnetite (Fe_3O_4) nanoparticles were synthesized by co-precipitation reaction. Lys coating was also performed through the synthesis reaction via one-put reaction approach [32]. In brief, ferrous sulfate heptahydrate ($FeSO_4 \cdot 7H_2O$, 0.6 g) and ferric chloride hexahydrate ($FeCl_3 \cdot 6H_2O$, 1.7 g) were dissolved in 50 mL distilled water. The mixture was stirred under N_2 atmosphere at 70 °C for 30 min and then Lys solution (1.6 g in 6 mL distilled water) was added. After another 30 min, 5 mL NH_4OH (32%) was suddenly injected into the reaction mixture and stirring continued for 1.5 h. Resulted suspension of Lys coated magnetite ($Lys@Fe_3O_4$) nanoparticles were harvested and washed with distilled water. The dark black precipitate was dried in an oven at 55 °C.

4.3. Characterization of Lys Coated Magnetite Nanoparticles

The crystallinity of $Lys@Fe_3O_4$ nanoparticles was determined by X-ray diffraction (XRD, Siemens D5000, Siemens, Aubrey, TX, USA). The samples were scanned with 45 kV and 40 mA current strength in a 2θ range of 10–60° with a step size of 0.02°, and the rate of $2° \, min^{-1}$. Particle size analysis and morphology investigations were done by transmission electron microscopy (TEM; Philips, Amsterdam, Netherlands, CM 10, HT 100 Kv). Samples were prepared by drying a drop of nanoparticle suspension (~100 µg/mL) on a carbon-coated copper grid. Resulting micrographs were subjected to image analysis by using an open platform for scientific image analysis (ImageJ 1.47v software). Statistical analysis was also performed by SPSS Software v.20 (IBM Analytics, Chicago, IL, USA). Fourier transformed infrared spectroscopy (FT-IR; Bruker, Germany, Vertex 70) was performed with KBr pellets to determine the chemical status of the resulted nanoparticles.

4.4. Preparation of PCL Nanofibers

Nanoparticles doped PCL nanofibers were fabricated by dispersion of $Lys@Fe_3O_4$ nanoparticles in the PCL solution prior to the electrospinning process. Briefly, $Lys@Fe_3O_4$ nanoparticles were dispersed in a co-solvent of chloroform/methanol (3:1) and stirred for 2 h. Then, PCL pellets (9% *w/w*) were added to the suspension and stirring continued for 24 h. The concentrations of nanoparticles were calculated to be 3, 5 and 10 wt% of PCL. Resulted fibers were labeled as 3MNP, 5MNP, and 10MNP, respectively. Consequently, the mixture was loaded into a 5 mL syringe with 20-gauge needle and connected to a high-voltage (11.5 kV) power supply of an electrospinning device (Nanoazma, Tehran, Iran). The distance of tip-to-collector was set to be 12 cm, the flow rate was 0.8 mL/h, and the drum speed was 250 rpm. The process was performed at ambient atmosphere.

4.5. Characterization of Fibers

Scanning electron microscopy (SEM, TESCAN-Vega 3, Tescan, Brno, Czech Republic) at an accelerating voltage of 20 kV was used to analyze 3D morphology of the PCL fibers. The image analysis was done by ImageJ software. Energy-dispersive X-ray spectroscopy (EDX) mapping evaluations were performed to visualize the pattern of elemental dispersion in nanofibers. Additionally, TEM (ZEISS EM10C-100 KV, Oberkochen, Germany) was used to visualize spread of nanoparticles in the fibers. Attenuated total reflection (ATR)-FTIR spectroscopy (Bruker, Tensor II, Germany) was used to determine the nanofibers' chemical profile. Water affinity was measured by water contact angle using the CA-500A analyzer (Sharifsolar, Iran). Tensile mechanical properties were measured by a universal testing instrument (SANTAM, STM-20; Navard, Iran). Small pieces (30×10 mm) of fibers with a thickness of 150–300 μm were prepared and used for the mechanical test. Crystallinity of the fibers was characterized by XRD (Bruker, D8-ADVANCE) and samples were scanned in 2θ range of 10–60° at the rate of $2°$ min^{-1}. Studies about thermal behavior and compositional fraction of the fibers were accomplished by thermogravimetric analysis (TGA; Mettler Toledo, Switzerland).

4.6. Cells Cultured and In Vitro Tests

The Hep-G2 cells were cultured under the standard incubation conditions (5% CO_2, 37 °C) using RPMI-1640 supplemented with 10% FBS and 1% pen/strep. Before any in vitro tests, scaffold specimens were sterilized by ultraviolet irradiation (30 min for each side of the film). Afterward, the specimens were attached to the bottom of the wells (48-well-plate) using agarose solution (0.5% *w/w*). Hep-G2 cells were incorporated in scaffold specimens and plates were incubated for 24 and 72 h. Attached cells were fixed on the scaffolds by using 2.5% glutaraldehyde at room temperature. Specimens were then rinsed by PBS and dehydrated by 25, 50, 70, 95, and 100% ethanol (three times and 5 min for each concentration). Afterwards, the specimens were evaluated via SEM.

MTT assay was used to evaluate cell viability and proliferation. After 24, 48, and 96 h incubation, attached wells were washed by PBS and incubated in MTT solution for 4 h. Resulted blue formazan crystals were dissolved in 100 μL DMSO and transferred into virgin wells for absorption measurement at 570 nm using a plate reader (Epoch, BioTek, Winooski, VT, USA).

5. Conclusions

Lys@Fe_3O_4 nanoparticles were successfully employed as an additive to fabricate iron-enriched soft tissue PCL scaffolds. Incorporation of magnetite nanoparticles in the electrospun PCL fibers has an immense impact on the physical and chemical characteristics of the scaffolds. The particles made PCL scaffolds less hydrophobic and also increased the tensile strength of the fibers. On the other hand, composite nanofibers were more thermal sensitive than PCL nanofibers. TEM and EDX mapping analysis showed that nanoparticles were well dispersed in the nano fibers. Magnetite nanoparticles that were embedded in the nanofibers can act as valuable sources of iron to promote cell growth and proliferation. It is interesting that in contrast to the PCL scaffold, enriched scaffolds can improve cell viability up to ~150%. However, no considerable difference was observed in the cell attachment process. These data indicated that iron-enriched scaffolds can be considered as the scaffolds with increased biological performance for liver tissue engineering.

Author Contributions: Funding acquisition, A.E.; Investigation, V.R. and S.-M.T.; Methodology, V.R., E.M., S.-M.T. and A.E.; Project administration, E.M. and A.E.; Software, S.-M.T.; Supervision, E.M. and A.E.; Visualization, V.R. and A.E.; Writing—original draft, V.R.; Writing—review & editing, A.B. and A.E. All authors have read and agreed to the published version of the manuscript.

Funding: This research was funded by the Shiraz University of Medical Sciences, Shiraz, Iran. The APC was funded by the University of Waikato, Hamilton, New Zealand.

Institutional Review Board Statement: The study was conducted according to the guidelines of the Declaration of Helsinki, and approved by the Ethics Committee of Shiraz University of Medical Sciences (approval ID: IR.SUMS.REC.1395.680, date of approval: 2019-07-23).

Informed Consent Statement: Not applicable.

Conflicts of Interest: The authors declare no conflict of interest.

References

1. Edgar, L.; McNamara, K.; Wong, T.; Tamburrini, R.; Katari, R.; Orlando, G. Heterogeneity of Scaffold Biomaterials in Tissue Engineering. *Materials* **2016**, *9*, 332. [CrossRef] [PubMed]
2. Asvar, Z.; Mirzaei, E.; Azarpira, N.; Geramizadeh, B.; Fadaie, M. Evaluation of electrospinning parameters on the tensile strength and suture retention strength of polycaprolactone nanofibrous scaffolds through surface response methodology. *J. Mech. Behav. Biomed. Mater.* **2017**, *75*, 369–378. [CrossRef]
3. Fadaie, M.; Mirzaei, E.; Geramizadeh, B.; Asvar, Z. Incorporation of nanofibrillated chitosan into electrospun PCL nanofibers makes scaffolds with enhanced mechanical and biological properties. *Carbohydr. Polym.* **2018**, *199*, 628–640. [CrossRef]
4. Hammond, J.S.; Beckingham, I.J.; Shakesheff, K.M. Scaffolds for liver tissue engineering. *Expert Rev. Med. Devices* **2006**, *3*, 21–27. [CrossRef]
5. Hasan, A.; Morshed, M.; Memic, A.; Hassan, S.; Webster, T.J.; Marei, H.E.-S. Nanoparticles in tissue engineering: Applications, challenges and prospects. *Int. J. Nanomed.* **2018**, *13*, 5637–5655. [CrossRef]
6. Singh, R.K.; Patel, K.D.; Lee, J.H.; Lee, E.J.; Kim, J.H.; Kim, T.H.; Kim, H.W. Potential of magnetic nanofiber scaffolds with mechanical and biological properties applicable for bone regeneration. *PLoS ONE* **2014**, *9*, e91584. [CrossRef]
7. Bhattarai, D.P.; Aguilar, L.E.; Park, C.H.; Kim, C.S. A Review on Properties of Natural and Synthetic Based Electrospun Fibrous Materials for Bone Tissue Engineering. *Membranes* **2018**, *8*, 62. [CrossRef] [PubMed]
8. Demir, D.; Güreş, D.; Tecim, T.; Genç, R.; Bölgen, N. Magnetic nanoparticle-loaded electrospun poly (ε-caprolactone) nanofibers for drug delivery applications. *Appl. Nanosci.* **2018**, *8*, 1461–1469. [CrossRef]
9. Jun, I.; Han, H.-S.; Edwards, J.R.; Jeon, H. Electrospun fibrous scaffolds for tissue engineering: Viewpoints on architecture and fabrication. *Int. J. Mol. Sci.* **2018**, *19*, 745. [CrossRef]
10. Lu, T.; Li, Y.; Chen, T. Techniques for fabrication and construction of three-dimensional scaffolds for tissue engineering. *Int. J. Nanomed.* **2013**, *8*, 337–350. [CrossRef]
11. Dulnik, J.; Denis, P.; Sajkiewicz, P.; Kołbuk, D.; Choińska, E. Biodegradation of bicomponent PCL/gelatin and PCL/collagen nanofibers electrospun from alternative solvent system. *Polym. Degrad. Stab.* **2016**, *130*, 10–21. [CrossRef]
12. Daňková, J.; Buzgo, M.; Vejpravova, J.; Kubíčková, S.; Sovková, V.; Vysloužilová, L.; Mantlíková, A.; Nečas, A.; Amler, E. Highly efficient mesenchymal stem cell proliferation on poly-ε-caprolactone nanofibers with embedded magnetic nanoparticles. *Int. J. Nanomed.* **2015**, *10*, 7307–7317. [CrossRef] [PubMed]
13. Manea, L.; Hristian, L.; Leon, A.; Popa, A. Recent advances of basic materials to obtain electrospun polymeric nanofibers for medical applications. *IOP Conf. Ser. Mater. Sci. Eng.* **2016**, *145*, 032006. [CrossRef]
14. Soares, R.M.; Siqueira, N.M.; Prabhakaram, M.P.; Ramakrishna, S. Electrospinning and electrospray of bio-based and natural polymers for biomaterials development. *Mater. Sci. Eng. C* **2018**, *92*, 969–982. [CrossRef] [PubMed]
15. Ding, J.; Zhang, J.; Li, J.; Li, D.; Xiao, C.; Xiao, H.; Yang, H.; Zhuang, X.; Chen, X. Electrospun polymer biomaterials. *Prog. Polym. Sci.* **2019**, *90*, 1–34. [CrossRef]
16. De Santis, R.; Gloria, A.; Russo, T.; d'Amora, U.; Zeppetelli, S.; Dionigi, C.; Sytcheva, A.; Herrmannsdörfer, T.; Dediu, V.; Ambrosio, L. A basic approach toward the development of nanocomposite magnetic scaffolds for advanced bone tissue engineering. *J. Appl. Polym. Sci.* **2011**, *122*, 3599–3605. [CrossRef]
17. Zhang, H.; Xia, J.; Pang, X.; Zhao, M.; Wang, B.; Yang, L.; Wan, H.; Wu, J.; Fu, S. Magnetic nanoparticle-loaded electrospun polymeric nanofibers for tissue engineering. *Mater. Sci. Eng. C* **2017**, *73*, 537–543. [CrossRef]
18. Hu, Y.; Feng, B.; Zhang, W.; Yan, C.; Yao, Q.; Shao, C.; Yu, F.; Li, F.; Fu, Y. Electrospun gelatin/PCL and collagen/PCL scaffolds for modulating responses of bone marrow endothelial progenitor cells. *Exp. Ther. Med.* **2019**, *17*, 3717–3726. [CrossRef]
19. Park, J.S.; Woo, D.G.; Sun, B.K.; Chung, H.M.; Im, S.J.; Choi, Y.M.; Park, K.; Huh, K.M.; Park, K.H. In vitro and in vivo test of PEG/PCL-based hydrogel scaffold for cell delivery application. *J. Control. Release* **2007**, *124*, 51–59. [CrossRef]
20. Ren, K.; Wang, Y.; Sun, T.; Yue, W.; Zhang, H. Electrospun PCL/gelatin composite nanofiber structures for effective guided bone regeneration membranes. *Mater. Sci. Eng. C* **2017**, *78*, 324–332. [CrossRef]
21. Zhou, Z.X.; Chen, Y.R.; Zhang, J.Y.; Jiang, D.; Yuan, F.Z.; Mao, Z.M.; Yang, F.; Jiang, W.B.; Wang, X.; Yu, J.K. Facile Strategy on Hydrophilic Modification of Poly(ε-caprolactone) Scaffolds for Assisting Tissue-Engineered Meniscus Constructs In Vitro. *Front. Pharmacol.* **2020**, *11*, 471. [CrossRef]
22. Kim, J.-J.; Singh, R.K.; Seo, S.-J.; Kim, T.-H.; Kim, J.-H.; Lee, E.-J.; Kim, H.-W. Magnetic scaffolds of polycaprolactone with functionalized magnetite nanoparticles: Physicochemical, mechanical, and biological properties effective for bone regeneration. *RSC Adv.* **2014**, *4*, 17325–17336. [CrossRef]
23. Ballesteros, C.A.; Correa, D.S.; Zucolotto, V. Polycaprolactone nanofiber mats decorated with photoresponsive nanogels and silver nanoparticles: Slow release for antibacterial control. *Mater. Sci. Eng. C* **2020**, *107*, 110334. [CrossRef] [PubMed]
24. Fahimirad, S.; Abtahi, H.; Satei, P.; Ghaznavi-Rad, E.; Moslehi, M.; Ganji, A. Wound healing performance of PCL/Chitosan based electrospun nanofiber electrosprayed with curcumin loaded chitosan nanoparticles. *Carbohydr. Polym.* **2021**, *259*, 117640. [CrossRef] [PubMed]
25. Rezk, A.I.; Bhattarai, D.P.; Park, J.; Park, C.H.; Kim, C.S. Polyaniline-coated titanium oxide nanoparticles and simvastatin-loaded poly (ε-caprolactone) composite nanofibers scaffold for bone tissue regeneration application. *Colloids Surf. B* **2020**, *192*, 111007. [CrossRef]

26. Gounani, Z.; Pourianejad, S.; Asadollahi, M.A.; Meyer, R.L.; Rosenholm, J.M.; Arpanaei, A. Polycaprolactone-gelatin nanofibers incorporated with dual antibiotic-loaded carboxyl-modified silica nanoparticles. *J. Mater. Sci.* **2020**, *55*, 17134–17150. [CrossRef]
27. Akbarzadeh, R.; Yousefi, A.M. Effects of processing parameters in thermally induced phase separation technique on porous architecture of scaffolds for bone tissue engineering. *J. Biomed. Mater. Res. Part B* **2014**, *102*, 1304–1315. [CrossRef]
28. Ebrahiminezhad, A.; Rasoul-Amini, S.; Kouhpayeh, A.; Davaran, S.; Barar, J.; Ghasemi, Y. Impacts of amine functionalized iron oxide nanoparticles on HepG2 cell line. *Curr. Nanosci.* **2015**, *11*, 113–119. [CrossRef]
29. Ebrahiminezhad, A.; Varma, V.; Yang, S.; Ghasemi, Y.; Berenjian, A. Synthesis and application of amine functionalized iron oxide nanoparticles on menaquinone-7 fermentation: A step towards process intensification. *Nanomaterials* **2015**, *6*, 1. [CrossRef]
30. Raee, M.J.; Ebrahiminezhad, A.; Gholami, A.; Ghoshoon, M.B.; Ghasemi, Y. Magnetic immobilization of recombinant *E. coli* producing extracellular asparaginase: An effective way to intensify downstream process. *Sep. Sci. Technol.* **2018**, *53*, 1397–1404. [CrossRef]
31. Ebrahiminezhad, A.; Ghasemi, Y.; Rasoul-Amini, S.; Barar, J.; Davaran, S. Impact of amino-acid coating on the synthesis and characteristics of iron-oxide nanoparticles (IONs). *Bull. Korean Chem. Soc.* **2012**, *33*, 3957–3962. [CrossRef]
32. Raee, M.J.; Ebrahiminezhad, A.; Ghoshoon, M.B.; Gholami, A.; Ghasemi, Y. Synthesis and characterization of L-lysin coated iron oxide nanoparticles as appropriate choices for cell immobilization and magnetic separation. *Nanosci. Nanotechnology-Asia* **2019**, *9*, 462–466. [CrossRef]
33. Durmus, Z.; Kavas, H.; Toprak, M.S.; Baykal, A.; Altincekic, T.G.; Aslan, A.; Bozkurt, A.; Cosgun, S. l-lysine coated iron oxide nanoparticles: Synthesis, structural and conductivity characterization. *J. Alloy. Compd.* **2009**, *484*, 371–376. [CrossRef]
34. Patel, D.; Chang, Y.; Lee, G.H. Amino acid functionalized magnetite nanoparticles in saline solution. *Curr. Appl. Phys.* **2009**, *9*, S32–S34. [CrossRef]
35. Bortolassi, A.C.C.; Nagarajan, S.; de Araújo Lima, B.; Guerra, V.G.; Aguiar, M.L.; Huon, V.; Soussan, L.; Cornu, D.; Miele, P.; Bechelany, M. Efficient nanoparticles removal and bactericidal action of electrospun nanofibers membranes for air filtration. *Mater. Sci. Eng. C* **2019**, *102*, 718–729. [CrossRef] [PubMed]
36. Chen, X.; Zhang, Y.; He, X.; Li, H.; Wei, B.; Yang, W. Electrospinning on a plucked string. *J. Mater. Sci.* **2019**, *54*, 901–910. [CrossRef]
37. Petras, D.; Slobodian, P.; Pavlínek, V.; Sáha, P.; Kimmer, D. The effect of pvac solution viscosity on diameter of pvac nanofibres prepared by technology of electrospinning. *AIP Conf. Proc.* **2011**, *1375*, 312–319.
38. Bagheri, M.; Mahmoodzadeh, A. Polycaprolactone/Graphene Nanocomposites: Synthesis, Characterization and Mechanical Properties of Electrospun Nanofibers. *J. Inorg. Organomet. Polym. Mater.* **2020**, *30*, 1566–1577. [CrossRef]
39. Ramírez-Cedillo, E.; Ortega-Lara, W.; Rocha-Pizaña, M.R.; Gutierrez-Uribe, J.A.; Elías-Zúñiga, A.; Rodríguez, C.A. Electrospun Polycaprolactone Fibrous Membranes Containing Ag, TiO_2 and $Na_2Ti_6O(13)$ Particles for Potential Use in Bone Regeneration. *Membranes* **2019**, *9*, 12. [CrossRef]
40. Huang, Z.-M.; Zhang, Y.; Ramakrishna, S.; Lim, C. Electrospinning and mechanical characterization of gelatin nanofibers. *Polymer* **2004**, *45*, 5361–5368. [CrossRef]
41. Taghizadeh, S.-M.; Berenjian, A.; Taghizadeh, S.; Ghasemi, Y.; Taherpour, A.; Sarmah, A.K.; Ebrahiminezhad, A. One-put green synthesis of multifunctional silver iron core-shell nanostructure with antimicrobial and catalytic properties. *Ind. Crop. Prod.* **2019**, *130*, 230–236. [CrossRef]
42. Meng, Z.; Zheng, W.; Li, L.; Zheng, Y. Fabrication and characterization of three-dimensional nanofiber membrance of PCL–MWCNTs by electrospinning. *Mater. Sci. Eng. C* **2010**, *30*, 1014–1021. [CrossRef]
43. Daniel, C.; Zhovner, D.; Guerra, G. Thermal stability of nanoporous crystalline and amorphous phases of poly (2, 6-dimethyl-1, 4-phenylene) oxide. *Macromolecules* **2013**, *46*, 449–454. [CrossRef]
44. Lian, H.; Meng, Z. Melt electrospinning vs. solution electrospinning: A comparative study of drug-loaded poly (ε-caprolactone) fibres. *Mater. Sci. Eng. C* **2017**, *74*, 117–123. [CrossRef] [PubMed]
45. Balakrishnan, P.B.; Gardella, L.; Forouharshad, M.; Pellegrino, T.; Monticelli, O. Star poly (ε-caprolactone)-based electrospun fibers as biocompatible scaffold for doxorubicin with prolonged drug release activity. *Colloids Surf. B* **2018**, *161*, 488–496. [CrossRef]
46. Rychter, M.; Baranowska-Korczyc, A.; Milanowski, B.; Jarek, M.; Maciejewska, B.M.; Coy, E.L.; Lulek, J. Cilostazol-loaded poly (ε-Caprolactone) electrospun drug delivery system for cardiovascular applications. *Pharm. Res.* **2018**, *35*, 32. [CrossRef]
47. Diao, H.J.; Low, W.C.; Milbreta, U.; Lu, Q.R.; Chew, S.Y. Nanofiber-mediated microRNA delivery to enhance differentiation and maturation of oligodendroglial precursor cells. *J. Control. Release* **2015**, *208*, 85–92. [CrossRef]
48. Liao, I.-C.; Chen, S.; Liu, J.B.; Leong, K.W. Sustained viral gene delivery through core-shell fibers. *J. Control. Release* **2009**, *139*, 48–55. [CrossRef]
49. Nogueira, F.; Karumidze, N.; Kusradze, I.; Goderdzishvili, M.; Teixeira, P.; Gouveia, I.C. Immobilization of bacteriophage in wound-dressing nanostructure. *Nanomedicine* **2017**, *13*, 2475–2484. [CrossRef] [PubMed]

Article

Preparation of Slow-Release Insecticides from Biogas Slurry: Effectiveness of Ion Exchange Resin in the Adsorption and Release of Ammonia Nitrogen

Quanguo Zhang [1,2], Zexian Liu [1], Francesco Petracchini [3], Chaoyang Lu [1], Yameng Li [1], Zhiping Zhang [1,2], Valerio Paolini [3] and Huan Zhang [1,*]

[1] Key Laboratory of New Materials and Equipment for Renewable Energy, Ministry of Agriculture and Rural Affairs, Henan Agricultural University, Zhengzhou 450002, China; zquanguo@163.com (Q.Z.); xnl421683@163.com (Z.L.); lu@henau.edu.cn (C.L.); liyameng2017@163.com (Y.L.); zhangzhiping715@163.com (Z.Z.)

[2] Institute of Modern Agricultural Engineering, Huanghe S&T University, Zhengzhou 450006, China

[3] Institute of Atmospheric Pollution Research, National Research Council of Italy, 29300 Rome, Italy; petracchini@iia.cnr.it (F.P.); valerio.paolini@iia.cnr.it (V.P.)

* Correspondence: zhangying@henau.edu.cn

Citation: Zhang, Q.; Liu, Z.; Petracchini, F.; Lu, C.; Li, Y.; Zhang, Z.; Paolini, V.; Zhang, H. Preparation of Slow-Release Insecticides from Biogas Slurry: Effectiveness of Ion Exchange Resin in the Adsorption and Release of Ammonia Nitrogen. *Processes* **2021**, *9*, 1461. https://doi.org/10.3390/pr9081461

Academic Editors: Pau Loke Show, Kit Wayne Chew and Aydin Berenjian

Received: 6 July 2021
Accepted: 16 August 2021
Published: 21 August 2021

Abstract: The insecticidal ingredient in a biogas solution being fully utilized by cation exchange resin to produce slow-release insecticide is of great social value. In this work, the feasibility of ammonia nitrogen in a biogas slurry loaded on resin as a slow-release insecticide was evaluated by studying the effect of adsorption and the slow release of ammonia nitrogen by resin. The effects of the ammonia nitrogen concentration, resin dosage, adsorption time and pH value on the ammonia nitrogen adsorption by the resin were studied. The results showed that the ion exchange resin had a good adsorption effect on the ammonia nitrogen. With the increase of the resin dosage, time and ammonia nitrogen concentration, the adsorption capacity increased at first and then stabilized. The ammonia nitrogen adsorption capacity reached its maximum value (1.13 mg) when the pH value was 7. The adsorption process can be fitted well by the Langmuir isothermal adsorption equation and quasi-second-order kinetic model. Additionally, the release rate of the ammonia nitrogen increased with the increasing sodium chloride concentration. The adsorption capacity of ammonia nitrogen by the D113 (resin type) resin decreased by 15.8% compared with the ammonium chloride solution. The report shows that the ion exchange resin has a good adsorption effect on ammonia nitrogen, which is of guiding significance for expanding the raw materials for slow-release insecticides, improving the utilization rate of biogas slurry and cleaner production of slow-release insecticides from biogas slurry. Additionally, all variables showed statistical differences ($p < 0.05$).

Keywords: biogas slurry; ion exchange resin; ammonia nitrogen; adsorption; slow release

1. Introduction

Agriculture is a basic industry for national development, and it concerns the fundamental interests of the people [1]. Since the twentieth century, pesticides have been used in pursuit of crop yield, which has made great contributions to agricultural production, yet many of its shortcomings have been found in the process of use, such as causing the serious destruction of the ecological environment [2], the deterioration of soil properties [3], the pollution of water resources, the increase in pest resistance [4] and the damage to the health of people [5]. In the meantime, 90% of pesticides are lost due to osmotic action [6], resulting in a significant reduction of the pesticide utilization rate. It is extremely important to solve these problems in the development of agriculture, and the search for economical, environmentally friendly and highly effective insecticides has attracted more attention [7]. Recently, the insecticidal effect of biogas slurry has been studied.

Biogas slurry is the residue tail liquid after anaerobic fermentation of biomass, which has the function of promoting the growth and reproduction of animals and plants [8] and killing insects because it is rich in a variety of nutrients and ammonia nitrogen [9]. Biogas slurry has a remarkable effect on the control of aphids and red spiders in fruits and vegetables [10]. The full and rational utilization of biogas slurry can produce huge economic benefits. Until now, many areas have begun to directly use clean and low-cost biogas slurry to kill insects, which results in wasting some nutrients and is less efficient than other synthetic pesticides [11]. The common practice is to mix biogas slurry with other pesticides to further improve insecticidal efficiency [12]. This still does not solve the problem of nutrient waste. Slow-release pesticides are currently the more popular method [13] to improve the utilization efficiency of pesticides and reduce the pollution to the environment via reducing pesticide dosages, pesticide spraying frequency and prolonging the effective duration [14]. The combination of biogas slurry and slow-release technology is an interesting research direction, and there have been no such reports yet. To date, a variety of slow-release types have been developed, including microcapsule, inclusion, homogeneous and adsorption [15]. Among these, adsorption is widely used because of its low cost and easy operation.

A lot of slow-release carriers have been developed, such as nanoparticles [16], polymer material [17], bentonite and zeolite. Cation exchange resin can effectively adsorb cations from a solution and slowly release the adsorbed substances under certain conditions as a common adsorbent [18]. Chen et al. found that the maximum exchange capacity of lead (II) adsorbed by lignin-based cation exchange resin was 2.26 mmol/g [19], which was comparable to a phenol cation exchange resin. Daisuke et al. found a new method to remove Se(IV) using an Fe^{3+} ion exchange resin [20], and the Se was effectively removed using a column packed with 12.8 g (10.4 cm^3) of adsorbent. Leng et al. used a cation exchange resin to remove iron from phosphoric acid and determined that the quasi-second-order kinetic model perfectly described the adsorption process [21]. Goutham R. et al. developed a polymer matrix film loaded with a combination of free diclofenac sodium and an ion exchange resin complex to achieve an immediate and slow release [22]. Atyabi et al. developed an effective sustained release system in which bicarbonate and other drugs are loaded on ion exchange resins [23], and the cation exchange resin can also efficiently adsorb ammonia nitrogen such that the efficiency is more than 90% [24]. To sum up, cation exchange resin has great potential for adsorbing ammonia nitrogen from biogas slurry to produce sustained release insecticides due to the high-efficiency adsorption of ammonia nitrogen.

In this study, D113 (resin type) macroporous weak acid acrylic cation exchange resin was selected, which is easier to regenerate than strongly acidic cation exchange resin and has good ion exchange performance (the schematic diagram of adsorption and release is shown in Figure 1). The adsorption capacity of the resin for ammonia nitrogen under different conditions was studied by the static adsorption method. The adsorption process and type were determined by fitting, and the effects of adsorption of ammonia nitrogen in the ammonium chloride solution and biogas slurry were compared. Slow-release studies were carried out. These revealed the correlation between the resin dosage, ammonia nitrogen concentration, adsorption time, adsorption temperature, pH value and ammonia nitrogen adsorption capacity and the characteristics of a slow release. The adsorption process parameters of the D113 macroporous weak acid acrylic cation exchange resin were optimized, which could provide scientific reference for the further research and development of technologies of efficient and environmentally friendly slow-release ecological insect repellent. The experimental results provide technical support for the harmless treatment of biogas slurry, and the use of pesticides will be reduced in the process of agricultural production, which is beneficial to the clean production of green agriculture and the health of humans and livestock.

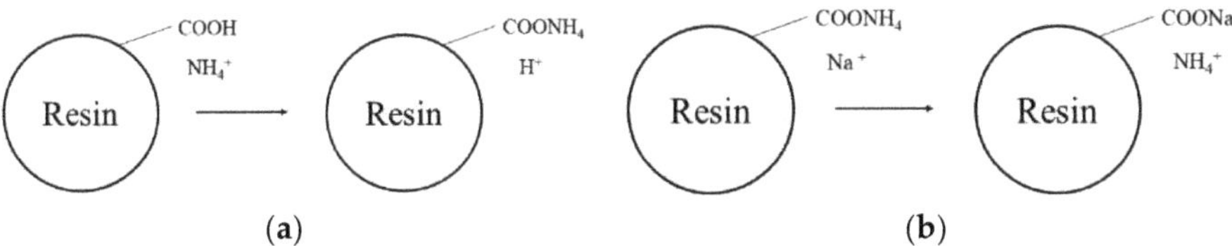

Figure 1. (**a**) Adsorption process. (**b**) Slow-release process.

2. Material and Methods

2.1. Experimental Materials

The D113 macroporous weak acid acrylic cation exchange resin (Shanxi Lanshen Special Resin Co. Ltd., Xian, China) was selected for this study, and the molecular structure diagram is shown in Figure 2. The resin was soaked in a 1.8 mol/L NaCl solution for 24 h and washed with ammonia-free water. Then, the resin was soaked in a 0.5 mol/L HCl solution for 10 h to remove inorganic impurities and repeatedly washed with ammonia-free water to neutrality. After being soaked in a 1 mol/L NaOH solution for 10 h to remove organic impurities, the resin was washed repeatedly with ammonia-free water to neutrality and dried in a drying oven (Beijing Zhongxi Yuanda Technology Co., Ltd., Shenyang, China) at 55 °C for use.

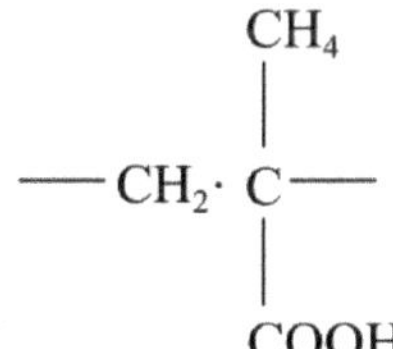

Figure 2. Schematic diagram of the molecular structure of the D113 resin.

The biogas slurry was obtained from the anaerobic reactor using straw and pig manure as raw materials, and the ammonia nitrogen concentration was 432.43 mg/L.

2.2. Standard Curve Drawing

The colorimetric tubes were filled with different volumes of an ammonia nitrogen standard solution (0 mL, 0.5 mL, 1 mL, 2 mL, 4 mL, 6 mL, 8 mL, and 10 mL) and ammonia-free water to 50 mL. a potassium sodium tartrate solution (Tianjin Yongda Chemical Reagent Company Limited, Tianjin, China) (1 mL) and Nessler's reagent (Merck KGaA, Darmstadt, Germany) (1 mL) were added and then shaken well. The absorbance was measured at 420 nm with a spectrophotometer after developing color for 15 min. The standard curve regression equation was observed. As shown in Figure 3.

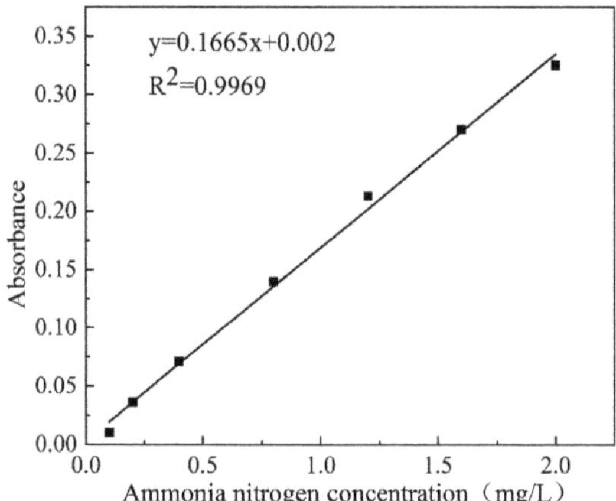

Figure 3. Standard curve.

2.3. Experimental Scheme

2.3.1. Resin Static Adsorption Test

In order to study the effects of the resin addition, ammonia nitrogen concentration, time, temperature and pH value on the ammonia nitrogen adsorption, a certain quality of ammonium chloride (Tianjin Kemiou Chemical Reagent Co., Ltd., Tianjin, China) was weighed with an electronic balance according to the experimental requirements (100, 500, 1000, 1500, 2000, 2500, and 3000 mg/L) and placed in a 100-mL Erlenmeyer flask. Ammonia-free water (50 mL) was added to fully dissolve it. The pH value of the solution was adjusted as required by the experiment (4, 5, 6, 7, 8, 9, and 10), and the required resin (0.1, 0.2, 0.3, 0.4, 0.5, 0.6, and 0.7 g) was weighed into the Erlenmeyer flask. After being sealed with a rubber stopper, the Erlenmeyer flask was shaken at 150 r/min in a constant temperature shaker (Shanghai Changken Test Equipment Co., Ltd., Shanghai, China) with different temperatures (20, 30, 40, 50, and 60 °C) and adsorption times (5, 10, 15, 20, 25, 30, 35, 40, 45, 50, 55, and 60 min).

2.3.2. Effect of Different NaCl Concentrations on the Slow Release of Ammonia Nitrogen

A sodium chloride solution (200 mg/L, 1000 mg/L, 2000 mg/L, 4000 mg/L, and 10,000 mg/L; 50 mL) was added to 100-mL Erlenmeyer flasks, and then 0.5 g of resin containing ammonia nitrogen (10 mg / g) was put into the Erlenmeyer flasks. All reactors remained still under room temperature. Samples were taken at 10 min, 30 min, 60 min, 120 min, 180 min, 240 min, 300 min, 360 min, 420 min, 480 min, 540 min, and 600 min, respectively, and the absorbance was measured.

2.3.3. Adsorption of Ammonia Nitrogen from the Biogas Slurry by the Resin

The biogas slurry was centrifuged, and the concentration of ammonia nitrogen without adsorption was measured. The supernatant (50 mL) was taken into the Erlenmeyer flask, and another Erlenmeyer flask contained the ammonium chloride solution (50 mL) with the same concentration of ammonia nitrogen as the biogas slurry. The adsorption experiments were conducted at a resin addition amount of 0.5 g, adsorption temperature of 30 °C, adsorption pH value of 7 and shaker speed of 150 r/min. Samples were taken at 5, 10, 15, 20, 25, 30, 35, 40, 45, and 50 min. The absorbance was measured. All experiments were performed in triplicate, and the average value of the experimental data was used as the result. The experimental procedures are shown in Figure 4.

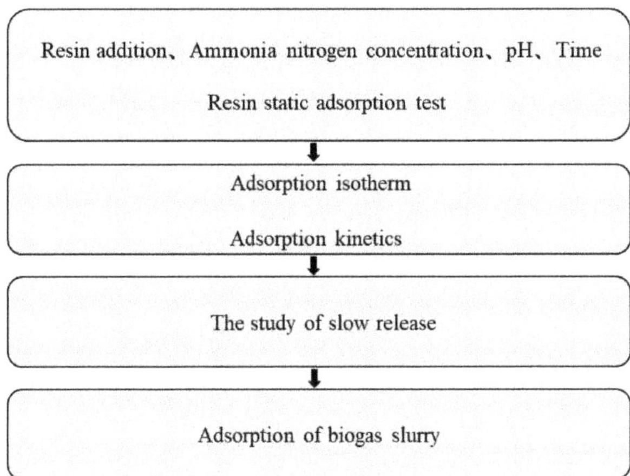

Figure 4. Schematic of the experimental procedures.

2.4. Determination of Ammonia Nitrogen by Nessler's Reagent Spectrophotometer

The supernatant of the ammonium chloride solution adsorbed by the resin was put into a 50-mL colorimetric tube by a liquid transfer gun. Ammonia-free water was added to 50 mL to dilute the solution. A potassium sodium tartrate solution (1 mL) and Nessler's reagent (1 mL) were added to the colorimetric tube and then shaken well. Ammonia nitrogen would react with Nessler's reagent to form a reddish complex. After developing color for 15 min, the absorbance of the solution was measured at 420 nm with a spectrophotometer, and the ammonia nitrogen concentration was calculated.

2.5. Isotherm Curve Fitting

The Langmuir and Freundlich adsorption isotherm equations [25] (Equations (1) and (2)) were used to simulate the data of ammonia nitrogen adsorption at different temperatures:

$$\frac{C_e}{Q_e} = \frac{1}{K_b Q_m} + \frac{C_e}{Q_m} \tag{1}$$

where Q_e is the adsorption capacity of the resin at equilibrium in mg/g, Q_m is the theoretical maximum adsorption capacity of the ion exchange resin in mg/g, C_e is the ammonia nitrogen concentration at the adsorption equilibrium in mg/mL and K_b is the equilibrium constant:

$$lnQ_e = lnK_f + \frac{1}{n}lnC_e \tag{2}$$

where Q_e is the adsorption capacity of the resin at equilibrium in mg/g, C_e is the ammonia nitrogen concentration at the adsorption equilibrium in mg/mL, K_f is the Freundlich constant and n is the empirical constant.

2.6. Adsorption Kinetic Model Fitting

Two kinetic equations (Equations (3) and (4)) were used to fit the data of the ammonia nitrogen adsorption under different adsorption times:

$$ln(Q_e - Q_t) = lnQ_e - K_1 t \tag{3}$$

(quasi-first-order kinetic equation)

where Q_e is the adsorption capacity of the resin at equilibrium in mg/g, Q_t is the adsorption capacity of the resin at moment t in mg/g and K_1 is the quasi-first-order kinetic rate constant:

$$\frac{t}{Q_t} = \frac{1}{K_2 Q_e^2} + \frac{t}{Q_e}$$

(4)

(quasi-second-order kinetic equation) where Q_e is the ammonia nitrogen adsorption capacity by the resin at equilibrium in mg/g, Q_t is the ammonia nitrogen adsorption capacity by the resin at moment t in mg/g and K_2 is the quasi-second-order kinetic rate constant.

2.7. Statistical Analysis

The experimental data were analyzed by ANOVA (α = 0.05) using Mintab19 software (Mintab19, Pennsylvania State University, State College, PA, USA, 2019).

3. Results and Discussion

3.1. Effect of Resin Addition on Ammonia Nitrogen Adsorption

The adsorption test was performed with resin addition amounts of 0.05 g, 0.1 g, 0.2 g, 0.3 g, 0.4 g, 0.5 g, and 0.6 g under an ammonium chloride concentration of 100 mg/L, adsorption time of 30 min, adsorption temperature of 30 °C, pH value of 7 and shaker speed of 150 r/min. The absorbance was measured using the Nessler's reagent spectrophotometer method, and then the adsorption capacity of ammonia nitrogen and resin unit adsorption capacity were calculated. The results are shown in Figure 5a.

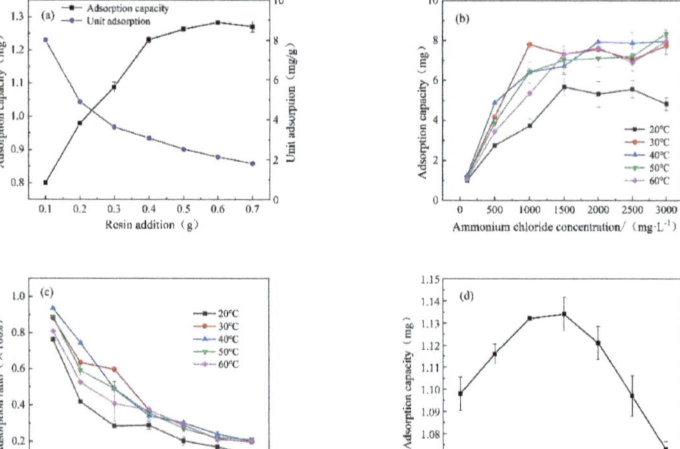

Figure 5. *Cont.*

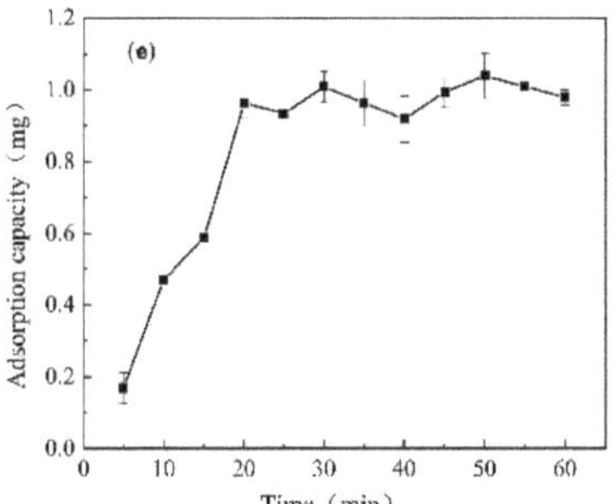

Figure 5. The effect of (**a**) resin addition on the ammonia nitrogen adsorption capacity, (**b**) the initial ammonia nitrogen concentration on the ammonia nitrogen adsorption capacity and (**c**) the ammonia nitrogen adsorption ratio, (**d**) pH and (**e**) time on the ammonia nitrogen adsorption capacity.

As shown in Figure 5a, the ammonia nitrogen adsorption capacity increased with the increase in the resin addition, but the unit adsorption capacity decreased. When 0.1 g of resin was added, a total of 0.79 mg of ammonia nitrogen was adsorbed, accounting for 61% of the total ammonia nitrogen, and the unit adsorption capacity of the resin reached its highest value (7.98 mg/g). The efficiency was higher than those of other slow-release carriers [26]. When the resin addition increased to 0.4 g, 1.22 mg of ammonia nitrogen was adsorbed, accounting for 94% of the total ammonia nitrogen, which reached a higher value, but the unit adsorption capacity dropped to 3.07 mg/g. When the amount of resin continued to increase, the adsorption effect slowed down significantly. At the resin addition amount of 0.6 g, 1.28 mg of ammonia nitrogen was adsorbed, accounting for 98% of the total ammonia nitrogen. However, the unit adsorption capacity decreased to 2.13 mg/g. The unit adsorption capacity decreased to 1.79 mg/g when the resin addition was 0.7 g. This adsorption law was consistent with the research results by Zhuang [27]. This may be due to the number of ammonia nitrogen exchange sites increasing with the increasing resin amount [28], and the unit utilization efficiency of the resin decreased with the increase in the resin content, which may have been due to the fact that the ammonia nitrogen could not completely contact the resin. After comprehensive consideration, the amount of resin selected was 0.5 g, which showed a significant ($p < 0.05$) difference from the other addition amounts.

3.2. Effect of the Ammonia Nitrogen Concentration on Ammonia Nitrogen Adsorption by the Resin at Different Temperatures

The ammonium chloride solution concentration was set to 100 mg/L, 500 mg/L, 100 mg/L, 1500 mg/L, 2000 mg/ L, 2500 mg/L, and 3000 mg/L, and the adsorption tests were performed at 20 °C, 30 °C, 40 °C, 50 °C, and 60 °C, respectively, under a resin addition amount of 0.5 g, adsorption time of 30 min, pH value of 7 and shaker speed 150 r/min. The absorbance was measured, and then the ammonia nitrogen adsorption capacity and adsorption ratio were calculated. The results are shown in Figure 5b,c.

As shown in Figure 5b, when the concentration of ammonium chloride was 100 mg/L, the adsorption capacity of the resin reached its minimum value. The adsorption capacity increased gradually and then tended to be stable with the increase in the ammonium chloride concentration because the number of adsorption sites of the resin was constant [29]. As the concentration of the ammonia nitrogen increased, the adsorption sites decreased and finally reached saturation and adsorption equilibrium. When the temperature was 20 °C, the resin had the lowest adsorption capacity, and the maximum adsorption capacity was 5.67 mg. The adsorption capacities at 30 °C, 40 °C, 50 °C and 60 °C were very close; the maximum adsorption capacities were 7.80 mg, 8.04 mg, 8.34 mg and 7.89 mg, respectively. There was a significant ($p < 0.05$) difference was observed between 20 °C and the other temperatures. Because the resin adsorption process is an endothermic reaction [30], when the temperature increases, the adsorption capacity of the resin increases and finally tends to be stable. According to Figure 5c, the adsorption rate fell off gradually from 88% to 19% with the increase in the ammonia nitrogen concentration at 30 °C.

3.3. Effect of the pH Value on Ammonia Nitrogen Adsorption

The adsorption test was performed at pH values of 4, 5, 6, 7, 8, 9 and 10 with the condition of a resin addition amount of 0.5 g, ammonium chloride concentration of 100 mg/L, adsorption time of 30 min, adsorption temperature of 30 °C and shaker speed of 150 r/min. The absorbance was measured using the Nessler's reagent spectrophotometer method, and the adsorption capacity was calculated. The results are shown in Figure 5d.

As shown in Figure 5d, the ammonia nitrogen adsorption capacity by the resin increased first and then decreased with the increase in the pH value. The initial pH value had a significant influence ($p < 0.05$) on the ammonia nitrogen adsorption capacity. The ammonia nitrogen adsorption capacity reached its maximum value (1.13 mg) when the pH value was 7. This was because when the solution was acidic, the concentration of H^+ increased, which competed with NH_4^+ for the adsorption sites on the resin. The higher the concentration of the H^+ was, the fiercer the competition with the NH_4^+ was, resulting in a reduction of the ammonia nitrogen adsorption capacity. When the solution was alkaline, NH_4^+ would combine with OH^- to become free ammonia [NH_3]. The method of resin adsorption became physical adsorption, and the adsorption capacity was weak. Hence, with the increase in the pH value, free ammonia [NH_3] would increase, leading to a lower ammonia nitrogen adsorption capacity for the resin [31]. The optimal pH value of the resin for ammonia nitrogen adsorption was seven.

3.4. Effect of the Adsorption Time on Ammonia Nitrogen Adsorption

The adsorption experiment was carried out under the condition of a resin addition amount of 0.5 g, ammonium chloride concentration of 100 mg/L, adsorption temperature of 30 °C, pH value of 7 and shaker speed of 150 r/min. Samples were taken at 5, 10, 15, 20, 25, 30, 35, 40, 45 50, 55 and 60 min. After the absorbance was measured using the Nessler's reagent spectrophotometer method, the ammonia nitrogen adsorption capacity was calculated. The results are shown in Figure 5e.

As shown in Figure 5e, when the time increased, the adsorption capacity gradually increased within 5–20 min. The maximum adsorption capacity reached 0.96 mg at 20 min, and the adsorption capacity was in a dynamic equilibrium at 20–60 min. This result from our study was consistent with the result of Zhu [32]. This may be because the number of adsorption sites decreased when the time increased, and the concentration of Na^+ gradually increased. After 20 min, the NH_4^+ and Na^+ competed with each other for adsorption, which caused the ammonia nitrogen concentration to form a stable fluctuation phenomenon. In order to ensure that the adsorption capacity reached the maximum, 30 min was selected as the adsorption time.

3.5. Adsorption Isotherm

The data from the experiment of the effect of the ammonia nitrogen concentration on the ammonia nitrogen adsorption at different temperatures were fitted by the Langmuir and Freundlich adsorption isotherm equations (Equations (1) and (2)) [33]. The isotherm adsorption curve was drawn, and the relevant parameters were calculated. The results are shown in Figure 6 and Table 1.

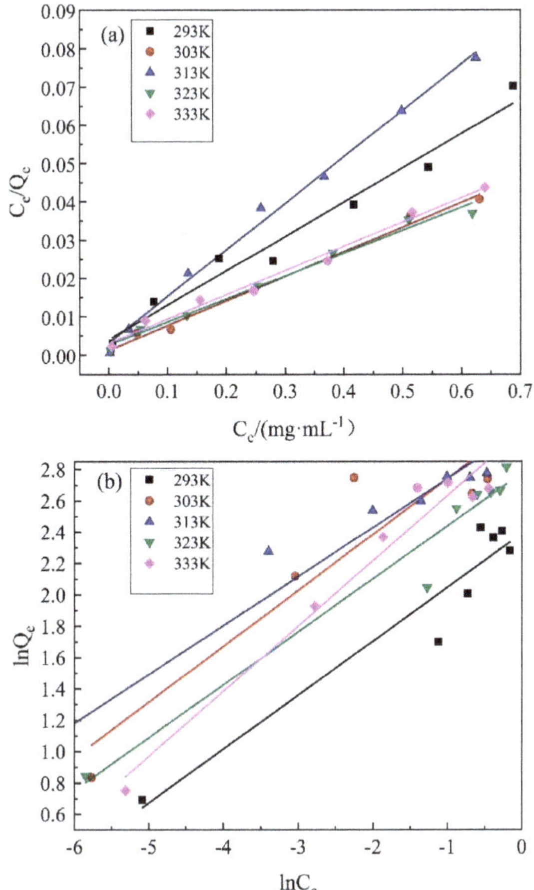

Figure 6. Adsorption isotherm. (**a**) Langmuir isotherms. (**b**) Freundlich isotherms.

Table 1. Fitting parameters of the adsorption isotherm.

Temperature (K)	Langmuir			Freundlich		
	Q_m (mg/g)	K_b (mL/mg)	R^2	K_f	n	R^2
293	11.1732	21.8292	0.9704	10.9091	2.9121	0.9182
303	15.5521	49.4615	0.9932	22.1447	2.8066	0.8836
313	16.5563	36.5294	0.9949	21.1915	3.2041	0.9462
323	16.7224	22.1482	0.9856	16.0691	2.9630	0.9605
333	16.7224	16.1622	0.9811	21.6932	2.3629	0.9682

According to Figure 6 and Table 1, the correlation coefficients R^2 of the Langmuir isothermal equation fitting at temperatures of 293 K, 303 K, 313 K, 323 K and 333 K were 0.9704, 0.9932, 0.9949, 0.9856 and 0.9811, respectively, and the correlation coefficients R^2 of the Freundlich isothermal equation fitting were 0.9182, 0.8836, 0.9462, 0.9605 and 0.9682, respectively. The Langmuir isothermal equation was more suitable for describing the adsorption process, which was single-molecule layer adsorption.

3.6. Adsorption Kinetics

According to the data obtained from the test of the effect of the adsorption time on ammonia nitrogen adsorption, two kinetic equations were used for the fitting curve (Equations (3) and (4)) [34]. Then, the fitting curve was drawn, and the relevant parameters were calculated. The results are shown in Figure 7 and Table 2.

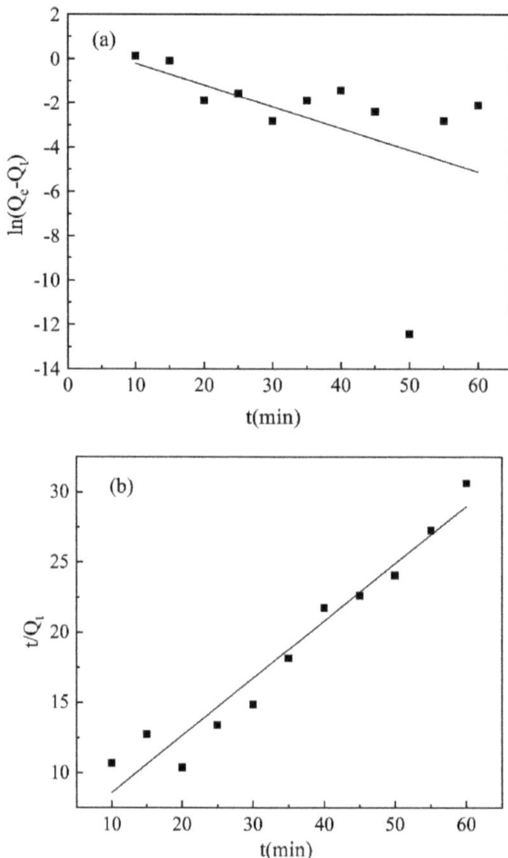

Figure 7. Adsorption kinetics. (**a**) Fitting curve of the quasi-first order kinetic equation. (**b**) Fitting curve of the quasi-second-order kinetic equation.

Table 2. Fitting parameters of the kinetic equations.

Quasi-First Order Kinetic Equation			Quasi-Second-Order Kinetic Equation		
Q_e mg/g	K_1 min^{-1}	R^2	Q_e mg/g	K_2 g/(mg·min)	R^2
2.1578	0.0981	0.2325	2.4510	0.03699	0.9496

According to Figure 7 and Table 2, the correlation coefficient R^2 fitted by the quasi-first-order kinetic equation was 0.2325, and the correlation coefficient R^2 fitted by the quasi-second-order kinetic equation was 0.9496. The ammonia nitrogen adsorption process by the resin was more in line with the quasi-second-order kinetic equation.

3.7. Effect of Different NaCl Concentrations on the Slow Release of Ammonia Nitrogen

The adsorption test was performed under the conditions of a resin addition amount of 3 g, ammonium chloride concentration of 1500 mg/L, adsorption temperature of 30 °C, adsorption pH value of 7, adsorption time of 30 min and shaker speed of 150 r/min. After the adsorption was completed, the resin was separated by filtration. The resin was placed in a drying box for drying to prepare a resin loaded with ammonia nitrogen. The NaCl solutions with concentrations of 200 mg/L, 1000 mg/L, 2000 mg/L, 4000 mg/L and 10,000 mg/L (50 mL) were configured in Erlenmeyer flasks of 100 mL. The resin loaded with ammonia nitrogen (0.5 g) was placed in an Erlenmeyer flask and remained still. Samples were taken at 10 min, 30 min, 60 min, 120 min, 180 min, 240 min, 300 min, 360 min, 420 min, 480 min, 540 min and 600 min, respectively. The absorbance was measured by the Nessler's reagent spectrophotometer method before the ammonia nitrogen slow-release amount was calculated. The results are shown in Figure 8.

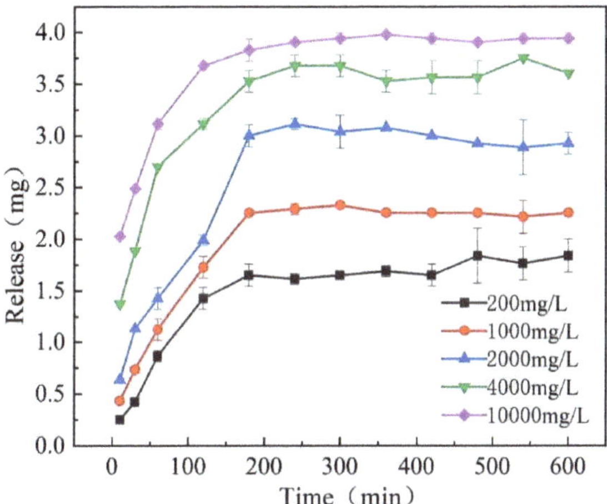

Figure 8. The effect of different NaCl concentrations on the slow release of ammonia nitrogen.

Figure 8 shows that when the concentration of the NaCl solution was 200 mg/L, the slow-release amount of the resin was the lowest, reaching 2 mg in the stable stage and accounting for 40% of the total. At 10,000 mg/L, the slow-release amount of the resin reached 4 mg at the stable stage and accounted for 80% of the total. The slow-release amount had a significant difference ($p < 0.05$) between each concentration. The slow-release amount increased with the increase in concentration of the NaCl solution. In the initial stage of slow release, the efficiency was the highest, and then it gradually decreased to a stabilized value. The reason for this was that during the initial stage of slow release, more Na^+ could be exchanged with the NH_4^+ on the resin. As the exchange continued, the exchange sites continued to decrease, and the exchange rate gradually slowed down. The concentration of the NaCl solution was flexibly adjusted according to the specific slow-release requirements and the salinity tolerance of the crops to achieve better results.

3.8. Comparison of the Adsorption Ammonium Nitrogen Effect of Resin from Biogas Slurry and the Ammonia Chloride Solution

The biogas slurry was centrifuged before 50 mL of supernatant was taken. Then, the ammonia nitrogen concentration without adsorption was scaled. The other Erlenmeyer flask contained 50 mL of the ammonium chloride solution with the same concentration of ammonia nitrogen as the biogas slurry. The adsorption test was performed under the conditions of a resin addition amount of 0.5 g, adsorption temperature of 30 °C, adsorption pH value of 7 and shaker speed of 150 r/min. Samples were taken at 5, 10, 15, 20, 25, 30, 35, 40, 45 and 50 min. After the absorbance was measured by Nessler's reagent spectrophotometer method, the ammonia nitrogen adsorption capacity was calculated. The results are shown in Figure 9.

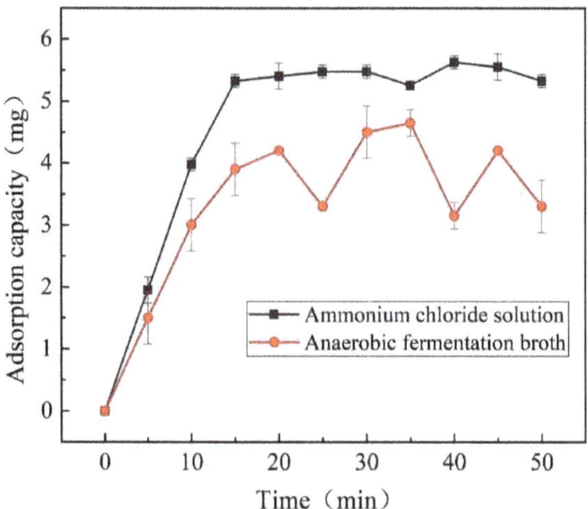

Figure 9. Comparison of resin adsorption capacities from the biogas slurry and ammonia chloride solution.

Figure 9 shows that the maximum adsorption capacity by the resin from the biogas slurry was 4.8 mg, and the maximum adsorption capacity in the ammonia chloride solution was 5.5 mg. There was a significant difference ($p < 0.05$) in the adsorption capacity. The ability of resin to adsorb ammonia nitrogen from the biogas slurry significantly reduced. In the process of adsorption of ammonia nitrogen in biogas slurry, the ammonia nitrogen concentration fluctuated greatly in the later period. This was because the biogas slurry contained a large number of interfering ions such as K ions, which would preempt the exchange site with NH_4^+, resulting in a decrease in the adsorption capacity of ammonia nitrogen [35]. When the ammonia nitrogen concentration in the solution dropped to a certain value, the relative concentration of K^+ increased, and the exchange capacity of K^+ was better than that of NH_4^+. The ammonia nitrogen adsorbed by the resin was exchanged, resulting in an increase of the ammonia nitrogen concentration. Then, the adsorption capacity of the resin for NH_4^+ was better than K^+ again, and from there, the concentration of ammonia nitrogen entered a state of fluctuation.

4. Conclusions

The effects and characteristics of adsorption and the slow release of ammonia nitrogen by D113 resin under different conditions were studied. The results showed that increasing the initial ammonia concentration, resin addition amount, temperature and time could improve the adsorption capacity of ammonia nitrogen, and the optimum conditions for adsorption were comprehensively analyzed. Under acidic conditions, hydrogen ions would compete with ammonium ions for adsorption sites, and under alkaline conditions, hydroxide ions would combine with ammonium ions in the solution to form free nitrogen, which greatly affected the adsorption capacity of the resin for ammonia nitrogen. Therefore, the optimal pH value is seven. The adsorption process for ammonia nitrogen is an endothermic process. The adsorption capacity will increase with the increase in temperature and become stable after reaching 30 °C.

The process of ammonia nitrogen adsorption was in accordance with the Langmuir isothermal curve and the quasi-second-order kinetic model. The process of adsorbing ammonia nitrogen is monolayer adsorption and a combination of multiple diffusions. The composition of biogas slurry is very complex, containing a large amount of organic matter and some metal ions, which will affect the adsorption process. Thus, the adsorption efficiency and adsorption capacity of resin for ammonia nitrogen in biogas slurry were lower than those in the NH_4Cl solution. Aside from that, the release of ammonia nitrogen can be controlled by adjusting the solubility of the sodium chloride solution. In the process of application, the concentration of the sodium chloride solution can be selected according to the actual situation and the properties of the crop itself. The aim of this study was to solve the problem of a low utilization rate of biogas slurry. The results provide theoretical and practical support for the harmless treatment of biogas slurry and agricultural safety production. The focus of the next research is to further improve the adsorption efficiency of the resin for ammonia nitrogen in biogas slurry and to optimize the slow-release effect of slow-release insecticides.

Author Contributions: Conceptualization, Q.Z. and Z.L.; methodology, Z.L.; software, Z.L.; formal analysis, Z.L.; investigation, Z.L.; data curation, Z.L.; writing—original draft preparation, Z.L.; writing—review and editing, C.L., Y.L. and Z.Z.; supervision, V.P. and F.P.; project administration, H.Z.; funding acquisition, Q.Z. All authors have read and agreed to the published version of the manuscript.

Funding: The present study was financed by the National Key R&D Program of China (2018YFE0206600), National Natural Science Foundation of China (52076068), and China Postdoctoral Science Foundation (2020M681069).

Institutional Review Board Statement: Not applicable.

Informed Consent Statement: Not applicable.

Data Availability Statement: The data are not publicly available due to the need for further research work.

Conflicts of Interest: We declare that we have no conflict of interest to this work. We declare that we do not have any commercial or associative interest that represents a conflict of interest in connection with the work submitted.

References

1. Perlatti, B.; de Souza Bergo, P.L.; da Silva, M.F.D.G.F.; Batista, J.; Rossi, M. Polymeric Nanoparticle-Based Insecticides: A Controlled Release Purpose for Agrochemicals. *Insectic. Dev. Safer More Eff. Technol.* **2013**, 523–550. [CrossRef]
2. Kaonga, C.C.; Takeda, K.; Sakugawa, H. Concentration and degradation of alternative biocides and an insecticide in surface waters and their major sinks in a semi-enclosed sea, Japan. *Chemosphere* **2016**, *145*, 256–264. [CrossRef] [PubMed]
3. Mattos, B.; Rojas, O.; Magalhaes, W. Biogenic silica nanoparticles loaded with neem bark extract as green, slow-release biocide. *J. Clean. Prod.* **2017**, *142*, 4206–4213. [CrossRef]
4. Ghormade, V.; Deshpande, M.V.; Paknikar, K. Perspectives for nano-biotechnology enabled protection and nutrition of plants. *Biotechnol. Adv.* **2011**, *29*, 792–803. [CrossRef] [PubMed]
5. Colson, A.O.; Besler, B.; Close, D.M.; Sevilla, M.D. Ab initio molecular orbital calculations of DNA bases and their radical ions in various protonation states: Evidence for proton transfer in GC base pair radical anions. *J. Phys. Chem.* **1992**, *96*, 661–668. [CrossRef]
6. Bollag, J.-M.; Myers, C.J.; Minard, R.D. Biological and chemical interactions of pesticides with soil organic matter. *Sci. Total. Environ.* **1992**, *123–124*, 205–217. [CrossRef]
7. Joshi, P.P.; Van Cleave, A.; Held, D.W.; Howe, J.A.; Auad, M.L. Preparation of slow release encapsulated insecticide and fertilizer based on superabsorbent polysaccharide microbeads. *J. Appl. Polym. Sci.* **2020**, *137*, 49177. [CrossRef]
8. Wang, P.; Zhang, X.; Gouda, S.G.; Yuan, Q. Humidification-dehumidification process used for the concentration and nutrient recovery of biogas slurry. *J. Clean. Prod.* **2019**, *247*, 119142. [CrossRef]
9. Shi, M.; He, Q.; Feng, L.; Wu, L.; Yan, S. Techno-economic evaluation of ammonia recovery from biogas slurry by vacuum membrane distillation without pH adjustment. *J. Clean. Prod.* **2020**, *265*, 121806. [CrossRef]
10. Ma, H. A Brief Analysis on Comprehensive Utilization Strategies of Biogas Slurry. *NEBEFS* **2012**, *36*, 5.
11. Baştabak, B.; Koçar, G. A review of the biogas digestate in agricultural framework. *J. Mater. Cycles Waste Manag.* **2020**, *22*, 1318–1327. [CrossRef]

12. You, L.; Yu, S.; Liu, H.; Wang, C.; Zhou, Z.; Zhang, L.; Hu, D. Effects of biogas slurry fertilization on fruit economic traits and soil nutrients of Camellia oleifera Abel. *PLoS ONE* **2019**, *14*, e0208289. [CrossRef]
13. Wang, L.; Yu, G.; Li, J.; Feng, Y.; Peng, Y.; Zhao, X.; Tang, Y.; Zhang, Q. Stretchable hydrophobic modified alginate double-network nanocomposite hydrogels for sustained release of water-insoluble pesticides. *J. Clean. Prod.* **2019**, *226*, 122–132. [CrossRef]
14. Margulis-Goshen, K.; Magdassi, S. Nanotechnology: An Advanced Approach to the Development of Potent Insecticides. *Adv. Technol. Manag. Insect Pests* **2012**, 295–314. [CrossRef]
15. Pang, W.; Hou, D.; Wang, H.; Sai, S.; Wang, B.; Ke, J.; Wu, G.; Li, Q.; Holtzapple, M. Preparation of Microcapsules of Slow-Release NPK Compound Fertilizer and the Release Characteristics. *J. Braz. Chem. Soc.* **2018**, *29*, 2397–2404. [CrossRef]
16. Bramhanwade, K.; Shende, S.; Bonde, S.; Gade, A.; Rai, M. Fungicidal activity of Cu nanoparticles against Fusarium causing crop diseases. *Environ. Chem. Lett.* **2015**, *14*, 229–235. [CrossRef]
17. Zhuang, C.; Shi, C.; Tao, F.; Cui, Y. Honeycomb structural composite polymer network of gelatin and functional cellulose ester for controlled release of omeprazole. *Int. J. Biol. Macromol.* **2017**, *105*, 1644–1653. [CrossRef] [PubMed]
18. Li, Q.; Fu, L.; Wang, Z.; Li, A.; Shuang, C.; Gao, C. Synthesis and characterization of a novel magnetic cation exchange resin and its application for efficient removal of Cu^{2+} and Ni^{2+} from aqueous solutions. *J. Clean. Prod.* **2017**, *165*, 801–810. [CrossRef]
19. Chen, X.; Sun, S.; Wang, X.; Wen, J.; Wang, Y.; Cao, X.; Yuan, T.; Wang, S.; Shi, Q.; Sun, R. One-pot preparation and characterization of lignin-based cation exchange resin and its utilization in Pb (II) removal. *Bioresour. Technol.* **2019**, *295*, 122297. [CrossRef] [PubMed]
20. Kawamoto, D.; Yamanishi, Y.; Ohashi, H.; Yonezu, K.; Honma, T.; Sugiyama, T.; Kobayashi, Y.; Okaue, Y.; Miyazaki, A.; Yokoyama, T. A new and practical Se(IV) removal method using Fe^{3+} type cation exchange resin. *J. Hazard. Mater.* **2019**, *378*, 120593. [CrossRef]
21. Leng, X.; Zhong, Y.; Xu, D.; Wang, X.; Yang, L. Mechanism and kinetics study on removal of Iron from phosphoric acid by cation exchange resin. *Chin. J. Chem. Eng.* **2018**, *27*, 1050–1057. [CrossRef]
22. Adelli, G.R.; Balguri, S.P.; Bhagav, P.; Raman, V.; Majumdar, S. Diclofenac sodium ion exchange resin complex loaded melt cast films for sustained release ocular delivery. *Drug Deliv.* **2017**, *24*, 370–379. [CrossRef] [PubMed]
23. Atyabi, F.; Sharma, H.; Mohammad, H.; Fell, J. Controlled drug release from coated floating ion exchange resin beads. *J. Control. Release* **1996**, *42*, 25–28. [CrossRef]
24. Tarpeh, W.A.; Udert, K.M.; Nelson, K.L. Comparing Ion Exchange Adsorbents for Nitrogen Recovery from Source-Separated Urine. *Environ. Sci. Technol.* **2017**, *51*, 2373–2381. [CrossRef] [PubMed]
25. Aljerf, L. High-efficiency extraction of bromocresol purple dye and heavy metals as chromium from industrial effluent by adsorption onto a modified surface of zeolite: Kinetics and equilibrium study. *J. Environ. Manag.* **2018**, *225*, 120–132. [CrossRef] [PubMed]
26. Cheng, H.; Zhu, Q.; Xing, Z. Adsorption of ammonia nitrogen in low temperature domestic wastewater by modification bentonite. *J. Clean. Prod.* **2019**, *233*, 720–730. [CrossRef]
27. Zhuang, H.; Zhong, Y.; Yang, L. Adsorption equilibrium and kinetics studies of divalent manganese from phosphoric acid solution by using cationic exchange resin. *Chin. J. Chem. Eng.* **2020**, *28*, 2758–2770. [CrossRef]
28. Kim, J.; Park, C.W.; Lee, K.-W.; Lee, T.S. Adsorption of Ethylenediaminetetraacetic Acid on a Gel-Type Ion-Exchange Resin for Purification of Liquid Waste Containing Cs Ions. *Polymers* **2019**, *11*, 297. [CrossRef] [PubMed]
29. Jatoi, A.S.; Baloch, H.A.; Mazari, S.A.; Mubarak, N.M.; Sabzoi, N.; Aziz, S.; Soomro, S.A.; Abro, R.; Shah, S.F. A review on extractive fermentation via ion exchange adsorption resins opportunities, challenges, and future prospects. *Biomass Convers. Biorefin.* **2021**, 1–12. [CrossRef]
30. Qiu, M.; Hu, C.; Liu, J.; Chen, C.; Lou, X. Removal of High Concentration of Ammonia from Wastewater by the Ion Exchange Resin. *Nat. Environ. Pollut. Technol.* **2017**, *16*, 261–264.
31. Ren, Z.; Jia, B.; Zhang, G.; Fu, X.; Wang, Z.; Wang, P.; Lv, L. Study on adsorption of ammonia nitrogen by iron-loaded activated carbon from low temperature wastewater. *Chemosphere* **2020**, *262*, 127895. [CrossRef] [PubMed]
32. Zhu, X.; Li, W.; Zhang, C. Extraction and removal of vanadium by adsorption with resin 201 * 7 from vanadium waste liquid. *Environ. Res.* **2019**, *180*, 108865. [CrossRef] [PubMed]
33. Bhattacharya, A.; Naiya, T.K.; Mandal, S.; Das, S. Adsorption, kinetics and equilibrium studies on removal of Cr(VI) from aqueous solutions using different low-cost adsorbents. *Chem. Eng. J.* **2008**, *137*, 529–541. [CrossRef]
34. Tang, X.; Li, Z.; Chen, Y. Adsorption behavior of Zn(II) on calcinated Chinese loess. *J. Hazard. Mater.* **2009**, *161*, 824–834. [CrossRef] [PubMed]
35. Ranjan, R.; Thust, S.; Gounaris, C.; Woo, M.; Floudas, C.A.; von Keitz, M.; Valentas, K.J.; Wei, J.; Tsapatsis, M. Adsorption of fermentation inhibitors from lignocellulosic biomass hydrolyzates for improved ethanol yield and value-added product recovery. *Microporous Mesoporous Mater.* **2009**, *122*, 143–148. [CrossRef]

processes

Article

Response Surface Methodology Routed Optimization of Performance of Hydroxy Gas Enriched Diesel Fuel in Compression Ignition Engines

Muhammad Usman [1,*], Saifuddin Nomanbhay [2,*], Mei Yin Ong [2], Muhammad Wajid Saleem [1], Muneeb Irshad [3], Zain Ul Hassan [1], Fahid Riaz [4], Muhammad Haris Shah [1], Muhammad Abdul Qyyum [5,*], Moonyong Lee [5,*] and Pau Loke Show [6,*]

[1] Department of Mechanical Engineering, University of Engineering and Technology Lahore, Lahore 54890, Pakistan; wajidsaleem@uet.edu.pk (M.W.S.); 2016me1@student.uet.edu.pk (Z.U.H.); 2018me162@student.uet.edu.pk (M.H.S.)
[2] Institute of Sustainable Energy (ISE), Universiti Tenaga Nasional (UNITEN), Jalan Ikram-Uniten, Kajang 43000, Selangor Darul Ehsan, Malaysia; me089475@hotmail.com
[3] Department of Physics, University of Engineering and Technology Lahore, Lahore 54890, Pakistan; muneeb_irshad@uet.edu.pk
[4] Department of Mechanical Engineering, National University of Singapore, Singapore 117575, Singapore; fahid.riaz@u.nus.edu
[5] School of Chemical Engineering, Yeungnam University, Gyeongsan 712-749, Korea
[6] Department of Chemical and Environmental Engineering, Faculty Science and Engineering, University of Nottingham, Semenyih 43500, Selangor Darul Ehsan, Malaysia
* Correspondence: muhammadusman@uet.edu.pk (M.U.); saifuddin@uniten.edu.my (S.N.); maqyyum@yu.ac.kr (M.A.Q.); mynlee@yu.ac.kr (M.L.); PauLoke.Show@nottingham.edu.my (P.L.S.)

Citation: Usman, M.; Nomanbhay, S.; Ong, M.Y.; Saleem, M.W.; Irshad, M.; Hassan, Z.U.; Riaz, F.; Shah, M.H.; Qyyum, M.A.; Lee, M.; et al. Response Surface Methodology Routed Optimization of Performance of Hydroxy Gas Enriched Diesel Fuel in Compression Ignition Engines. *Processes* 2021, 9, 1355. https://doi.org/10.3390/pr9081355

Academic Editor: Hector Budman

Received: 28 June 2021
Accepted: 27 July 2021
Published: 1 August 2021

Publisher's Note: MDPI stays neutral with regard to jurisdictional claims in published maps and institutional affiliations.

Abstract: In this study, the response surface methodology (RSM) optimization technique was employed for investigating the impact of hydroxy gas (HHO) enriched diesel on performance, acoustics, smoke and exhaust gas emissions of the compression ignition (CI) engine. The engine was operated within the HHO flow rate range of 0–10 L/min and engine loads of 15%, 30%, 45%, 60% and 75%. The results disclosed that HHO concentration and engine load had a substantial influence on the response variables. Analysis of variance (ANOVA) results of developed quadratic models indicated the appropriate fit for all models. Moreover, the optimization of the user-defined historical design of an experiment identified an optimum HHO flow rate of 8 L/min and 41% engine load, with composite desirability of 0.733. The responses corresponding to optimal study factors were 25.44%, 0.315 kg/kWh, 117.73 ppm, 140.87 ppm, 99.37 dB, and 1.97% for brake thermal efficiency (BTE), brake specific fuel consumption (BSFC), CO, HC, noise, and smoke, respectively. The absolute percentage errors (APEs) of RSM were predicted and experimental results were below 5%, which vouched for the reliable use of RSM for the prediction and optimization of acoustics and smoke and exhaust emission characteristics along with the performance of a CI engine.

Keywords: CI engine; HHO; response surface methodology; prediction; noise; smoke; optimization

1. Introduction

The oil reserves are depleting rapidly and are only sufficient to meet the drastically increasing power demand for the next fifty years [1]. Energy demand is soaring at an unprecedented pace and the available sources are too meagre to satisfy the needs [2,3]. In this scenario, the consumption of diesel as a transportation fuel has also increased by about 40% over the last decade [4]. The agriculture sector makes a major contribution to diesel consumption [5,6] as heavy machinery uses diesel as a fuel [7]. Moreover, automotive diesel engines share 26% of total greenhouse gas emissions into the environment, which is an unignorable threat to the stability of the Earth [8–10]. This has motivated researchers to investigate alternative fuels, such as hydroxy gas (HHO) for the versatile dual-fuel compression ignition engines [11,12], to find clean, economical, and sustainable energy resources [13–15].

The emission and combustion characteristics of internal combustion engines are mainly governed by the chemical and physical properties of the burning fuel [16,17]. Hydroxy gas (HHO), also termed brown gas, has only hydrogen and oxygen in its structure, and provides clean-burning to control CO_2 emissions and produces pure water when employed in CI engines [17]. Although hydrogen has a favorable high-octane rating, specific energy content, and autoignition temperature, it alone is not an appropriate option as a primary fuel due to the high safety risks of pressurized storage tanks in vehicles [18–21]. Consequently, an onboard HHO generating unit can mitigate the operational and safety problems concerning hydrogen generation, transportation and storage [22,23]. However, it can only be used as an additive because the same amount of energy as released from combustion (240 kJ/mol) is mandatory for HHO production [24].

The use of HHO as an alternative fuel has been the focus of researchers for quite a long time. In this regard, Pushpendra Kumar Sharma et al. explored the influence of varying flow rates of HHO with varying engine loads and observed a maximum increase of 6.5% for BTE along with a reduction of 58%, 60%, and 49% in CO and HC emissions, and smoke for 0.75 L/min and a 10 kg load, respectively [25]. Conversely, Subramanian et al. reported a 7% decrease in BTE at 36 L/min of flow rate owing to the higher auto-ignition temperature of hydrogen, a non-homogeneous air–diesel–HHO mixture and incomplete combustion at higher flow rates [26]. Usman et al. conducted a comparative assessment of gasoline, LPG, and LPG–HHO blends and reported improved performance and emissions for HHO blended LPG as compared to neat LPG [27]. Similarly, in another study, CNG–HHO blend showed 15.4% higher brake power and reduced CO and HC emissions [28]. In addition, Hydroxy gas can also be used as a secondary additive with the blends of other liquid and gaseous fuels [22], such as bio-fuels, to improve the performance (BP and BSFC) and emissions (CO_2 and HC) of IC engines [26,29,30].

Over the last two decades, several studies have investigated the use of RSM optimization to improve engine operation along with pollutant reduction of diesel-fueled engines [31–33]. Samet Uslu defined RSM models of the emission and performance of an engine operated with a palm oil diesel blend. He found that the correlation coefficients of all models were 0.90. Moreover, the optimum palm oil percentage of 17.88% was identified by multi-response optimization [31]. Milind Yadav et al. used RSM for the prediction and optimization of the performance characteristics of an oxy–hydrogen blended gasoline fueled SI engine [34]. Moreover, the use of RSM for the prediction of the emission and performance of the biodiesel–diesel blend was conducted by Mustafa Aydın et al. They reported a 32% biodiesel ratio, engine load of 816 W, and 470 bar injection pressure for the best performance and minimal emissions [35].

The cited literature reveals that the use of RSM for the optimization of diesel engines has been extensively carried out. However, one of the important aspects of emissions, that is, acoustic emissions, has not been given attention. This study addresses this very issue. Further, CI engines have not been optimized by employing the RSM technique for performance, acoustic, smoke, and exhaust emissions of the diesel–HHO blend. In this study, HHO was introduced with diesel at a flow rate of 0–10 L/min at varying loads. Later, RSM was used for studying the individual interactions between the study factors along with the statistical significance of developed models. The optimization identified the use of HHO with diesel as an effective alternative fuel that promises improved performance and reduced emissions.

Section 2 describes the detailed experimental approach of this study utilizing hydroxy gas. The results are discussed in detail utilizing ANOVA analysis in Section 3. The work is optimized utilizing RSM and validated in Sections 4 and 5, respectively. The study is concluded with recommendations and future research directions in Section 6.

2. Materials and Methods

This study used a 30 kW Perkins (AD 3.152), 91.4 mm bore length, four stroke diesel engine for experimentation. The specifications of the test engine are shown in Table 1. The experiments were performed for different flow rates of HHO, ranging from 0 L/min to 10 L/min while loads were varied from 0–75%, with an equal increment of 15% for each strategic test run. The physicochemical properties of diesel and hydrogen are presented in Table 2. Diesel fuel was directly supplied to the engine through the fuel injectors. However, HHO gas was supplied to the test engine intake manifold at varying flow rates for the diesel HHO mixture. The schematic of the experimental setup, which includes HHO generator, noise measuring meter, smoke meter, emission analyzer and electric heaters, is displayed in Figure 1. Brake thermal efficiency and brake specific fuel consumption were calculated numerically by utilizing calorific value (CV), brake power (BP), and fuel consumption (FC). The brake power was measured from the integrated control panel with heaters, which indicates the value of voltage and current at different load conditions, varied through electrical switches to turn on/off the heaters. In the experimental setup, heaters acted as a resistance load. The test engine was equipped with three phase AC generator having five breakers. Each breaker was equivalent to a load of 15%, which was applied to the engine through the generator. Simply, all the breakers turned on means the engine is operating on a load of 75%. Fuel consumption was determined by measuring time for the consumption of 100 mL of liquid fuel indicated by a gauged cylinder fixed adjacent to diesel containing tank while calorific value was obtained from Pakistan State Oil (PSO). HHO flow rate was ascertained from the rotameter connected at the output of the HHO generator. An emission analyzer (TESTO 350) was employed as a CO and HC emission content recorder, with a measuring sensitivity of 1 ppm CO, and 10 ppm HC. A smoke (opacity) meter (Wager 6500 manufactured by GasTech), which was in full compliance with the requirements of the SAE J1667 test criteria, was used to notice the smoke within a range of 0.0–100.0%. The engine noise level was measured with a sound level meter (UNI-T UT353), which can accurately measure 30–130 dB sound at a frequency response of 31.5 Hz to 8KHz.

Table 1. Test engine specifications.

Factors	Narrative
Type/Make	AD 3.152/Perkins
Volumetric efficiency (percent)	85
Stroke (cm)	1.27
Bore (cm)	0.91
Number of nozzles	3
Diesel injection	17° before TDC

Table 2. Properties of test fuels.

Properties	Diesel	Hydrogen [28]
Physical state	Liquid	Gas
Specific gravity	0.83–0.86 @ 16 °C	0.000083
Stoichiometric A/F	14.5	34.2
Viscosity (at 40 °C) (mm^2/s)	2.42	N/A
Boiling Range	160 to 366 °C	N/A
Cetane Number	57.86	-
Flash Point (°C)	59	N/A
Calorific Value (kJ/kg)	44,000	120,000

HHO was generated by electrolysis of water, which is the most commonly used method. Alkaline hydroxides, for example, KOH and NaOH and so forth, were used for speeding up the reaction [36]. The HHO generation system included AC supply, load controller, transformer, rectifier, reactor, and bubbler. The maximum production capacity of the unit was 10 L/min. Potassium hydroxide (KOH) was used as a catalyst owing to its higher solubility and affinity for water [22,37]. The flow rate of the produced gas was controlled using a potentiometer which was directly dependent on the current passing through the cell.

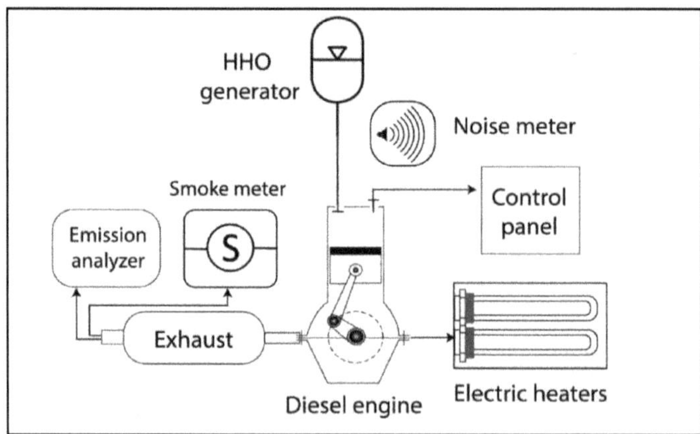

Figure 1. Schematic of the experimental Setup.

3. Results and Discussion

3.1. Response Surface Methodology

RSM is a statistical technique used for the estimation of relationships between input and response variables. It adopts linear, quadratic, or higher-order polynomial functions to investigate the statistical significance of the study factors and their interactions. Moreover, the regression concept is used for the prediction and optimization of responses. Over the years, the use of RSM in engineering fields has shown welcome results in the prediction of complex systems.

In the current study, the examined parameters were engine load and HHO concentration. Design-Expert version 11 was used for defining the multi-level historical design. The candidate set was created using user-defined discrete levels. Engine load and HHO concentration were assigned to four and six levels, respectively. The response variables measured were BTE, BSFC, CO, HC, noise, and smoke. The best fit model for each response was selected and analysis of variance (ANOVA) was applied for a better understanding of model attributes. In ANOVA, F is a probability distribution in different samplings, Df is degrees of freedom and the p-value is a statistical measure of variations in samples of a particular property. The decision rule for significance was benchmarked as a p-value less than 0.05. The percentage contribution (PC%) of each model term was calculated, which is a ratio of an aggregate of squared deviations to an individual sum of squares (SOS). PC% is a tool that provides a rough idea about the relative importance of study factors and the interactions.

3.2. ANOVA Results

The ANOVA results and fit statistics for BTE are presented in Table 3. The F-value of 1980.51 and p-value less than 0.0001 show that the model for BTE is significant. Moreover, the R^2 value of 0.9976 (refer to Table 4) is close to positive unity and there is sufficient agreement between predicted and adjusted R^2. The p values from the ANOVA table show that both load and concentration of HHO are significant. However, the load is significantly contributing to aggregated variations with a PC% of 84.4 compared to fuel concentration (3.5%). The best fitted quadratic model from the fit summary was selected owing to the poor fit and aliased nature of linear and cubic models, respectively. The actual regression equation for BTE is given by Equation (1).

$$BTE\ (\%) = 0.456285 + 0.793251 \times Load - 0.00680684 \times Concentration$$
$$+ 0.00680858 \times Load \times Concentration - 0.0059656 \times Load^2 + 0.00697741 \times Concentration^2. \tag{1}$$

Table 3. ANOVA results for BTE.

Source	Sum of Squares	Df	Mean Square	F-Value	*p*-Value	PC%
Model	1344.94	5	268.99	1980.51	<0.0001	99.7582
A-Load	1138.40	1	1138.40	8381.87	<0.0001	84.43851
B-HHO Concentration	47.75	1	47.75	351.56	<0.0001	3.541759
AB	7.30	1	7.30	53.76	<0.0001	0.541463
A^2	151.34	1	151.34	1114.29	<0.0001	11.22534
B^2	0.1454	1	0.1454	1.07	0.3111	0.010785
Residual	3.26	24	0.1358			0.241804
Cor Total	1348.20	29				

Table 4. Coefficient of determination for BTE.

Coefficient of Determination	Value
R^2	0.9976
Adjusted R^2	0.9971
Predicted R^2	0.9958

The contour plot (see Figure 2a) reveals the impact of load and fuel addition on BTE variation. The red color of the contour region advocates high engine BTE at high load and high HHO concentration. The color gradually shifted to red with an increase in HHO amount. The more explicit variation of response (BTE) is noticeable in Figure 2b. The 3D surface plot shows the rising curve of BTE with positive moments along the load and fuel axes. The maximum thermal efficiency is observable at an engine load of 75% and a 10 L/min flow rate of HHO. The improvement in BTE with HHO enrichment is due to the complete combustion of diesel in the presence of hydroxy gas which resulted from higher mean effective pressure near TDC, owing to the faster flame travel in the case of hydrogen. The dark and light circles above and below the surface represent the experimental and predicted values, respectively. Furthermore, the accuracy of the given models could be assessed using certain diagnostics tests and graphs. In general, small deviations between experimental and predicted results are desirable for efficient models. Figure 3 shows a comparison of actual and predicted BTE. The minimal deviations of predicted values from actual data sets are testimony to a good fit of the quadratic regression model.

ANOVA results for the second response variable, BSFC, are shown in Table 5. The model is significant owing to an F value of 169.80, a *p*-value less than the designated range, and R^2 (0.9725) close to 1, as indicated in Table 6. The ANOVA findings show the significant effect of both load and HHO on fuel consumption. However, compared on a comparative scale, the load variations were found to have a greater impact on an engine than HHO concentration, as evidenced by PCs of 72.9% and 1.4%, respectively. The quadratic regression equation for BSFC on an actual scale is shown by Equation (2).

$$BSFC\ (kg/KWh) = 1.01215 - 0.0241143 \times Load - 0.00582281 \times HHO\ Concentration$$
$$+ 2.20781 \times 10^{-5} \times Load \times HHO\ Concentration + 0.000197256 \times Load^2 \tag{2}$$
$$- 4.87589 \times 10^{-5} \times HHO\ Concentration^2.$$

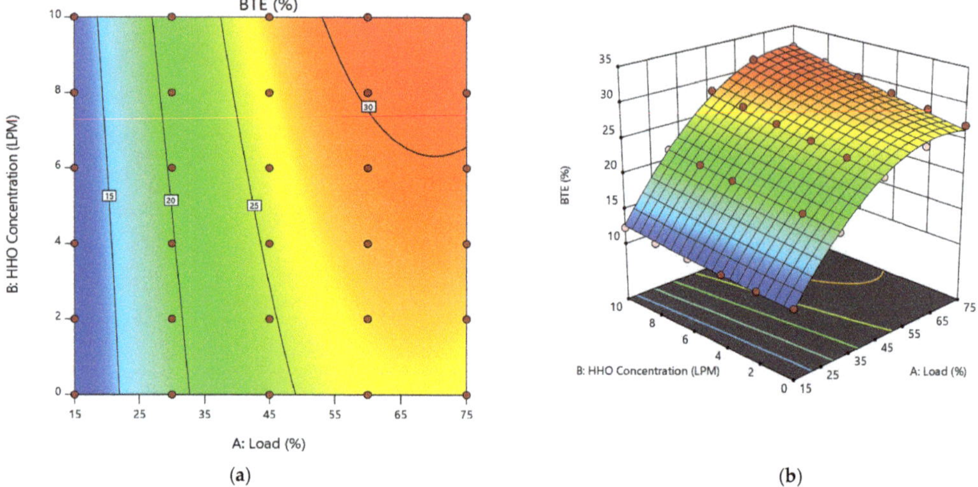

Figure 2. (**a**) Contour plot and (**b**) response surface of BTE.

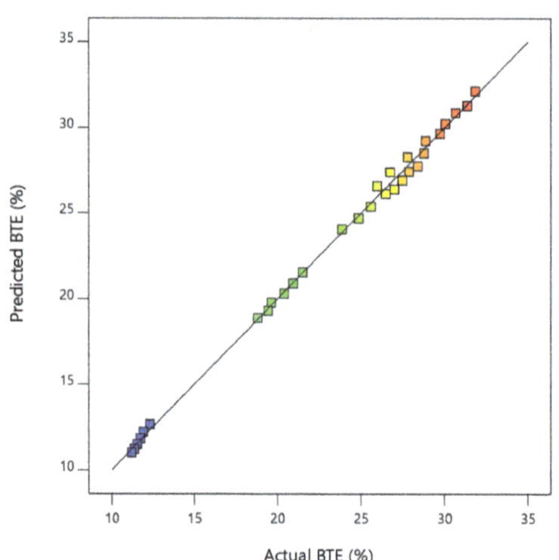

Figure 3. Comparison of actual and predicted BTE.

Table 5. ANOVA results for BSFC.

Source	Sum of Squares	Df	Mean Square	F-Value	*p*-Value	PC (%)
Model	0.7029	5	0.1406	169.80	<0.0001	97.24682
A-Load	0.5275	1	0.5275	637.11	<0.0001	72.98008
B-HHO Concentration	0.0099	1	0.0099	11.95	0.0020	1.369673
AB	0.0001	1	0.0001	0.0927	0.7634	0.013835
A^2	0.1655	1	0.1655	199.85	<0.0001	22.89707
B^2	7.101×10^{-6}	1	7.101×10^{-6}	0.0086	0.9270	0.000982
Residual	0.0199	24	0.0008			2.753182
Cor Total	0.7228	29				

Table 6. Coefficient of determination for BSFC.

Coefficient of Determination	Value
R^2	0.9725
Adjusted R^2	0.9668
Predicted R^2	0.9587

The effect of HHO addition and load on the fuel consumption trend of an engine is shown in Figure 4a,b. The contour plot in Figure 4a shows that fuel economy improved with successive addition of HHO to diesel at high loads. Moreover, it is also evident that, for the load range of 15–45%, there are more abrupt variations in BSFC compared to high loads, as indicated by a multi-color region. The response surface curve in Figure 4b shows the decreasing increasing trend of BSFC with load and fuel concentration. The sudden lift in the curve at the culmination is due to increased fuel demand at a high load owing to ample friction resistance. The improved fuel economy with HHO enrichment is primarily because of the higher calorific value of HHO and efficient combustion due to the lean diesel–HHO–air mixture [38,39]. The comparison of predicted and actual BSFC, as given in Figure 5, shows a bit of disorder data near the regression line. The disorderliness is due to the manual use of equipment in calculating BSFC. However, the deviations are not so large and therefore the model is acceptable.

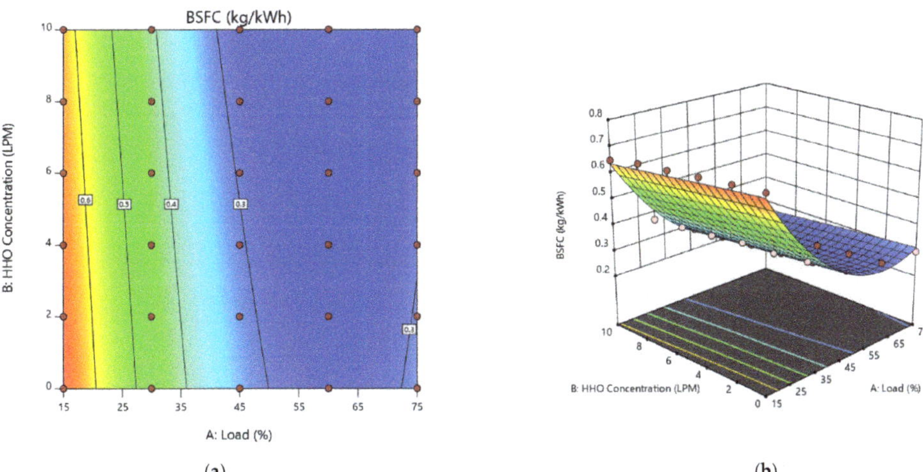

Figure 4. (**a**) Contour plot and (**b**) response surface of BSFC.

Similar to performance, quadratic models for emissions were also analyzed using ANOVA. The defined model for CO emissions was significant as shown in Table 7. The results revealed that both factors were significant; however, the percentage contribution of load to overall variations was greater compared to fuel concentration. Moreover, the R^2 value was 0.9819 (refer to Table 8) and there was a reasonable agreement between adjusted and predicted R^2. In an attempt to see the accuracy of the selected model, the actual versus predicted description in Figure 6 could be used as a model accuracy measuring tool. It is discernible from the figure that the data points are near to the linear regression line and deviations are negligible. The CO emission regression equation on a coded scale is given by Equation (3).

$$\text{CO (ppm)} = 140.804 - 1.66356 \times \text{Load} - 3.55691 \times \text{HHO Concentration}$$
$$- 0.0911124 \times \text{Load} \times \text{HHO Concentration} + 0.0622159 \times \text{Load}^2 - 0.0267857 \times \text{HHO Concentration}^2. \tag{3}$$

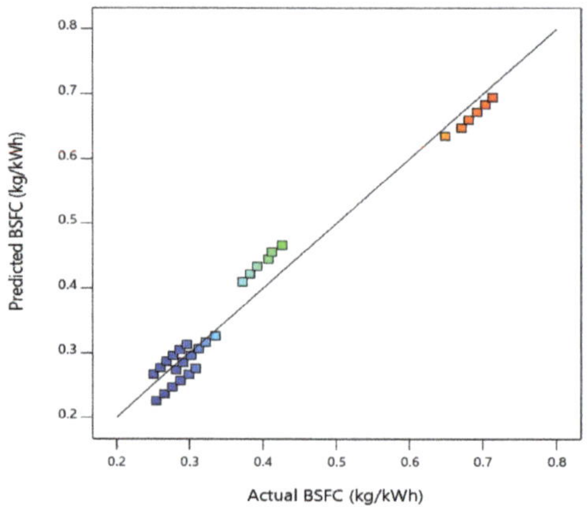

Figure 5. Comparison of actual and predicted BSFC.

Table 7. ANOVA results for CO.

Source	Sum of Squares	Df	Mean Square	F-Value	p-Value	PC%
Model	2.033×10^5	5	40,654.18	260.45	<0.0001	98.20
A-Load	1.635×10^5	1	1.635×10^5	1047.57	<0.0001	78.98
B-HHO Concentration	21,980.99	1	21980.99	140.82	<0.0001	10.62
AB	1307.48	1	1307.48	8.38	0.0080	0.63
A^2	16,460.64	1	16,460.64	105.45	<0.0001	7.95
B^2	2.14	1	2.14	0.0137	0.9077	0.00
Residual	3746.27	24	156.09			98.20
Cor Total	2.070×10^5	29				

Table 8. Coefficient of determination for CO.

Coefficient of Determination	Value
R^2	0.9819
Adjusted R^2	0.9781
Predicted R^2	0.9684

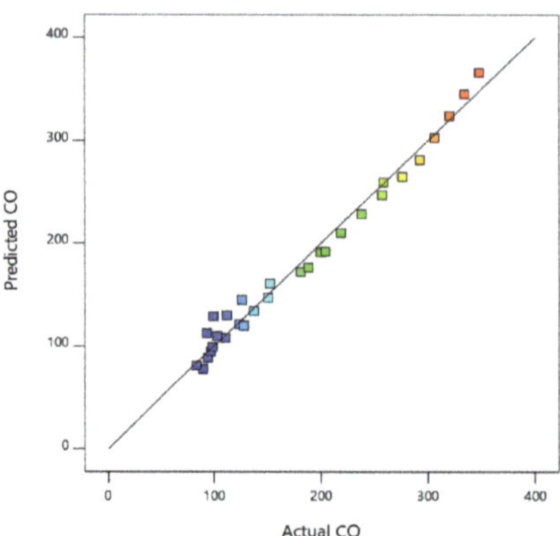

Figure 6. Comparison of actual and predicted CO.

The variations in emissions of carbon monoxide with load and HHO concentrations are shown in Figure 7a,b. The contour plot (Figure 7a) provides a general illustration of the

CO emission pattern of the engine subjected to various loads. The emissions are shown with the multi-color scheme, where blue stands for the minimum and red for the maximum. The response surface in Figure 7b depicts the CO variations with load and HHO. The main root of carbon monoxide generation is the partial burning of fuel inside the engine. The addition of hydroxy gas not only reduces the carbon content but also facilitates complete combustion which consequently reduces the emissions [40]. Therefore, a curve is seen to be following a decreasing trend in the presence of HHO.

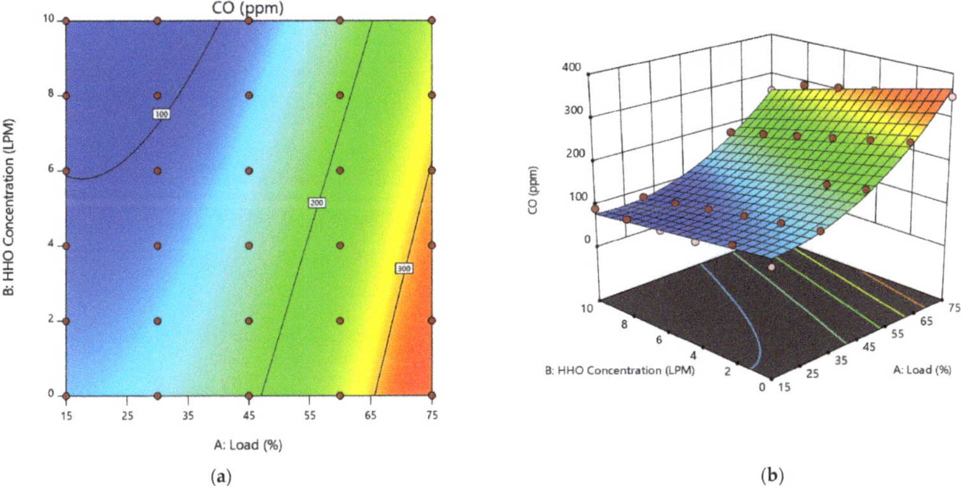

Figure 7. (**a**) Contour plot and (**b**) response surface of CO.

Similarly, Table 9 presents the ANOVA results of HC emission. The model selected and input variables are significant because of p values less than 0.005. The coefficient of determination, the R^2 value, however, is shown in Table 10. Engine load and HHO concentration had percentage contributions of 82.4% and 10.3% respectively. The comparison of actual and predicted HC emissions in Figure 8 shows that the selected model is accurate. Equation (4) gives the predicting regression equation of HC emissions.

$$\text{HC (ppm)} = 94.8363 + 1.16673 \times \text{Load} + 0.320067 \times \text{HHO Concentration}$$
$$- 0.134163 \times \text{Load} \times \text{HHO Concentration} + 0.0232022 \times \text{Load}^2 + 0.00446429 \times \text{HHO Concentration}^2. \tag{4}$$

Table 9. ANOVA results for HC.

Source	Sum of Squares	Df	Mean Square	F-Value	p-Value	PC (%)
Model	1.065×10^5	5	21,306.90	176.60	<0.0001	97.32231317
A-Load	90,147.78	1	90,147.78	747.17	<0.0001	82.3792533
B-HHO Concentration	11,262.46	1	11,262.46	93.35	<0.0001	10.29191229
AB	2834.94	1	2834.94	23.50	<0.0001	2.590637732
A^2	2289.29	1	2289.29	18.97	0.0002	2.092009374
B^2	0.0595	1	0.0595	0.0005	0.9825	5.43726×10^{-5}
Residual	2895.67	24	120.65			2.646134296

Table 10. Coefficient of determination for HC.

Coefficient of Determination	Value
R^2	0.9735
Adjusted R^2	0.9680
Predicted R^2	0.9592

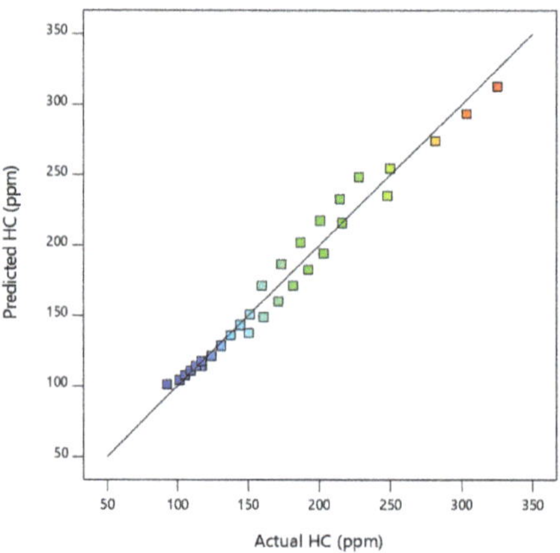

Figure 8. Comparison of actual and predicted HC.

The detailed effect of varying factors on hydrocarbon emissions is shown in Figure 9a,b. The addition of HHO reduced HC emissions for all concentrations and the minimum emissions were found to be for 10 L/min, as shown in Figure 9a. Similarly, the response surface shows the emission variations of each fuel combination and is seen following a decreasing trend. The presence of hydroxy gas reduces HC, while carbon present in lubricating oil and primary diesel fuel is oxidized by excessive oxygen and high combustion temperatures inside the cylinder. Moreover, a relatively short quenching distance and a wider flammability range in the case of gaseous fuel have improved the engine performance in this regard [41].

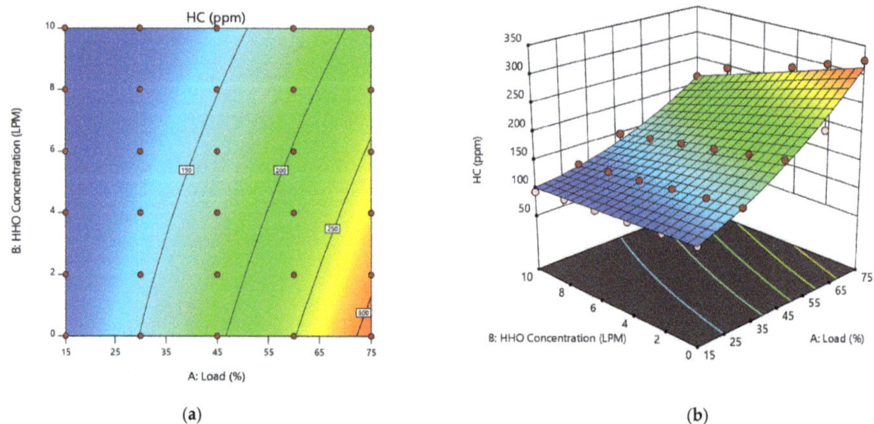

(a)   (b)

Figure 9. (a) Contour plot and (b) response surface of HC.

In addition to the performance and emission of an engine, factors of noise and smoke have also been considered. When the piston oscillates in the cylinder, it creates vibrations which consequently cause high noise levels. Moreover, when sudden ignition of fuel occurs inside the combustion chamber, it generates pressure waves that increase the intensity of the vibrations [42]. The smoke is produced as the result of a rich air–fuel mixture and lubricant burning in the combustion chamber [12]. Tables 11 and 12 present the ANOVA results for noise and smoke. The quadratic models and study factors for both responses were significant. Variations in noise would be more due to HHO concentration rather than load, as shown by percentage contributions of 13.19% and 74.30%. Similarly, the smoke model unveils that both load and fuel amount have a significant impact on smoke produced. Moreover, a model R^2 value close to one (based on Tables 13 and 14) and actual versus predicted diagnostic descriptions (Figures 10 and 11) evidenced the accuracy of the selected models. Equations (5) and (6) give the second-order regression equations of noise and smoke.

$$\text{Noise (dB)} = 96.9203 - 0.0315619 \times \text{Load} + 0.353839 \times \text{HHO Concentration}$$
$$+ 0.0018 \times \text{Load} \times \text{HHO Concentration} + 0.000519577 \times \text{Load}^2 - 0.0078125 \times \text{HHO Concentration}^2. \quad (5)$$

$$\text{Smoke (\%)} = 1.77495 - 0.0799333 \times \text{Load} - 0.0645821 \times \text{HHO Concentration}$$
$$- 0.00115714 \times \text{Load} \times \text{HHO Concentration} + 0.00239947 \times \text{Load}^2 + 0.0055625 \times \text{HHO Concentration}^2. \quad (6)$$

Table 11. ANOVA results for noise.

Source	Sum of Squares	Df	Mean Square	F-Value	*p*-Value	PC (%)
Model	54.28	5	10.86	46.00	<0.0001	90.5503968
A-Load	7.91	1	7.91	33.50	<0.0001	13.18848462
B-HHO Concentration	44.54	1	44.54	188.68	<0.0001	74.2915545
AB	0.5103	1	0.5103	2.16	0.1545	0.851247727
A^2	1.15	1	1.15	4.86	0.0373	1.915023405
B^2	0.1823	1	0.1823	0.7723	0.3882	0.304086551
Residual	5.66	24	0.236			9.449603204
Cor Total	59.95	29				

Table 12. ANOVA results for smoke.

Source	Sum of Squares	Df	Mean Square	F-Value	*p*-Value	PC (%)
Model	54.28	5	10.86	46.00	<0.0001	97.08694515
A-Load	7.91	1	7.91	33.50	<0.0001	87.15578304
B-HHO Concentration	44.54	1	44.54	188.68	<0.0001	0.496195224
AB	0.5103	1	0.5103	2.16	0.1545	0.080273582
A^2	1.15	1	1.15	4.86	0.0373	9.319517418
B^2	0.1823	1	0.1823	0.7723	0.3882	0.035175877
Residual	5.66	24	0.236			2.913054855
Cor Total	59.95	29				

Table 13. Coefficient of determination for noise.

Coefficient of Determination	Value
R^2	0.9055
Adjusted R^2	0.8858
Predicted R^2	0.8527

Table 14. Coefficient of determination for smoke.

Coefficient of Determination	Value
R^2	0.9709
Adjusted R^2	0.9468
Predicted R^2	0.9528

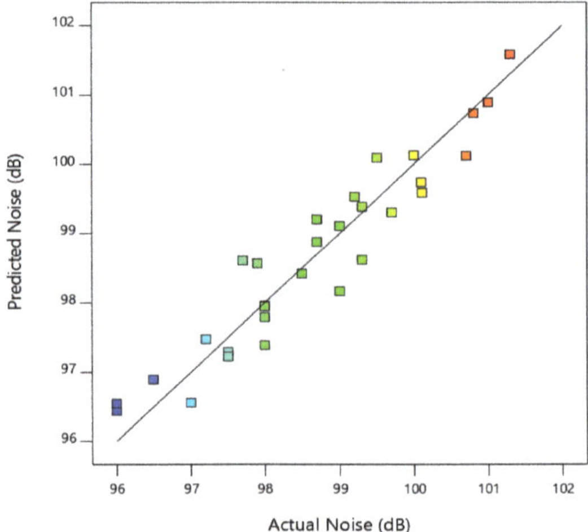

Figure 10. Comparison of actual and predicted noise.

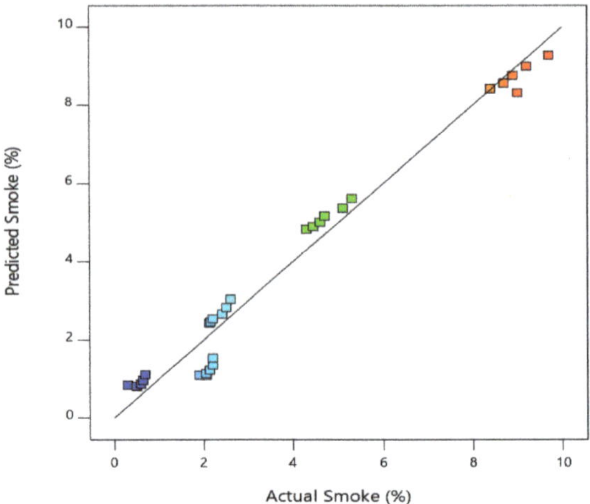

Figure 11. Comparison of actual and predicted smoke.

The effect of load and HHO on noise could be studied using the contour plots and response surface presented in Figure 12a,b. The red-colored region at the right top corner of Figure 12a indicates that, with the addition of HHO, the noise level increased and is at a maximum for 10 L/min HHO. The same trend could be seen more explicitly in Figure 12b where the response surface shows the gradual increase in noise level. The increased noise level with the addition of hydroxy gas could be apprehended by improved thermal efficiency and excessive combustion at high pressures inside the chamber [42,43]. The opacity is seen following a decreasing trend with a rise in fuel enrichment and load, as shown by Figure 13a,b. The contour plot and 3D response surface show that the least smoke is found for a blend of diesel with 10 L/min of HHO. The improved performance of an engine in terms of smoke emissions could be attributed to reduced HC emissions, high flame propagation, and high flame temperature of hydrogen [44].

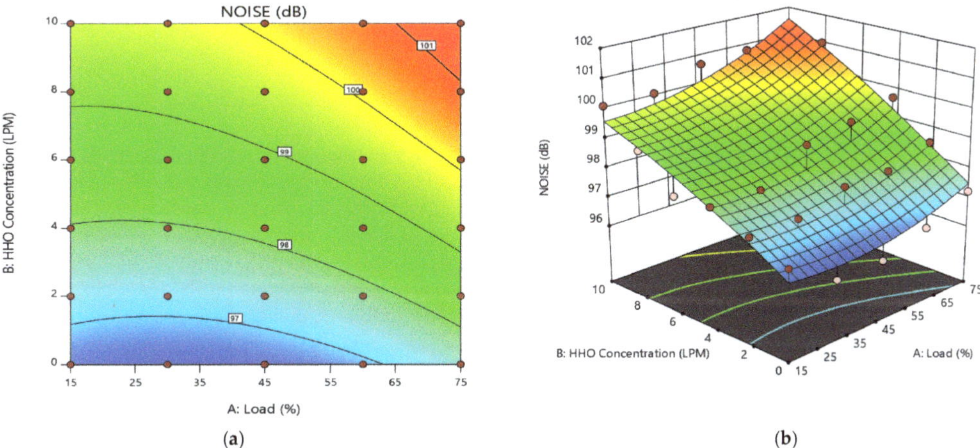

Figure 12. (**a**) Contour plot and (**b**) response surface of noise.

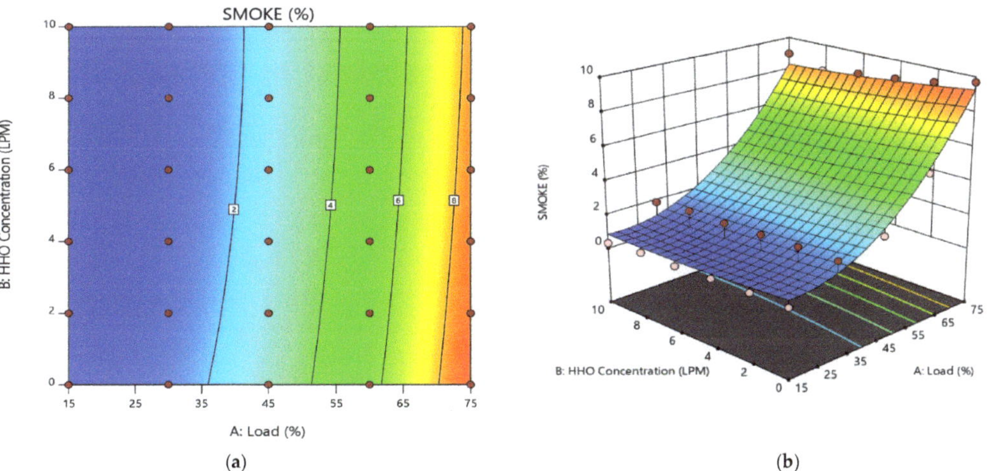

Figure 13. (**a**) Contour plot and (**b**) response surface of smoke.

4. RSM Based Optimization

Optimization is the study of maximizing output. An RSM-based optimization is a method to identify optimized conditions by maximizing or minimizing the study factors. In the current work, the emission and performance parameters of the engine are optimized using the numerical optimization feature of the Design Expert. In the optimization setup shown in Table 15, the goal of maximum was assigned to BTE only, while for smoke, noise, BSFC, CO, and HC the minimum criteria were selected. Moreover, the default in the range criterion for study factors was selected.

Table 15. Optimization setup.

Name	Goal	Lower Limit	Upper Limit	Lower Weight	Upper Weight	Importance
A: Load (%)	is in range	15	75	1	1	3
B: HHO Concentration (L/min)	is in range	0	10	1	1	3
BTE (%)	maximize	11.2216	31.8402	1	1	3
BSFC (kg/kWh)	minimize	0.25126	0.71291	1	1	3
Noise (dB)	minimize	96	101.3	1	1	3
HC (ppm)	minimize	92.812	325	1	1	3
Smoke (%)	minimize	0.3	9.7	1	1	3
CO (ppm)	minimize	84	348	1	1	3

The engine operating conditions identified by optimization were 41% engine load and blend of diesel with 8 L/min HHO, both rounded to the nearest whole number. The response variables, corresponding to optimized operating conditions, were 25.44% BTE, 0.315 kg/kWh BSFC, 117.7 3 ppm of CO, 140.86 ppm of HC, 99.4 dB of noise, and smoke of 1.97%. The optimum gained values of study factors and response variables are shown by the red and blue dots in Figure 14. The experimentation and RSM models in the previous sections advocated the use of 10 L/min blended diesel for boosted performance and reduced emissions. However, at the same time, all the blend percentages were unfavorable for noise and therefore an optimum concentration of 8 L/min sounds reasonable.

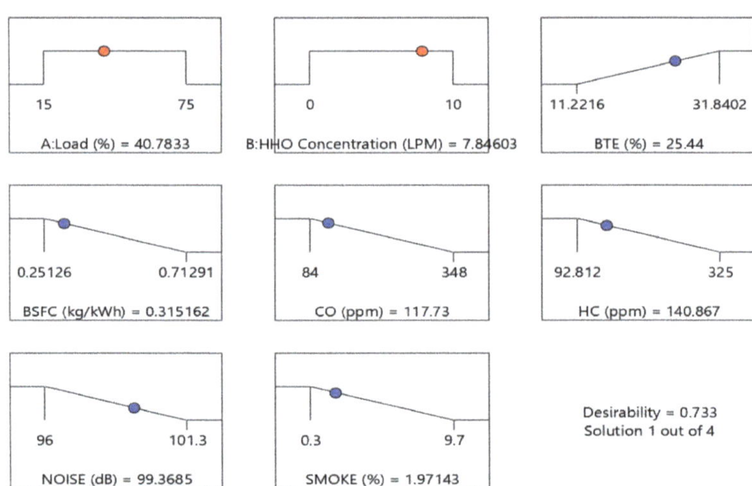

Figure 14. Design expert identified optimum points.

The statistical identification of how optimization involved the overall responses could be studied through composite desirability (D). It is a unitless value in the range of 0–1, with 1 for the best and 0 for the worst case. In the current study, composite desirability is 0.733, which is a clear indication that the optimization settings have achieved favorable outcomes for all responses. The contour plot of desirability is shown in Figure 15. Moreover, the impact of individual responses on the overall setting could be assessed through the individual desirability (d) of each response, as shown by the bar graph in Figure 16. It is evident that d is largest for CO (0.877) and lowest for noise (0.364). The numerical values show that minimizing carbon monoxide emissions would have the greatest impact on the overall settings compared, and minimizing noise would impart the least impact to the setting as a whole.

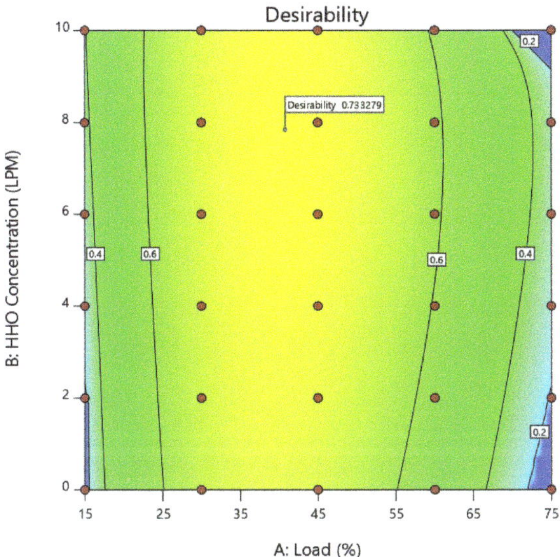

Figure 15. Contour plot of desirability.

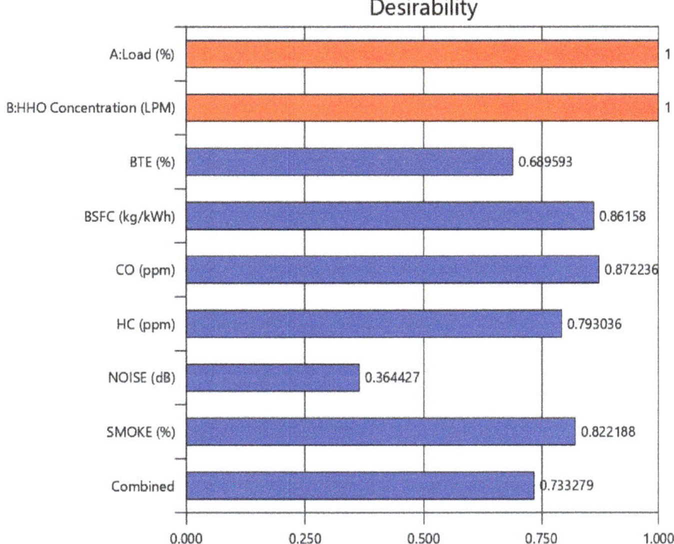

Figure 16. Desirability chart.

5. Validation of RSM Results

The obtained RSM multi-optimization results were validated using experimentation. The engine was operated on the optimal values of load and HHO concentration and the responses were recorded. The absolute percentage error (APE) between the RSM predicted and experimentally obtained results was calculated as shown in Table 16.

Table 16. Comparison of RSM and experimental values.

HHO Concentration (L/min)	Load (%)	Value	BTE (%)	BSFC kg/kWh	CO ppm	HC ppm	Noise dB	Smoke (%)
8	41	RSM Predicted	25.44	0.315	117.73	140.87	99.37	1.97
		Experimental	26.22	0.4	121	138.4	96.22	2
		APE	3.07	4.76	2.78	1.75	3.17	1.52

The APE shows that the developed RSM models and optimization results are accurate. The predicted results showed a reasonable agreement with the experimental results, with APE of all responses being below 5%. However, the maximum APE of 4.76% was evaluated for BSFC, which may be due to inefficient desirability resulting from manual recording during experimentation. Collectively, the predicted results of the developed models were efficient, which promised the simplification of complex performances with the least investment of time, effort and capital.

6. Conclusions

The purpose of the current investigation was to examine the impact of the blends of diesel with HHO on performance, noise, smoke and tailpipe emissions. Engine load and blend percentage of HHO were the varying factors. The following conclusions could be obtained from the research.

- ANOVA analysis of all the developed quadratic models indicated suitable fits.
- The 10 L/min HHO blended diesel proved valuable for improving performance, smoke, and for containing emissions.
- The noise increased for all the blended fuels and was maximum for 10 L/min HHO.
- The optimum blend flow rate among 0–10 L/min was 8 L/min for an engine load of 41%.
- Optimization revealed a composite desirability value of 0.733 with 25.44%, 0.315 kg/kWh, 117.73 ppm, 140.87 ppm, 99.37 dB, and 1.97% for BTE, BSFC, CO, HC, noise, and smoke respectively.
- In the optimization model, the most and least significant factors affecting desirability (D) were CO and noise, respectively.
- APE predicted that experimental results were below 5%.

The ANOVA and optimization results indicated the potential of hydroxy gas to be used as an alternative fuel in a CI engine. Thus, the use of HHO in blend percentages with diesel will help to save the stability of the Earth from deteriorating due to significantly reduced exhaust emissions compared to pure diesel. Moreover, the use of the RSM technique is beneficial and could save time and capital.

Author Contributions: Conceptualization, M.U., Z.U.H. and M.W.S.; methodology, M.H.S.; software, M.I.; validation, M.U. and M.I.; resources, P.L.S.; writing—original draft preparation, M.U.; writing—review and editing, S.N., M.Y.O., F.R., M.A.Q. and M.L.; supervision, M.U.; funding acquisition, S.N. All authors have read and agreed to the published version of the manuscript.

Funding: This work was funded by BOLD Grant 2021 under iRMC UNITEN (code: J510050002/2021092). The APC was funded by BOLD Refresh Publication Fund 2021 under iRMC, UNITEN (grant: J5100D4103-BOLDREFRESH2025-CENTER OF EXCELLENCE).

Institutional Review Board Statement: Not applicable.

Informed Consent Statement: Not applicable.

Acknowledgments: M.Y.O. and S.N. would like to acknowledge AAIBE Chair of Renewable Energy (Project code: 202006KETTHA) for the fellowship financial support. Besides, M.Y.O. would also like to thank UNITEN for the UNITEN Postgraduate Excellence Scholarship 2019 (YCU-COGS).

Conflicts of Interest: The authors declare no conflict of interest.

Nomenclature

APE	Absolute percentage error
A/F	Air fuel ratio
ANOVA	Analysis of variance
BSFC	Brake specific fuel consumption
BTE	Brake thermal efficiency
CO	Carbon monoxide
CI	Compression ignition
D	Composite desirability
d	Individual desirability
dB	Decibel
Df	Degrees of freedom
HC	Hydrocarbons
HHO	Hydroxy gas
Ppm	Parts per million
PC	Percentage contribution
R^2	Coefficient of determination
RSM	Response surface methodology
TDC	Top dead center

References

1. Höök, M.; Tang, X. Depletion of fossil fuels and anthropogenic climate change—A review. *Energy Policy* **2013**, *52*, 797–809. [CrossRef]
2. Abas, N.; Kalair, A.; Khan, N. Review of fossil fuels and future energy technologies. *Futures* **2015**, *69*, 31–49. [CrossRef]
3. Kverndokk, S. *Depletion of Fossil Fuels and the Impact of Global Warming*; Discussion Papers; Statistics Norway Research Department: Oslo, Norway, 1994.
4. Martins, F.; Felgueiras, C.; Smitkova, M.; Caetano, N. Analysis of fossil fuel energy consumption and environmental impacts in European countries. *Energies* **2019**, *12*, 964. [CrossRef]
5. Li, N.; Mu, H.; Li, H.; Gui, S. Diesel consumption of agriculture in China. *Energies* **2012**, *5*, 5126–5149. [CrossRef]
6. González-Marrero, R.M.; Lorenzo-Alegría, R.M.; Marrero, G.A. A dynamic model for road gasoline and diesel consumption: An application for Spanish regions. *Int. J. Energy Econ. Policy* **2012**, *2*, 201–209.
7. Agheli, L. Estimating the demand for diesel in agriculture sector of Iran. *Int. J. Energy Econ. Policy* **2015**, *5*, 660–667.
8. Nesamani, K.S. Estimation of automobile emissions and control strategies in India. *Sci. Total Environ.* **2010**, *408*, 1800–1811. [CrossRef] [PubMed]
9. Sequera, A.; Parthasarathy, R.; Gollahalli, S. Effect of Fuel Injection Timing in the Combustion of Biofuels in a Diesel Engine. In Proceedings of the 7th International Energy Conversion Engineering Conference, Denver, CO, USA, 2–5 August 2009.
10. Singh, P.; Chauhan, S.R.; Goel, V.; Gupta, A.K. Enhancing diesel engine performance and reducing emissions using binary biodiesel fuel blend. *J. Energy Resour. Technol.* **2020**, *142*, 01220. [CrossRef]
11. Castro, N.; Toledo, M.; Amador, G. An experimental investigation of the performance and emissions of a hydrogen-diesel dual fuel compression ignition internal combustion engine. *Appl. Therm. Eng.* **2019**, *156*, 660–667. [CrossRef]
12. Devarajan, Y. Experimental evaluation of combustion, emission and performance of research diesel engine fuelled Di-methyl-carbonate and biodiesel blends. *Atmos. Pollut. Res.* **2019**, *10*, 795–801. [CrossRef]
13. Elgarhi, I.; El-Kassaby, M.M.; Eldrainy, Y.A. Enhancing compression ignition engine performance using biodiesel/diesel blends and HHO gas. *Int. J. Hydrogen Energy* **2020**, *45*, 25409–25425. [CrossRef]
14. Hassan, Z.U.; Usman, M.; Asim, M.; Kazim, A.H.; Farooq, M.; Umair, M.; Imtiaz, M.U.; Asim, S.S. Use of diesel and emulsified diesel in CI engine: A comparative analysis of engine characteristics. *Sci. Prog.* **2021**, *104*, 368504211020930. [CrossRef] [PubMed]
15. Song, J.; Wang, G. An Experimental Study on Combustion and Performance of a Liquefied Natural Gas–Diesel Dual-Fuel Engine With Different Pilot Diesel Quantities. *J. Therm. Sci. Eng. Appl.* **2020**, *12*, 021011. [CrossRef]
16. Kemal, A.; Kahraman, N.; Çeper, B.A. Prediction of performance and emission parameters of an SI engine by using artificial neural networks. *Isi Bilimi Tek. Derg. J. Therm. Sci. Technol.* **2013**, *33*, 57–64.
17. Elkelawy, M.; Etaiw, S.E.-d.H.; Bastawissi, H.A.-E.; Marie, H.; Elbanna, A.; Panchal, H.; Sadasivuni, K.; Bhargav, H. Study of diesel-biodiesel blends combustion and emission characteristics in a CI engine by adding nanoparticles of Mn (II) supramolecular complex. *Atmos. Pollut. Res.* **2020**, *11*, 117–128. [CrossRef]
18. Arat, H.T.; Baltacioglu, M.K.; Özcanli, M.; Aydin, K. Effect of using Hydroxy—CNG fuel mixtures in a non-modified diesel engine by substitution of diesel fuel. *Int. J. Hydrogen Energy* **2016**, *41*, 8354–8363. [CrossRef]
19. Jakliński, P.; Czarnigowski, J. An experimental investigation of the impact of added HHO gas on automotive emissions under idle conditions. *Int. J. Hydrogen Energy* **2020**, *45*, 13119–13128. [CrossRef]

20. Kumar, V.; Gupta, D.; Kumar, N. Hydrogen use in internal combustion engine: A review. *Int. J. Adv. Cult. Technol.* **2015**, *3*, 87–99. [CrossRef]

21. Zohuri, B. Cryogenics and Liquid Hydrogen Storage. In *Hydrogen Energy*; Springer Nature Switzerland AG: Cham, Switzerland, 2019; pp. 121–139.

22. Kazim, A.H.; Khan, M.B.; Nazir, R.; Shabbir, A.; Abbasi, M.S.; Abdul Rab, H.; Shahid Qureishi, N. Effects of oxyhydrogen gas induction on the performance of a small-capacity diesel engine. *Sci. Prog.* **2020**, *103*, 36850420921685. [CrossRef]

23. Polverino, P.; D'Aniello, F.; Arsie, I.; Pianese, C. Study of the energetic needs for the on-board production of Oxy-Hydrogen as fuel additive in internal combustion engines. *Energy Convers. Manag.* **2019**, *179*, 114–131. [CrossRef]

24. Rajasekar, E.; Murugesan, A.; Subramanian, R.; Nedunchezhian, N. Review of NOx reduction technologies in CI engines fuelled with oxygenated biomass fuels. *Renew. Sustain. Energy Rev.* **2010**, *14*, 2113–2121. [CrossRef]

25. Sharma, P.K.; Sharma, D.; Soni, S.L.; Jhalani, A.; Singh, D.; Sharma, S. Characterization of the hydroxy fueled compression ignition engine under dual fuel mode: Experimental and numerical simulation. *Int. J. Hydrog. Energy* **2020**, *45*, 8067–8081. [CrossRef]

26. Subramanian, B.; Thangavel, V. Experimental investigations on performance, emission and combustion characteristics of Diesel-Hydrogen and Diesel-HHO gas in a Dual fuel CI engine. *Int. J. Hydrog. Energy* **2020**, *45*, 25479–25492. [CrossRef]

27. Usman, M.; Farooq, M.; Naqvi, M.; Saleem, M.W.; Hussain, J.; Naqvi, S.R.; Jahangir, S.; Jazim Usama, H.M.; Idrees, S.; Anukam, A. Use of gasoline, LPG and LPG-HHO blend in SI engine: A comparative performance for emission control and sustainable environment. *Processes* **2020**, *8*, 74. [CrossRef]

28. Usman, M.; Hayat, N.; Bhutta, M.M.A. SI Engine Fueled with Gasoline, CNG and CNG-HHO Blend: Comparative Evaluation of Performance, Emission and Lubrication Oil Deterioration. *J. Therm. Sci.* **2020**, *30*, 1199–1211. [CrossRef]

29. Baltacioglu, M.K.; Kenanoglu, R.; Aydın, K. HHO enrichment of bio-diesohol fuel blends in a single cylinder diesel engine. *Int. J. Hydrog. Energy* **2019**, *44*, 18993–19004. [CrossRef]

30. Kenanoğlu, R.; Baltacıoğlu, M.K.; Demir, M.H.; Özdemir, M.E. Performance & emission analysis of HHO enriched dual-fuelled diesel engine with artificial neural network prediction approaches. *Int. J. Hydrog. Energy* **2020**, *45*, 26357–26369.

31. Uslu, S. Optimization of diesel engine operating parameters fueled with palm oil-diesel blend: Comparative evaluation between response surface methodology (RSM) and artificial neural network (ANN). *Fuel* **2020**, *276*, 117990. [CrossRef]

32. Usman, M.; Naveed, A.; Saqib, S.; Hussain, J.; Tariq, M.K. Comparative assessment of lube oil, emission and performance of SI engine fueled with two different grades octane numbers. *J. Chin. Inst. Eng.* **2020**, *43*, 734–741. [CrossRef]

33. Xu, H.; Yin, B.; Liu, S.; Jia, H. Performance optimization of diesel engine fueled with diesel–jatropha curcas biodiesel blend using response surface methodology. *J. Mech. Sci. Technol.* **2017**, *31*, 4051–4059. [CrossRef]

34. Yadav, M.; Sawant, S. Effect of oxy-hydrogen blending with gasoline on vehicle performance parameters and optimization using response surface methodology. *J. Chin. Inst. Eng.* **2019**, *42*, 553–564. [CrossRef]

35. Aydın, M.; Uslu, S.; Çelik, M.B. Performance and emission prediction of a compression ignition engine fueled with biodiesel-diesel blends: A combined application of ANN and RSM based optimization. *Fuel* **2020**, *269*, 117472. [CrossRef]

36. Kehayov, D.; Komitov, G.; Ivanov, I. Influence of some factors on the gas flow produced by hho generator. *Land Reclam. Earth Obs. Surv. Environ. Eng.* **2019**, *8*, 188–191.

37. Nabil, T.; Dawood, M.M.K. Enabling efficient use of oxy-hydrogen gas (HHO) in selected engineering applications; transportation and sustainable power generation. *J. Clean. Prod.* **2019**, *237*, 117798. [CrossRef]

38. Al-Rousan, A.A.; Alkheder, S.; Musmar, S.e.A.; Al-Dabbas, M.A. Green transportation: Increasing fuel consumption efficiency through HHO gas injection in diesel vehicles. *Int. J. Glob. Warm.* **2018**, *14*, 372–384. [CrossRef]

39. Subramanian, B.; Ismail, S. Production and use of HHO gas in IC engines. *Int. J. Hydrog. Energy* **2018**, *43*, 7140–7154. [CrossRef]

40. Dahake, M.; Patil, S.; Patil, S. Effect of hydroxy gas addition on performance and emissions of diesel engine. *Int. Res. J. Eng. Technol.* **2016**, *3*, 756–760.

41. Momirlan, M.; Veziroglu, T.N. The properties of hydrogen as fuel tomorrow in sustainable energy system for a cleaner planet. *Int. J. Hydrog. Energy* **2005**, *30*, 795–802. [CrossRef]

42. Tüccar, G. Effect of hydroxy gas enrichment on vibration, noise and combustion characteristics of a diesel engine fueled with Foeniculum vulgare oil biodiesel and diesel fuel. *Energy Sources Part A RecoveryUtil. Environ. Eff.* **2018**, *40*, 1257–1265. [CrossRef]

43. Yıldırım, S.; Tosun, E.; Çalık, A.; Uluocak, İ.; Avşar, E. Artificial intelligence techniques for the vibration, noise, and emission characteristics of a hydrogen-enriched diesel engine. *Energy Sources Part A RecoveryUtil. Environ. Eff.* **2019**, *41*, 2194–2206. [CrossRef]

44. Rimkus, A.; Matijošius, J.; Bogdevičius, M.; Bereczky, Á.; Török, Á. An investigation of the efficiency of using O2 and H2 (hydrooxile gas-HHO) gas additives in a ci engine operating on diesel fuel and biodiesel. *Energy* **2018**, *152*, 640–651. [CrossRef]

 *processes* 

Comparative Study of a Life Cycle Assessment for Bio-Plastic Straws and Paper Straws: Malaysia's Perspective

Chun-Hung Moy [1], Lian-See Tan [1], Noor Fazliani Shoparwe [2], Azmi Mohd Shariff [3] and Jully Tan [4,*]

[1] Malaysia-Japan International Institute of Technology, Universiti Teknologi Malaysia, Jalan Sultan Yahya Petra, Kuala Lumpur 54100, Malaysia; hung9098703@rocketmail.com (C.-H.M.); tan.liansee@utm.my (L.-S.T.)

[2] Faculty of Bioengineering and Technology, Jeli Campus, Universiti Malaysia Kelantan, Jeli, Kelantan 17600, Malaysia; fazliani.s@umk.edu.my

[3] Chemical Engineering Department, Universiti Teknologi PETRONAS, Seri Iskandar, Perak 32610, Malaysia; azmish@utp.edu.my

[4] School of Engineering, Monash University Malaysia, Jalan Lagoon Selatan, Bandar Sunway, Selangor 47500, Malaysia

* Correspondence: tan.jully@monash.edu

Citation: Moy, C.-H.; Tan, L.-S.; Shoparwe, N.F.; Shariff, A.M.; Tan, J. Comparative Study of a Life Cycle Assessment for Bio-Plastic Straws and Paper Straws: Malaysia's Perspective. *Processes* **2021**, *9*, 1007. https://doi.org/10.3390/pr9061007

Academic Editors: Pau Loke Show, Aydin Berenjian, Kit Wayne Chew and Anil K. Bhowmick

Received: 30 March 2021
Accepted: 2 June 2021
Published: 7 June 2021

Abstract: Plastics are used for various applications, including in the food and beverage industry, for the manufacturing of plastic utensils and straws. The higher utilization of plastic straws has indirectly resulted in the significant disposal of plastic waste, which has become a serious environmental issue. Alternatively, bio-plastic and paper straws have been introduced to reduce plastic waste. However, limited studies are available on the environmental assessment of drinking straws. Life cycle assessment (LCA) studies for bio-plastic and paper straws have not been comprehensively performed previously. Therefore, the impact of both bio-plastic and paper straws on the environment are quantified and compared in this study. Parameters, such as the global warming potential (GWP), acidification potential (AP) and eutrophication potential (EP), were evaluated. The input–output data of the bio-plastic and paper straws processes from a gate-to-grave analysis were obtained from the literature and generated using the SuperPro Designer V9 process simulator. The results show that bio-plastic straws, which are also known as polylactic acid (PLA) straws, had reduced environmental impacts compared to paper straws. The outcomes of this work provide an insight into the application of bio-plastic and paper straws in effectively reducing the impact on the environment and in promoting sustainability, especially from the perspective of Malaysia.

Keywords: life cycle assessment; global warming potential; acidification potential; eutrophication potential; bio-plastic straws; paper straws

1. Introduction

Plastic pollution is a serious and long-standing issue that threatens human health at a global scale. The issue is alarming as plastic is non-biodegradable and does not completely disintegrate [1]. In fact, Malaysia has been listed as the eighth-worst country worldwide for the mismanagement of plastic waste [2]. It was estimated that there were almost one million tons of mismanaged plastic waste in Malaysia, of which 0.14 to 0.37 million tons may have been washed into the oceans in 2010 [3]. The incineration of plastic waste could emit dioxin, which is carcinogenic and a hormone disruptor, and, with persistent exposure, dioxin can accumulate in human body fat [4] and cause toxicity. Moreover, plastic packaging and straws that have been washed into the ocean and were disposed of in landfills also threaten the lives of the marine and land animals [5]. According to the United States National Oceanographic and Atmospheric Administration, plastic debris kills an estimated 100,000 marine mammals and millions of birds and fishes annually [6].

One of the strategies to reduce plastic waste and the resulting pollution issues is to replace conventional fossil-based plastics, such as polyethylene (PE) and polypropylene

(PP), with bio-plastics. Bio-plastics can be produced from renewable feedstocks without depleting natural resources, and they can biodegrade at a much faster rate than conventional plastics. Polylactic acid (PLA) is one of the most commonly used bio-plastics recently due to its versatility and biodegradable properties [7]. A previous study reported that the application of a polybutylene succinate (PBS) and PLA mixture to produce bio-plastic straws resulted in a lower carbon footprint compared to conventional PP straws [8].

In addition to bio-plastic, paper is also an alternative to plastic. In comparison to the conventional fossil-based plastics, paper is manufactured from logs wood, which is also a renewable source. Therefore, paper is usually claimed to be more environmentally friendly. However, the usage of either paper straws or bio-plastic straws can also pose some impacts on the environment [9].

To date, the study of environmental impacts has been conducted using different analytical tools, such as material flow analysis (MFA), environmental impact assessment (EIA), and life cycle assessment (LCA). LCA is defined as the compilation and evaluation of inputs, outputs, and environmental impacts of a product system throughout its life cycle according to the ISO 14,040 standard [10]. It is a structured step-by-step framework, which defines the goal and functional unit and leads to impact assessment [11].

Among the environmental impact categories, global warming potential (GWP) is an important indicator [12]. GWP represents the amount of carbon dioxide (CO_2) and other greenhouse gases (GHGs) emitted over a full life cycle of a process or a product. Meanwhile, acidification potential (AP) is associated with atmospheric pollution arising from anthropogenically derived sulfur (S) and nitrogen (N) as nitrogen oxides (NO_x) or ammonia (NH_3). Anthropogenically derived pollutant deposition was found to enhance the rate of acidification and increase the natural neutralizing capacity of soils [13]. Soil acidification is also one of the major contemporary environmental issues globally. Acidification potential is usually calculated in sulfur dioxide equivalents (SO_2-eq) [14].

On the other hand, eutrophication potential (EP) is linked to the release of macronutrients, such as nitrogen (N) and phosphorus (P), into the air, water, and land, which can affect both aquatic and terrestrial environments. A high level of nutrients can cause a deplorable composition shift in species living in a polluted environment. The presence of macronutrients in water systems often causes algal blooms, which impact the aquatic ecosystem and domestic water quality.

About 300 million tons of plastic products are produced every year, and half of them are single-use types, such as cups, straws, and shopping bags. A preliminary investigation of marine litter pollution along a beach in India during the period of observation from January to March 2020 found plastic straws as the third most common debris at 9.3% [15] Meanwhile, straws and stirrers ranked fifth, representing 7.9% among the most common debris found in the International Coastal Ocean Cleanup in year 2019 [16].

In Malaysia, it was estimated that each person uses one straw daily, amounting to 30 million straws being used daily. This usage of straws per capita varies from country to country. For example, it was estimated that the straw usage in the United States of America was at an average daily rate of 1.6 straws per capita. This is a total of 500 million plastic straws being used every day in the United States alone [17]. In addition, collectively, up to 8.3 billion plastic straws were estimated to pollute the world's beaches [18]. Therefore, aligning with the global effort to reduce plastic waste, in 2019, the Malaysian government implemented a ban on single-use plastic straws [19]. This increased the awareness of sustainability and the use of alternative straws. While some may argue the effectiveness of banning plastic straw usage to promote a reduction in plastic waste, with the ban on single-use plastic straws, the shift towards alternative straws has been gaining momentum. To the best of our knowledge, there is limited environmental analysis of alternative drinking straws, especially from the Malaysian perspective. Thus far, there are limited studies available on the environmental assessment of drinking straws. A search of the keyword "drinking straws" in the Scopus database showed 334 available published documents. However, further filtering using the additional search parameter "life cycle assessment" showed seven published documents, of which only one paper was relevant. Limited results showed environmental assessment for different types of plastics and/or straws. LCA of PP, PLA, paper, glass, and stainless steel from cradle to grave was carried out by Chitaka et al. [20], and the results showed that paper straws had the least impact as compared to the other types of raw materials. However, Rana [21] found that stainless steel straws had a significantly lower overall environmental impact than that of other straws. LCA for PP, stainless steel, borosilicate glass, paper, bamboo, and wheat stem straws from cradle to grave was performed by Zanghelini et al. [22], and it showed that plastic drinking straws posed a lower environmental impact when compared to reusable straws. Meanwhile, Chang and Tan [23] developed an integrated sustainability assessment of drinking straws. The study was carried out to quantify the potential environmental impacts of different drinking straws based on the scenarios of different countries, such as South Africa [20] and Brazil [22]. However, no studies are found on the environmental assessments of drinking straws in a Malaysian scenario.

Information on the environmental impact of bio-plastic and paper is scarce. Hence, a significant research gap was noted in the evaluation of alternative natural drinking straws, especially from a Malaysian perspective. Therefore, this study aims to present the environmental impacts of both bio-plastic and paper straws from cradle to grave using the LCA approach. Specifically, the main environmental protection indicators, such as GWP, AP and EP, were evaluated. The work flow was arranged such that the methodology of the LCA was detailed, followed by the results and discussion. Finally, the conclusion and recommendations are presented.

2. Methodology

2.1. Life Cycle Assessment

There are four phases in a life cycle assessment, which are the goal and scope definition, inventory analysis, impact assessment and interpretation [11]. In this study, an LCA of bio-plastic and paper straws was carried out to evaluate their impacts on the environment.

2.2. Goal and Scope Definition

The goal of this study was to determine the overall environmental impact of bio-plastic and paper straws from the manufacturing of the raw materials (gate) to their end-of-life (grave). To compare the environmental impacts of bio-plastic and paper drinking straws, data were normalized to a functional unit of 100 units of drinking straws produced which was equivalent to 133 g of bio-plastic straws (given 1.33 g per straw) or 260 g of paper straws (given 2.60 g per straw).

Figures 1 and 2 illustrate the bio-plastic straws and paper straws system boundaries in the study, respectively. The input and output data presented included each process that releases environmental pollutants, such as carbon monoxide (CO), carbon dioxide (CO_2), nitrogen oxide (NO_x), nitrous oxide (N_2O), methane (CH_4), ammonia (NH_3), sulfur dioxide (SO_2) and volatile organic compounds (VOC) into the atmosphere. Based on Figure 1, raw materials (raw corns, NH_3 solution, sulfuric acid, and protease), electricity, and diesel fuel are fed into the system. Meanwhile, for the quantification of GWP, AP, and EP, the pollutants (CO, CO_2, NO_x, N_2O, CH_4, NH_3, SO_2, and VOC) were considered as the outputs.

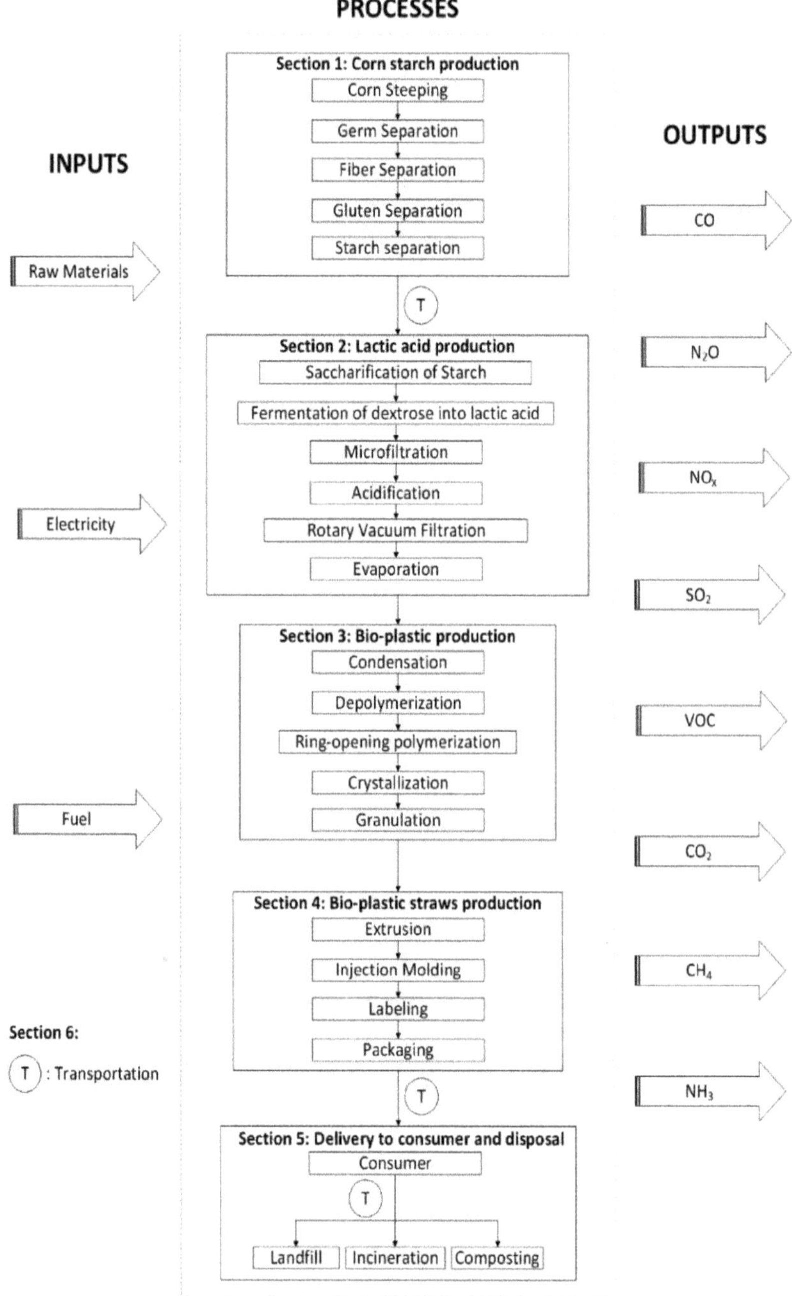

Figure 1. Overall system boundary of bio-plastic straws from gate to grave.

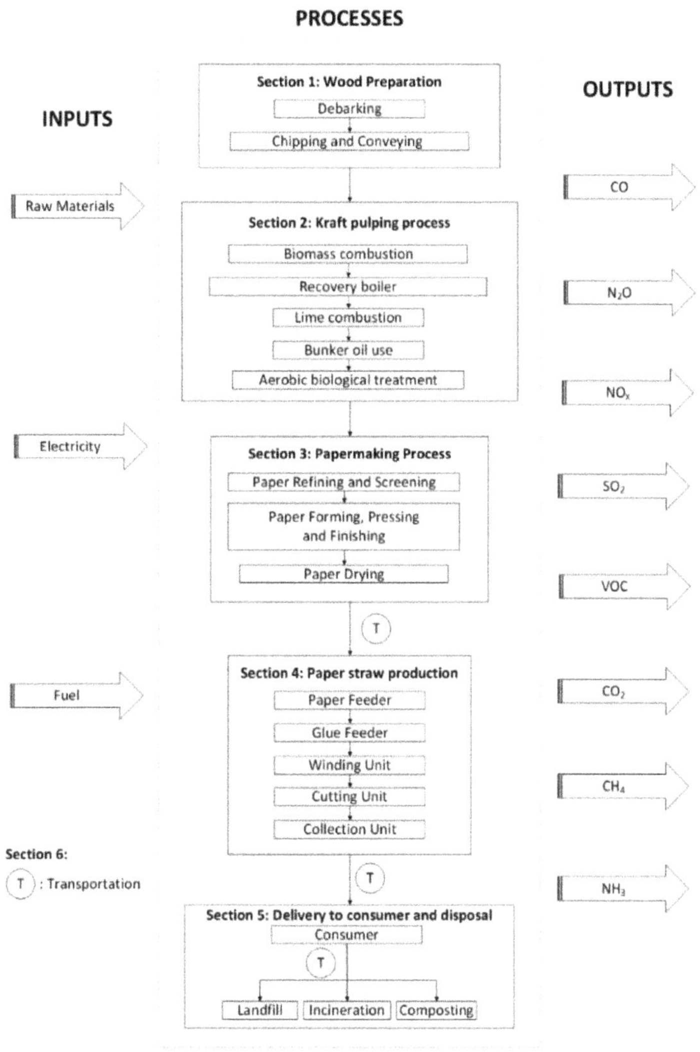

Figure 2. Overall system boundary of paper straws from gate to grave.

The manufacturing process of bio-plastic straws consists of six main steps: the production of corn starch, the production of lactic acid, the production of bio-plastic, the production of bio-plastic straws, delivery to consumers, disposal, and transportation. During the production of raw corn starch, processes involved include corn steeping and the separation of germ, fiber, gluten, and starch, which were considered for the input–output data analysis [24]. Next, lactic acid was produced from corn starch by saccharification of starch followed by the fermentation of dextrose into lactic acid, microfiltration, acidification, rotary vacuum filtration, and, finally, evaporation. The bio-plastic production also included processes such as condensation, depolymerization, ring-opening polymerization, crystallization and granulation. Before the final step (straw delivery), extrusion, injection molding, labeling, and packaging steps were carried out. As bio-plastic straws are utilized for a single use, different disposal methods (landfills, incineration and composting) were evaluated. The wastes were equally divided among three different disposal methods with

33% for each. Moreover, the transportation of raw materials to the production site and the transportation of products to the consumer and to the disposal site were also considered in this study. It should be noted that there is no transportation between Section 2 (lactic acid preparation), Section 3 (bio-plastic production), and Section 4 (bio-plastic straw production) as these sections are assumed to occur at the same manufacturing facility (as shown in the Figure 1).

In addition, Figure 2 illustrates the overall system boundary of paper straws from the gate to the grave. The setting of the system boundaries was performed in manner similar to that of the bio-plastic straw products. The boundaries consisted of six sections of processes that started from the wood preparation, kraft pulping, papermaking, paper straw production, delivery to consumer and disposal site and transportation. The first three steps are the extraction and manufacturing processes of raw materials for paper straw production. The first step was the wood preparation, which consisted of debarking, chipping and conveying. The next step was the kraft pulping process, which consisted of three main units (energy generation, chemical recovery, and wastewater treatment). The papermaking process also involved paper refining and screening, paper reforming, pressing, finishing and drying before the production of paper straws. In the production of paper straws, there are five main units, which are the paper feeder, glue feeder, winding unit, cutting, and the collection unit. Similar to bio-plastic straws, paper straws are utilized for a single use. Therefore, the disposal methods of the bio-plastic straws, such as landfills, incineration, and composting, were included. The transportation of raw materials to the production site and the transportation of products to the consumer and the disposal site were also considered in this study. Similarly, for the bio-plastic straw, it should be noted that there was no transportation between Section 1 (wood preparation), Section 2 (kraft pulping process), and Section 3 (papermaking process) as these sections are assumed to occur at the same manufacturing facility (as shown in Figure 2).

2.3. Inventory Analysis

A process simulation model was developed using a SuperPro Designer V9.0 based on Figures 1 and 2. The process simulation flowsheet of bio-plastic and paper straws is shown in Figures S1–S7 (Supplementary Data). The following assumptions and limitations were made for the inventory analysis:

- All calculations were based on 100 units of drinking straws produced, which were equal to 133 g of bio-plastic straws and 260 g of paper straws.
- Corn starch production was adapted from the corn refinery simulation [24].
- Similar physical properties in the injection molding of the PLA and the PP were assumed as the PLA straws are very flexible and perform similarly to conventional plastic straws made of PP [25].
- For the kraft pulping process, biomass combustion was used in the energy generation, which is commonly used in the pulp and paper industry [26].
- The disposal of bio-plastic and paper straws was equally divided between a composite facility, landfill and incineration. A similar amount of bio-plastic paper straws for each disposal method was ensured. The equal division was assumed for different disposal methods in order to analyze how each of the processes contributes to the GWP and AP [27].
- The landfill sites of bio-plastic and paper straws are located in Malaysia. Thus, both landfill sites have similar site characteristics, i.e., weather, humidity and temperature.
- The transportation of raw materials to the manufacturing site, the transportation of the product to the customer and the transportation of used bio-plastic and paper straws to disposal sites were based on the actual location of the supply chain in Peninsular Malaysia as a case study.

The equipment set-up in the process simulator SuperPro Designer v9.0 (by Intelligen Inc., Scotch Plains, NJ, USA) is illustrated in the Supplementary Data (Tables S1–S7). Based on the input as stipulated in Tables S1–S7, SuperPro Designer performs thorough material and energy balances and calculates each of the process's environmentally significant stream properties. Then, the data are tabulated for further analysis. Details of the process data inventory for bio-plastic and paper straws can be obtained from the Supplementary Data (Tables S8 to S10). The quantity of the power consumption and emission of pollutants (mass of pollutant, m_i) from each of the unit processes of the bio-plastic and paper straw production are essential to calculate the environmental impact categories.

2.4. Impact Assessment

The purpose of the impact assessment is to convert and aggregate the inventory analysis findings into the relevant environmental indicator. This can be explained as the transformation of the inventory results into the number of contributions to environmental impact categories (GWP, AP and EP). The main environmental effects identified by the European Commission in the Economics and Cross-Media Effects document includes global warming, acidification, and eutrophication [28].

All the identified environmental potential indexes were evaluated using the expressions summarized in Table 1. The main parameters used in the formula are the mass (m_i) in kilograms (kg) of a specific pollutant released to the air and pollutant specific weighting factors (GWP_i, AP_i and EP_i). These factors are representative of potential environmental effects per mass unit of the specific pollutant.

Table 1. Potential index definitions of the considered environmental effects and respective units of measurement [10,28].

Index	Formula	Unit of Measure
Global Warming Potential	$GWP = \sum GWP_i \times m_i$	kg CO_2 equivalent (kg CO_2-eq)
Acidification Potential	$AP = \sum AP_i \times m_i$	kg SO_2 equivalent (kg SO_2-eq)
Eutrophication Potential	$EP = \sum EP_i \times m_i$	kg PO_4 equivalent (kg PO_4-eq)

Each of the environmental potential indexes was evaluated as the sum of the effects of several pollutants based on the data tabulated in Tables S8 to S10. Each pollutant mass (m_i) was weighted by a specific weighting factor, which was expressed based on a reference substance. This allows a direct comparison and summation of the effects of several unrelated pollutants according to a cross-media effect assessment approach.

The specific weighting factor values for selected pollutants are listed in Table 2. The weight of various pollutants can be different as displayed in Table 2. In general, the sum of each considered potential index can be calculated once the pollutant mass levels are specified using the expression and specific factors reported in Tables 1 and 2, respectively. For instance, to calculate the GWP for Section 1 of bio-plastic straws production, all the pollutants related to GWP were taken into the calculation, as shown in Table 3. The total GWP of individual process (such as corn steeping) can be calculated ($0.26 \times 2 + 0.03 \times 3 + 1100)/1000 = 1.1$). The total GWP of Section 1 is the summation of the GWP of each individual process.

Table 2. List of specific weighting factors for the pollutants [10,28,29].

Item of Measurement	GWP_i	AP_i	EP_i
	kg CO_2-eq/kg	kg SO_2-eq/kg	kg PO_4-eq/kg
CO	2	0	0
NO_x	0	0.7	0.13
SO_x	0	1	0
VOC	3	0	0
NH_3	0	1.88	0.35
CO_2	1	0	0
CH_4	23	0	0
N_2O	310	0	0

Table 3. Example illustration of calculation for environmental impact category versus pollutants mass.

Pollutants	GWP_i	Corn Starch Production (Section 1)				
		Corn Steeping	Germ Separation	Fiber Separation	Gluten Separation	Starch Separation
CO (g)	2	2.6×10^{-1}	1.5	7.7×10^{-1}	6.0×10^{-1}	8.0×10^{-1}
VOC (g)	3	3.0×10^{-2}	1.7×10^{-1}	8.9×10^{-2}	6.9×10^{-2}	9.2×10^{-2}
CO_2 (g)	1	1.1×10^{3}	6.3×10^{3}	3.4×10^{3}	2.6×10^{3}	3.5×10^{3}
Total GWP		1.1	6.3	3.4	2.6	3.5
Overall GWP		16.9 kg CO_2-eq/100 straws				

The generation of 1 MJ of electricity emits a specific amount of pollutants [30,31] and further contributes to the GWP, AP and EP. The GWP is impacted by CO, CO_2 and VOC, while the AP is impacted by SO_2 and NO_x. The EP is impacted by NO_x. The GWP is impacted by CO, CO_2 and VOC, while the AP is impacted by SO_2 and NO_x. The EP is impacted by NO_x. These data were obtained from the power consumption of Tables S8–S10 (Supplementary Data).

The transportation distances from the corn plantation site to the corn starch production site, corn starch production site to straw manufacturing plant, straw manufacturing plant to consumer and, finally, consumer to the disposal site (i.e., incinerator, composting facility, and landfill) were taken into consideration based on the sites in Malaysia. The total distance of bio-plastic straws is tabulated in Table 4. Similarly, Table 5 illustrates the transportation details of the paper straws, which include the transportation of raw wood to the paper mill, paper to paper straws, paper straws to the consumers, and, lastly, the used paper straws transferred for end-of-life processing.

Table 4. Transportation distance for bio-plastic straws.

Starting Point	Destination	Distance (km)
Raw Corn	Corn Starch	187
Corn Starch	Bio-Plastic Straws	202
Bio-Plastic Straws	Consumer	50
	Incineration Plant	562
Consumer	Composting Facility	315
	Landfill	27
Total distance (km)		1343

Table 5. Transportation distance for paper straws.

Starting Point	Destination	Distance (km)
Wood Supplier	Paper Mill	59
Paper Mill	Paper Straw	119
Paper Straw	Consumer	50
	Incineration Plant	252
Consumer	Composting Facility	31
	Landfill	33
Total distance (km)		544

The details of the different locations are referenced based on the existing sites located in Malaysia (Supplementary Data Figures S9 and S10 for bio-plastic and paper straws, respectively). It was assumed that a medium- and heavy-duty truck was employed throughout the transportation of the raw materials and products. The emission factors of the transport used are shown in Table 6 to calculate the environmental impact based on the total distances in Tables 4 and 5.

Table 6. Product transport emission factors [32].

Vehicle Type	Medium- and Heavy-Duty Truck
CO_2 Factor (kg/km)	0.904716
CH_4 Factor (g/km)	0.011185
N_2O Factor (g/km)	0.006835

2.5. Data Interpretation

The final step in the LCA according to the ISO 14,044 standard on environmental management is interpretation. There are three main objectives of LCA interpretation. The first objective is to identify significant issues based on the LCA results. Next is the evaluation of completeness, sensitivity and consistency. The results were interpreted according to the goal and scope of the study, which includes an assessment and a sensitivity evaluation of the significant inputs, outputs and methodological choices to understand the uncertainty of the results. According to ISO 14044, the process of the completeness check identifies any missing or incomplete information which is related to the goal and scope of the LCA, and, if there is any, it shall be recorded and justified. The consistency check can also be defined as a process to determine whether the assumptions, methods and data are consistent with the goal and scope. The process addressed the data quality, regional and/or temporal differences, system boundary, and consistency of impact assessment. The findings from this research were compared to those of other similar studies to detect any incomplete or erroneous data. Note that the data input of the equipment setup during the initial simulation is based on the judgement of the researchers after considering the expert input and literature review. As such, the value may vary if the type of the equipment is varied and the efficiency of the equipment is improved. In summary, data validation was conducted and compared with the published research, and then the conclusions were drawn in line with the study objectives.

3. Results and Discussion

3.1. Overall Result of Bio-Plastic Straws

Figure 3 shows the GWP, AP and EP of bio-plastic straws. The initial sections of the whole process, which are Sections 1 and 2, emitted a high amount of pollutants which contributed to the three impacts. These processes included the production of starch from raw corn followed by the production of lactic acid from the starch, which required a relatively high amount of electricity. The extraction of the lactic acid from the raw corn involves a series of energy-intensive process such as corn steeping, starch separation, saccharification of the starch, fermentation and purification [24]. Therefore, a huge amount of carbon dioxide was emitted to the environment and resulted in a high GWP for the overall process. The amount of AP and EP in Section 1 was much higher than that in Section 2 due to the release of NH_3 during germ, gluten and starch separation. Overall, the electricity consumption greatly reduced starting from Section 3 onwards where bio-plastic was produced, until Section 5 of the system boundary (Figure 1) where delivery to the consumer and the disposal of bio-plastic straws took place. This could be attributed to the lower energy-intensive processes in the later sections of the overall manufacturing process of bio-plastic straws.

In Section 4, where the bio-plastic straw production was carried out, small amounts of pollutants contributing to GWP, AP and EP were captured in the inventory list. This could be due to the continuous high amount of heat required to melt the PLA pellets in the extrusion process. Therefore, a high amount of electricity supply was required in Section 4. Lower AP and EP were noted as Section 5 of the system boundary involves delivery to the consumer and the disposal of straws. These were mainly due to a smaller amount of NO_x and SO_x emitted during the processes in Section 5 of the system boundary. In Section 6 of the system boundary (Figure 1), AP and EP were observed to be zero because CO_2, CH_4, and N_2O were the only pollutants considered during the transportation section.

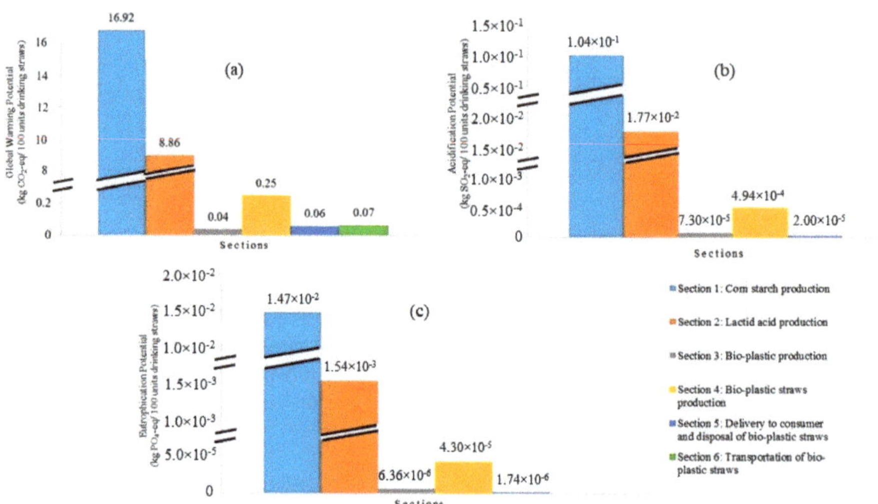

Figure 3. (**a**) GWP, (**b**) AP, and (**c**) EP of 100 units of bio-plastic straws.

3.2. Overall Result of Paper Straws

Based on Figure 4, it can be observed that Section 2 contributed to the highest GWP, AP and EP compared to other sections. Section 2 involves the kraft pulping process where wood chips were converted into pulp, which was further processed into paper. There are three main units in the kraft pulping process section, which are the energy generation, chemical recovery and wastewater treatment. The energy generation unit contributed to the highest GWP, AP and EP due to the biomass combustion, which released high amounts of carbon dioxide, methane, nitrous oxide and nitrogen oxides [26].

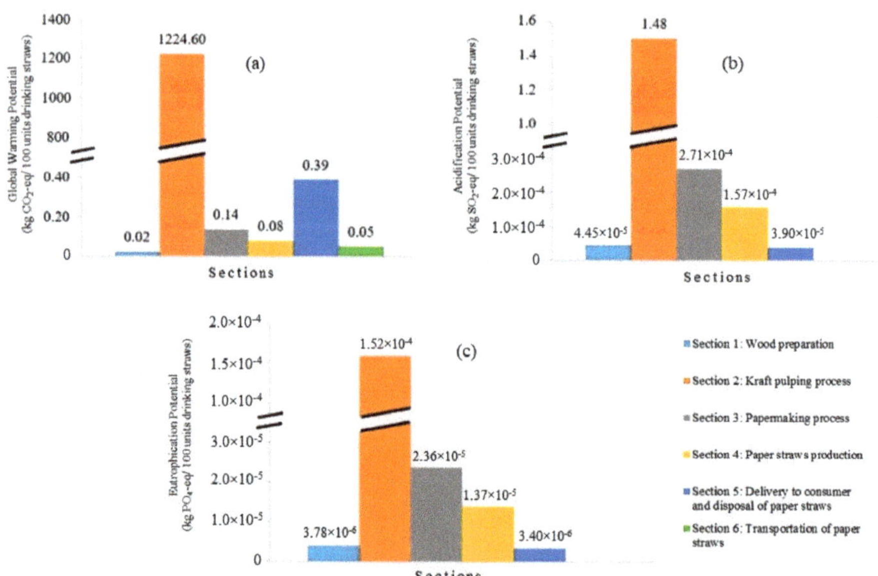

Figure 4. (**a**) GWP, (**b**) AP, and (**c**) EP of 100 units of paper straws.

Section 5 of the system boundary, which includes disposal of paper straws by composting, landfill or incineration, contributed to the highest GWP, followed by Sections 3 and 4. The highest GWP noted in Section 5 of the system boundary could be attributed to the emission of methane from the landfill. On the contrary, Section 1 (the wood preparation process) contributed to the least GWP. No emission of pollutants was involved in the wood preparation process except a small amount of electricity that was used for the debarking, chipping, and conveying process [33]. Section 3 (the papermaking process, which consumed a high amount of electricity) contributed to the highest AP compared to other sections. There were several processes in Section 3, which included paper refining and screening, paper forming, pressing, finishing and paper drying. Paper forming, pressing, and finishing processes required the most electricity annually and therefore resulted in a high AP. By contrast, the disposal of paper straws (Section 5 of the system boundary) contributed the least to AP.

3.3. Overall Comparison of Bio-Plastic Straws and Paper Straws

Six sections of the life cycle of bio-plastic and paper straws were studied regarding the GWP, AP and EP. The first three sections were the extraction and production of raw materials, which were the PLA for bio-plastic straws and paper for paper straws. Therefore, Sections 1–3 were grouped as one for comparison, as shown in Figure 5. Based on Figure 5a,b, paper straws contributed to a higher GWP and AP compared to bio-plastic straws concerning the extraction and production of raw materials. Section 1 for the preparation of corn starch for bio-plastic straws contributed to the higher GWP and AP compared to paper straws (Section 1: wood preparation). The production of starch from raw corn in Section 1 of bio-plastic straws generated a large portion of GWP and AP. This was due to the extraction process of starch, which involved corn steeping and germ, fiber, gluten and starch separation. On the contrary, the kraft pulping process in Section 2 of paper straws generated a large portion of GWP and AP due to higher energy generation that involved the combustion of biomasses. Moreover, the papermaking process for paper straws in Section 3 contributed to a higher GWP and AP than the polymerization of lactic acid for bio-plastic straws due to the higher electricity consumption. However, based on Figure 5c, bio-plastic straws contributed to a higher EP than paper straws, which could be ascribed to the release of NH_3 during germ, gluten, and starch separation in Section 1.

As for Section 4 where the production of drinking straws takes place, based on the results shown in Figure 6, bio-plastic straws contributed to higher GWP, AP and EP compared to that of paper straws. This could be explained by the energy-intensive production of bio-plastic straws compared to paper straws which involved the process of extrusion and injection moulding. In contrast, paper straw production does not require high electricity consumption. In Section 5 of the system boundary, delivery to the consumer and the disposal of drinking straws and paper straws contributed to the higher amount of GWP as compared to bio-plastic straws. This was due to the high emission of methane and carbon dioxide as a result of anaerobic decomposition in the landfill for paper straws. However, both paper and bio-plastic straws contributed to the same amount of AP due to similar electricity consumption for composting, landfill and incineration. For Section 6 of the system boundary (transportation), bio-plastic straws contributed to a higher GWP as compared to paper straws. This was due to longer distance travelled in the transport of bio-plastic straws than paper straws, which was affected by the different locations of the plants in the supply chain of the respective drinking straws. There was no AP and EP for either bio-plastic or paper straws due to zero AP pollutants for the transportation section.

The AP findings of the present study are consistent with those reported by Chaffee and Yaros (2007) where an LCA study was conducted to evaluate the environmental impact of grocery bags. However, the authors stated that the GWP of paper grocery bags was lower than that of bio-plastic grocery bags. This was due to the different system boundary as their study involved the extraction of fuels and feedstocks from the earth which included the tree growing process. Therefore, the CO_2 emissions, which are one of the main GWP contributors, were greatly reduced, as most of the CO_2 was absorbed during the photosynthesis process that takes place during the development of trees. Moreover, the raw material of paper bags consisted of a mixture of paper and recycled paper instead of pure paper that was used for paper straw processing, which also reduced the overall GWP. In contrast, the raw material of bio-plastic does not only consist of PLA but also other compostable plastics [34]. Therefore, the GWP trend for grocery bags is different from that of drinking straws; however, the AP pattern is similar.

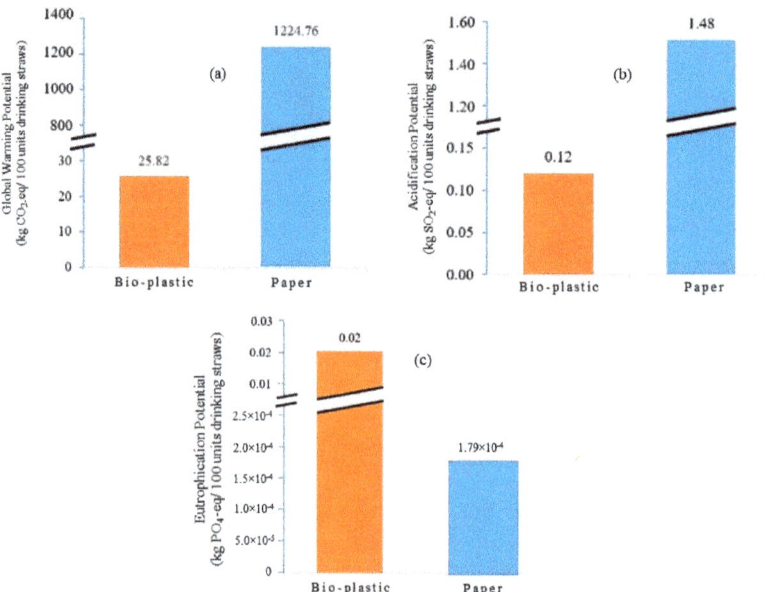

Figure 5. (a) GWP, (b) AP, and (c) EP of bio-plastic straws and paper straws for extraction and production of raw materials.

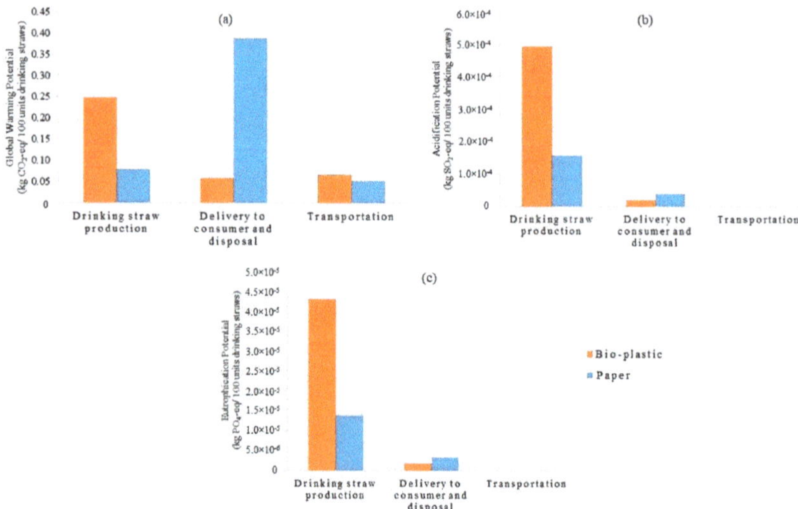

Figure 6. (a) GWP, (b) AP, and (c) EP of bio-plastic straws and paper straws for drinking straw production, delivery to the consumer and production and transportation sections.

Overall, as shown in Figure 7, paper straws have a higher GWP and AP compared to bio-plastic straws. However, bio-plastic straws have a slightly higher EP than paper straws, but, generally, the EP for both types of straw is very small and insignificant compared to the GWP and AP. The GWP and AP were mainly impacted by the extraction and production of raw materials for the production of drinking straws. This included corn starch production and wood preparation for Section 1, lactic acid production and kraft pulping process for Section 2, and polylactic acid production and papermaking process for Section 3. Overall, based on the simulated case study, bio-plastic straws are found to be a better option compared to paper straws for a milder impact on the environment concerning the GWP and AP.

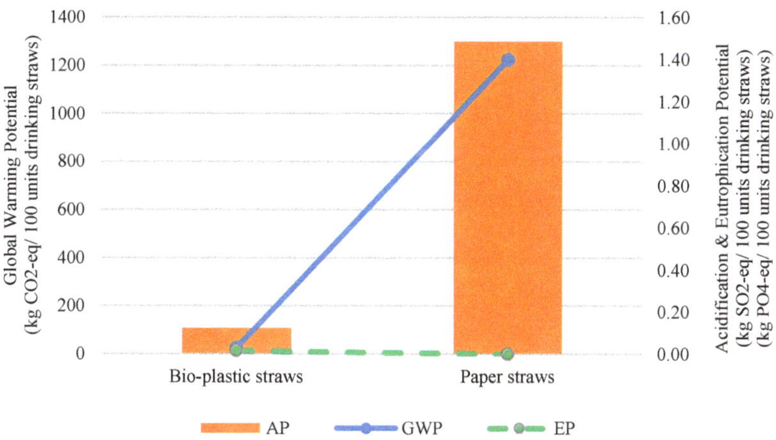

Figure 7. Overall grand total of GWP, AP, and EP for bio-plastic straws and paper straws.

3.4. Sensitivity Analysis

Sensitivity analysis was performed to demonstrate the effects of changing process variables, i.e., power consumption, on the fluctuation of GWP, AP and EP values of bio-plastic straws and paper straws. The results would enable the analysis of uncertainty propagation in an LCA calculation. The results could also indicate how well the process coped with uncertainty under different conditions [35]. The sensitivity analysis was done by switching the simulation model to rating mode and tested for its robustness. The probabilities were calibrated for the LCA outcomes arising from uncertainty in the inventory and from data variation characteristics.

The power consumption for one of the unit operations, which was used for the extrusion process in bio-plastic straws production (Section 4), was varied using SuperPro Designer for sensitivity analysis. The effect of the power consumption variation towards the GWP, AP and EP values of bio-plastic straws is shown in Figure 8a. On the other hand, the power consumption for paper drinking straw machine in paper straws production (Section 4) was also varied using SuperPro Designer for similar sensitivity analysis. The result is shown in Figure 8b. The results shown in Figure 8 indicated that the LCA conducted in this study experienced a slight increase of its GWP, AP and EP values (at the range of 0.0004–0.500%) when the power consumption increased with an interval of 5–10%, respectively. Hence, the sensitivity analysis conducted in this study showed minimal fluctuations of GWP, AP, and EP values with the variation in the power consumption.

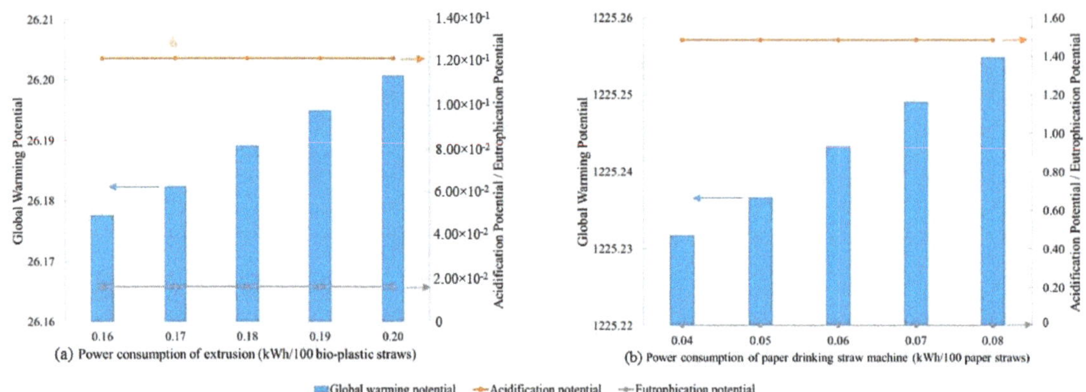

Figure 8. Overall GWP, AP, and EP values of (a) bio-plastic straws and (b) paper straws versus power consumption.

4. Conclusions

The GWP, AP, and EP of bio-plastic straws were successfully evaluated using the LCA with data obtained from process simulator. It was found that the corn starch production contributed to the highest GWP, AP and EP, which was ascribed to energy-intensive processes such as corn steeping and the separation of germ, fiber, gluten and starch. Moreover, the GWP, AP, and EP of paper straws were also successfully investigated using the LCA with data obtained from both the literature and the process simulator. The kraft pulping process contributed to the highest GWP, AP, and EP due to the energy generation unit, which involved biomass combustion. The GWP, AP, and EP for both bio-plastic and paper straws were compared for an indication of straws with less environmental impact. It was found that bio-plastic straws have a lower GWP of 26 kg CO_2-eq per 100 units of drinking straws, an AP of 0.12 kg SO_2-eq per 100 units of drinking straws and an EP of 0.016 kg PO_4-eq per 100 units of drinking straws. Additionally, paper straws have a GWP of 1225 kg CO_2-eq per 100 units of drinking straws, an AP of 1.5 kg SO_2-eq per 100 units of drinking straws and an EP of 0.0002 kg PO_4-eq per 100 units of drinking straws. Therefore, concerning the GWP and AP, bio-plastic straws are a better option than paper straws for a milder impact on the environment. Previous studies have shown that plastic straws posed a lower environmental impact compared to reusable straws, and bio-plastic straws showed a lower carbon footprint compared to conventional plastic straws. From this point of view, bio-plastic straws could be a feasible replacement for conventional plastic straws. The outcome of this study is able to serve as a benchmark in selecting alternative straws in efforts towards zero-plastic straws in Malaysia. However, the decision to switch to bio-plastic straws should not be rushed. Comprehensive information related to their biodegradability as well as the water and land footprints could be included for consideration to provide a more holistic sustainability assessment of drinking straws.

Supplementary Materials: The following are available online at https://www.mdpi.com/article/10.3390/pr9061007/s1, Table S1: Equipment details of corn starch production (Section 1) based on SuperPro Designer, Table S2: Equipment details of lactic acid production (Section 2) based on SuperPro Designer, Table S3: Equipment details of bio-plastic production (Section 3) based on SuperPro Designer, Table S4: Equipment details of bio-plastic straws production (Section 4) based on SuperPro Designer, Table S5: Equipment details of delivery to consumer and disposal of bio-plastic straws (Section 5) based on SuperPro Designer, Table S6: Equipment details of paper straws production (Section 4) based on SuperPro Designer, Table S7: Equipment details of delivery to consumer and disposal of paper straws (Section 5) based on SuperPro Designer, Table S8: Process data inventory of bio-plastic straws production obtained from SuperPro Designer simulator (based on 100 drinking straws unit functional), TableS9: Process data inventory of bio-plastic straws production obtained from SuperPro Designer simulator (based on 100 drinking straws unit functional) (continue),

Table S10: Process data inventory of paper straws production obtained from SuperPro Designer simulator (based on 100 drinking straws unit functional), Figure S1: Overall process flow diagram of the corn starch production (Section 1), Figure S2: Process flow diagram of the lactic acid production (Section 2), Figure S3: Process flow diagram of the bio-plastic production (Section 3), Figure S4: Process flow diagram of the bio-plastic straws production (Section 4), Figure S5: Process flow diagram of the delivery consumer and disposal of bio-plastic straws (Section 5), Figure S6: Process flow of the production of paper straws (Section 2), Figure S7: Process flow diagram of the delivery to consumer and disposal of paper straws (Section 5), Figure S8: Transportation detail for the overall process for bio-plastic straws from gate to grave, Figure S9: Transportation detail for the overall process for bio-plastic straws from gate to grave.

Author Contributions: C.-H.M.: designed the LCA model and the simulation of the processes and analyzed and interpreted the data. He also wrote the manuscript with input from all authors. L.-S.T.: was involved in planning and supervised the work, reviewed the results and gave final approval of the version to be published. N.F.S.: provided a critical review of the manuscript. A.M.S.: was involved in impact assessment and interpretation. J.T.: contributed to the study design and conceptual framework, reviewed the results and gave final approval of the version to be published. All authors have read and agreed to the published version of the manuscript.

Funding: This research was funded by Universiti Teknologi PETRONAS via the Joint Research Project (JRP8) funding and Universiti Teknologi Malaysia via Matching Grant (PY/2021/00347).

Institutional Review Board Statement: Not applicable.

Informed Consent Statement: Not applicable.

Data Availability Statement: The data presented in this study are available in the article and Supplementary Materials.

Acknowledgments: Inputs and suggestions from Yeap Swee Pin from UCSI University are gratefully appreciated.

Conflicts of Interest: The authors declare no conflict of interest.

References

1. Pavani, P.; Rajeswari, T.R. Impact of Heavy Metals on Environmental Pollution. *J. Chem. Pharm. Sci.* **2014**, *94*, 87–93.
2. Tibbetts, J. Managing Marine Plastic Pollution. *Environ. Health Perspect.* **2015**, *123*, A90–A93. [CrossRef]
3. Malaysian Plastics Manufacturers Association. *An Advanced Plastics Recycling Industry for Malaysia*; Malaysian Plastics Manufacturers Association: Petalin Jaya, Malaysia, 2019; pp. 1–31.
4. Darbre, P.D. Overview of air pollution and endocrine disorders. *Int. J. Gen. Med.* **2018**, *11*, 191–207. [CrossRef] [PubMed]
5. Wilcox, C.; van Sebille, E.; Hardesty, B.D.; Estes, J.A. Threat of plastic pollution to seabirds is global, pervasive, and increasing. *Proc. Natl. Acad. Sci. USA* **2015**, *112*, 11899–11904. [CrossRef] [PubMed]
6. Butterworth, A.; Clegg, I.; Bass, C. *Untangled. Marine Debris: A Global Picture of the Impact on Animal Welfare and of Animal-Focused Solutions*; World Society for the Protection of Animals: London, UK, 2012; pp. 1–78.
7. Prendiz, J.; Mena, M.; Vega-Baudrit, J.R. Polylactic Acid (PLA) As A Bioplastic And Its Possible Applications in the Food Industry. *Food Sci. Nutr.* **2019**, *5*, 048.
8. Boonniteewanich, J.; Pitivut, S.; Tongjoy, S.; Lapnonkawow, S.; Suttiruengwong, S. Evaluation of carbon footprint of bioplastic straw compared to petroleum based straw products. *Energy Procedia* **2014**, *56*, 518–524. [CrossRef]
9. Gironi, F.; Piemonte, V. Bioplastics and petroleum-based plastics: Strengths and weaknesses. *Energy Sources A Recover. Util. Environ. Eff.* **2011**, *33*, 1949–1959. [CrossRef]
10. Guinée, J.B. *Handbook on Life Cycle Assessment*; Kluwer Academic Publishers: Dordrecht, The Netherlands, 2002; Volume 7.
11. Heijungs, R.; Sangwon, S. *The Computational Structure of Life Cycle Assessment*; Kluwer Academic Publishers: Dordrecht, The Netherlands, 2002; Volume 11.
12. Wu, W. Carbon Footprint A Case Study on the Municipality of Haninge. Master's Thesis, School of Technology and Health (STH), Stockholm, Sweden, 2011.
13. Kunhikrishnan, A.; Thangarajan, R.; Bolan, N.; Xu, Y.; Mandal, S.; Gleeson, D.; Seshadri, B.; Zaman, M.; Barton, L.; Tang, C.; et al. Functional Relationships of Soil Acidification, Liming, and Greenhouse Gas Flux. *Adv. Agron.* **2016**, *139*, 1–71.
14. Hunter, K.A.; Liss, P.S.; Surapipith, L.V.; Dentener, F.; Duce, R.; Kanakidou, M.; Kubilay, N.; Mahowald, N.; Okin, G.; Sarin, M.; et al. Impacts of anthropogenic SOx, NOx and NH3 on acidification of coastal waters and shipping lanes. *Geophys. Res. Lett.* **2011**, *38*, 2–7. [CrossRef]
15. Behera, D.P.; Kolandhasamy, P.; Sigamani, S.; Devi, L.P.; Ibrahim, Y.S. A preliminary investigation of marine litter pollution along Mandvi beach, Kachchh, Gujarat. *Mar. Pollut. Bull.* **2021**, *165*, 112100. [CrossRef]

16. Statista. 2020. Available online: https://www-statista-com.iclibezp1.cc.ic.ac.uk/statistics/726724/weight-of-most-common-items-found-in-global-oceans/ (accessed on 25 May 2021).
17. The Star. Spare that Straw, Please. Available online: https://www.thestar.com.my/news/nation/2017/08/20/spare-that-straw-please-a-seemingly-harmless-tool-that-helps-you-enjoy-your-cold-drinks-is-sapping-t/ (accessed on 17 May 2021).
18. The Sun Daily. The Ban on Single-Use Plastic Straws. Available online: https://www.thesundaily.my/local/the-ban-on-single-use-plastic-straws-DL1836630 (accessed on 17 May 2021).
19. Wong, H.-Y.; Goh, Y.-N.; Kiumarsi, S. *Single-Use Straw Ban: Malaysians Behavioural Intentions Towards Reusable Straws*; Contemporary Issues in Business and Management, Graduate School of Business, Universiti Sains Malaysia: Penang, Malaysia, 2019.
20. Chitaka, T.Y.; Russo, V.; von Blottnitz, H. In pursuit of environmentally friendly straws: A comparative life cycle assessment of five straw material options in South Africa. *Int. J. Life Cycle Assess.* **2020**, *25*, 1818–1832. [CrossRef]
21. Rana, K. Plasticless: A Comparative Life-Cycle, Socio-Economic, and Policy Analysis of Alternatives to Plastic Straws. Master's Thesis, Michigan Technological University, Houghton, MI, USA, 2020.
22. Zanghelini, G.M.; Cherubini, E.; Dias, R.; Kabe, Y.H.O.; Delgado, J.J.S. Comparative life cycle assessment of drinking straws in Brazil. *J. Clean. Prod.* **2020**, *276*, 123070. [CrossRef]
23. Chang, L.; Tan, J. An Integrated Sustainability Assessment of Drinking Straws. *J. Environ. Chem. Eng.* **2021**, *9*, 105527. [CrossRef]
24. Ramírez, E.C.; Johnston, D.B.; McAloon, A.J.; Singh, V. Enzymatic corn wet milling: Engineering process and cost model. *Biotechnol. Biofuels* **2009**, *2*, 1–9. [CrossRef] [PubMed]
25. Mofokeng, J.P.; Luyt, A.S.; Tábi, T.; Kovács, J. Comparison of injection moulded, natural fibre-reinforced composites with PP and PLA as matrices. *J. Thermoplast. Compos. Mater.* **2012**, *25*, 927–948. [CrossRef]
26. Jawjit, W.; Kroeze, C.; Soontaranun, W.; Hordijk, L. An analysis of the environmental pressure exerted by the eucalyptus-based kraft pulp industry in Thailand. *Environ. Dev. Sustain.* **2006**, *8*, 289–311. [CrossRef]
27. Horan, E.; Norrish, M. Waste management options to control greenhouse gas emissions—Landfill, compost or incineration? In *Paper for the International Solid Waste Association (ISWA) Conference*; ISWA: Vienna, Austria, 2009.
28. Branchini, L.; Cagnoli, P.; de Pascale, A.; Lussu, F.; Orlandini, V.; Valentini, E. Environmental assessment of renewable fuel energy systems with cross-media effects approach. *Energy Procedia* **2015**, *81*, 655–664. [CrossRef]
29. Morawicki, R.O.; Hager, T. Energy and Greenhouse Gases Footprint of Food Processing. *Encycl. Agric. Food Syst.* **2014**, *3*, 82–99.
30. Poh, G.K.X.; Chew, I.M.L.; Tan, J. Life Cycle Optimization for Synthetic Rubber Glove Manufacturing. *Chem. Eng. Technol.* **2019**, *42*, 1771–1779. [CrossRef]
31. Kuan, C.K.; Foo, D.C.Y.; Tan, R.R.; Kumaresan, S.; Aziz, R.A. Streamlined life cycle assessment of residue utilization options in Tongkat Ali (Eurycoma longifolia) water extract manufacturing process. *Clean Technol. Environ. Policy* **2007**, *9*, 225–234. [CrossRef]
32. Perumal, A.; Timmons, D. Contextual Density and US Automotive Carbon Dioxide Emissions across the Rural–Urban Continuum. *Int. Reg. Sci. Rev.* **2017**, *40*, 590–615. [CrossRef]
33. U.S. Department of Environment. *Energy and Environmental Profile of the U.S. Pulp and Paper Industry*; U.S. Department of Environment: Washington, DC, USA, 2005.
34. Chaffee, C.; Yaros, B.R. Life Cycle Assessment for Three Types of Grocery Bags -Recyclable Plastic; Compostable, Biodegradable Plastic; and Recycled, Recyclable Paper. 2007. Available online: https://plastics.americanchemistry.com/Life-Cycle-Assessment-for-Three-Types-of-Grocery-Bags.pdf (accessed on 2 June 2021).
35. Ciroth, A.; Fleischer, G.; Steinbach, J. Uncertainty calculation in life cycle assessments: A combined model of simulation and approximation. *Int. J. Life Cycle Assess.* **2004**, *9*, 216–226. [CrossRef]

 *processes*

Article

Removal of Anionic and Cationic Dyes from Wastewater Using Activated Carbon from Palm Tree Fiber Waste

Basma G. Alhogbi [1,*], Shoruq Altayeb [1], Effat. A. Bahaidarah [1] and Mahmoud F. Zawrah [2]

[1] Department of Chemistry, Faculty of Science, King Abdulaziz University, Jeddah 21589, Saudi Arabia; sfateel0001@stu.kau.edu.sa (S.A.); ebahaidarah@kau.edu.sa (E.A.B.)

[2] National Research Centre, Center of Excellence for Advanced Sciences-Nano Group, 12622 Dokki, Cairo, Egypt; mzawrah@hotmail.com

* Correspondence: balhogbi@kau.edu.sa or alhogbib@gmail.com

Abstract: This study focuses on using a facile method for the green preparation of activated carbon (AC) from palm tree fiber (PTF) waste. The synthesized cost-effective AC was investigated for the removal of an anionic dye (Congo red, CR) and a cationic dye (Rhodamine B, RhB) from wastewater. The morphological and structural characterization of the synthesized AC were performed by scanning electron microscopy (SEM), transmission electron microscopy (TEM), surface area, Fourier transform infrared spectroscopy (FTIR), total pore volume, average pore diameter and pore size distribution, zeta potential, and zero-point charge. To investigate the adsorption efficiency, different parameters such as adsorbent dosage, solution pH, initial dye concentration, and duration were applied using the batch experiments. Various adsorption isotherm and kinetics models were applied to study the adsorption mechanism and dynamics. The results showed that chemical activation with a weak acid (H_3PO_4) at 400 °C for 30 min is a fast method for the activation of each precursor and produces a high yield. The result of analysis showed an increase in the adsorption capacity at pH 2. The maximum adsorption capacity was 9.79 and 26.58 mg g^{-1} at 30 min for CR dye and RhB dye, respectively. The optimum adsorbent dosage for the activated carbon from palm tree fiber (PTFAC) was 0.15 g with a high percentage removal of CR (98.24%) and RhB (99.86%) dyes. The adsorption isotherm and kinetic studies were found to be favorable and feasible for assessing the adsorption of dyes with the Langmuir model and pseudo-second-order reaction, respectively. In addition, the AC showed reusability up to five cycles. The results showed that the synthesized AC was environmentally friendly and successfully removed dyes from wastewater.

Keywords: activated carbon; adsorption; anionic and cationic dyes; palm tree fiber wastes; recycles

Citation: Alhogbi, B.G.; Altayeb, S.; Bahaidarah, E..A.; Zawrah, M.F. Removal of Anionic and Cationic Dyes from Wastewater Using Activated Carbon from Palm Tree Fiber Waste. *Processes* **2021**, *9*, 416. https://doi.org/10.3390/pr9030416

Academic Editors: Monika Wawrzkiewicz and Pau Loke Show

Received: 25 January 2021
Accepted: 19 February 2021
Published: 25 February 2021

Publisher's Note: MDPI stays neutral with regard to jurisdictional claims in published maps and institutional affiliations.

1. Introduction

The continuous growing population and their requirements have seriously contaminated water with several undesirable materials, for example, artificial dyes [1]. Synthetic dyes obtained from organic or inorganic compounds are mainly composed of two compounds—chromophores and auxochromes. The chromophore produces the color of dye, whereas the auxochrome is responsible for the intensity of color [2,3]. Before the invention of artificial dyes in 1956, the natural dyes obtained from plant sources as roots, berries, bark, leaves, wood, fungi, and lichens were used [4]. After 1956, artificial dyes derived from petrochemicals were used as coloring agents for several products in the market. Additional engaging colors have been introduced to various advanced industries such as textiles, paper, animal skin tanning, food processing, plastic, cosmetic, and dye-producing industries [5]. Synthetic dyes are usually carcinogenic, cause serious damage to water resources, and contribute to environmental pollution. The environmental and health concerns associated with the wastewater effluents have led scientists to search for simple, inexpensive, and rapid solutions. The putrefaction of dyes from wastewater can be done using three methods, namely physical, biological, and chemical methods [6]. These include precipitation, oxidation, coagulation, adsorption, and membrane separation [7–9]. Among these methods, adsorption is a cost-effective, simple, and feasible technique for the removal of organic dyes from wastewater. Many synthetic ingredients have been effectively used as adsorbents for the elimination of contaminants from water [10–12].

Among them, activated carbon (AC) is considered as the most commonly used adsorbent in the industry for the removal of contaminants [13]. The pore structure of the AC is quite developed with a high surface area and good adsorption capacity. The external chemical functional groups and structural properties of AC make it suitable for many applications used for the elimination of organic and inorganic contaminants from contaminated water [14,15]. However, the fabrication of marketable AC is costly; thus, research attention has been paid to develop a low-cost and competent AC using several sources. The production of AC from natural sources or agroindustry wastelands was considered because they are renewable, highly available, and cost-effective, and they have good properties. Thus, their disposal is required for a cleaner environment. In this regard, many researchers have reported on the preparation of AC from various industrial denim fabric waste [16] and agriculture waste (biomasses) such as coconut coir, apricot stones [17], Nigella sativa L [18], palm shell [19], pecan shell [20], pine cone [21], tapioca peel [22], macadamia nutshell [23], Mucuna pruriens and Manihot esculenta [24], snail shell [25], and tea leaves [26].

There are two methods to produce AC: physical and chemical activation. In physical activation, raw material is carbonized and then activated with steam and carbon dioxide. Chemical activation involves the impregnation of a chemical-activating agent in the precursor material followed by activation at temperatures between 400 and 700 °C under the nitrogen atmosphere [27]. The advantages of chemical activation are the following: (a) it occurs in a shorter time compared to that prepared by physical activation; (b) the produced AC has a high surface area with high yield; and (c) the activating chemical agent influences the breakdown of carbon and prevents the development of tar and volatile substance, thereby increasing the yield of AC. In addition, in the case of chemical activation, the process of dehydration and oxidation requires lower temperatures compared to that in the physical activation [28]. In the chemical activation method, $ZnCl_2$, KOH, and H_3PO_4 are extensively used as chemical-activating agents for the activation of lignocellulosic materials, which have not been carbonized. In addition, KOH is used for the activation of coal precursors or chars. The most common of activating agents include KOH and $ZnCl_2$ since they provide microporous features, which restricts these materials' applications for the adsorption of high molecular weight molecules in aqueous solution [16]. Thus, attempts were made toward the use of H_3PO_4 as a proper and low-cost activating agent for of AC as reported earlier for many biosorbent lignocellulosic material [16]. In the literature, various parameters were studied such as the concentration of activation agent, temperature, and activation time [29–31].

The present study focuses on the facile preparation of AC from biomass by chemical and thermal activation methods. The vast number of date palm trees in Saudi Arabia is about 23 million. Thus, palm tree fiber (PTF) is one of the greatest freely existing biomasses, and it is common agricultural waste in Saudi Arabia. Therefore, PTF has been chosen as the specific biomass in order to produce more economical Ac and easy accessibility collection [32]. The parameters such as the activation agent, activation temperature, and activation time were investigated for PTF activation in this study. Another goal of the study was to use various analytical techniques to examine the efficiency of the prepared AC to remove anionic dye Congo red (CR) and cationic dye rhodamine B (RhB) from polluted water. The influence of initial concentration, duration, adsorbent dosage, and pH on the adsorption of CR and RhB were studied. Performing equilibrium isotherm and kinetics of adsorption for CR and RhB as well as exploring the re-generation of used activated carbon from palm tree fiber (PTFAC) were also considered as the main objectives.

2. Materials and Experimental Procedures

2.1. Instruments

The pH values of dyes and adsorbent powder were measured using a pH meter (MP220; Mettler Toledo). A cyclone mill (CT 193 *Cyclotec^{TM}*; Mill Collection) was used to ground the sample, and the sample was sifted by a KimLab *ISO 3310* Std Test Sieve; the required size was a 0.43 mm mesh size. The mixtures were agitated using a reciprocating shaker (cat. no. 3006; GFL Shakers). The samples were carbonized in a muffle furnace (A-550; Vulcan), and after washing, they were dried in an oven (FN 055/120 Dry Heat Sterilizers). The spectroscopic reading was generated by using a UV-vis spectrophotometer (UV-1650PC, Shimadzu). Transmission electron microscopy (TEM) was performed using JEM-1230 (JEOL). The zeta potential of the PTFAC was measured by a Zetasizer (version 7.04, serial number MAL 1074157; Malvern Instruments). Scanning electron microscopy (SEM) was performed by Quanta FEG250. Fourier transform infrared spectroscopy (FTIR) was conducted by using a Spectrum Two FTIR Spectrometer (PerkinElmer) at the absorption spectra range of 4000–400 cm^{-1}.

2.2. Materials and Chemical Reagent

Palm tree fiber waste was collected from the locally available palm trees on the campus of King Abdulaziz University, Jeddah, Saudi Arabia. The PTF was washed multiple times with hot distilled water to eliminate dust and impurities and then dried in sunlight for a few days. The dried palm fiber was crushed to powder form, sifted to the desired particle size, and used to prepare the AC. The obtained granular particle size of the pulverized fiber was approximately 0.43 mm. Raw materials and the prepared AC are shown in Figure 1.

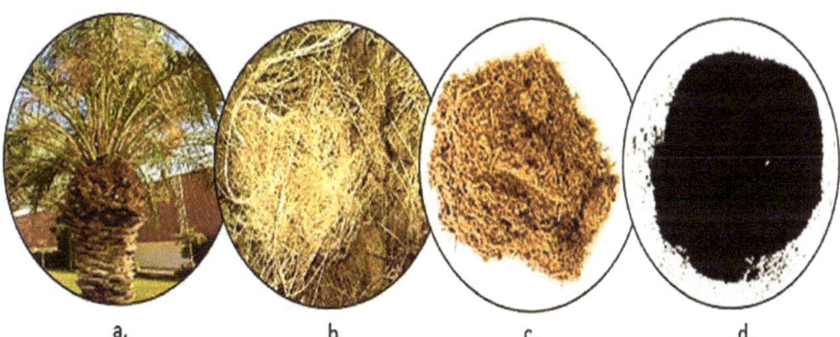

Figure 1. (**a**) Date palm tree, (**b**) Palm fibers, (**c**) Palm fibers powder, and (**d**) Activated carbon powder.

Congo red anionic dye (HiMedia) and RhB cationic dye (Sigma Chemical) were used as adsorbates (Table 1). Stock dyes solution (1000 mg L^{-1}) was made by dissolving a suitable amount of dye powder, which was accurately weighed on an electronic balance using deionized water. During the adsorption experiments, a diluted dye solution that had a concentration between 5 and 50 mg L^{-1} was prepared from the stock solution. The chemicals reagents such as H_3PO_4, H_2SO_4, HCl, and KOH were supplied by Fisher Scientific.

2.3. Chemical and Thermal Activation of Carbon from Palm Tree Fiber

The activation was carried out by impregnation of the PTF with various chemical agents as such as phosphoric acid (85%), sulfuric acid (98%), and potassium hydroxide (2 M) with an impregnation ratio of chemical/biomass being 3:1. The duration of the process was 24 h. The experiment was done under full safety precautions. Inside a fume hood, the chemical agent was added gradually to PTF to obtain a mixture. Then, the slurry was filtered, washed, and dehydrated in a dryer oven at 130 °C for 2 h. The designed samples of chemically activated carbon PTF were termed as CAPTF-H_3PO_4, CAPTF-H_2SO_4, and CAPTF-KOH. The pyrolysis process was carried out to increase the efficiency of the fibers activated by phosphoric acid (85%). The fibers were carbonized at 200, 300, 400, 500, 600, and 700 °C several times (30, 60, 120, and 180 min) in a muffle furnace using a covered porcelain crucible for reduced oxygen. Then, they were cooled at room temperature and soaked in 1.0 M NaOH for 2 h. After soaking, the obtained materials were washed thoroughly with hot distilled water followed by cold distilled water to remove all free acid until they attained a neutral pH. Then, they were oven-dried at 130 °C for 2 h. Finally, the dried materials were milled into a fine powder and kept in a plastic bag until used. The samples produced by activated carbon palm tree fiber were termed as PTFAC, as shown in Figure 1.

Table 1. Properties of synthetic dyes.

Dyes	Rhodamine B (RhB)	Congo Red (CR)
Molecular formula	$C_{28}H_{31}N_2O_3Cl$	$C_{32}H_{22}N_6Na_2O_6S_2$
Molecular structure		
Molecular weight (g/mol)	479.02	696.66
Chemical/Dye class	Xanthene dye	Diazo dye
λ_{max} (nm)	554 nm	496 nm

The adsorption process was performed by the batch-adsorption method. To select the fiscal chemical-activating agent (i.e., H_3PO_4, H_2SO_4, or KOH), the best pyrolysis temperature (200, 400, 600, or 800 °C), and suitable time (30, 60, 80, and 120 min) to remove CR and RhB dyes, the synthesized PTFAC were conducted under the following conditions: volume of dye solution = 20 mL, initial concentration = 25 mg L^{-1}, contact time = 120 min, temperature = 25 °C, adsorbent dose = 0.1 g, agitation speed = 300 rpm, pH = natural.

2.4. Characterization of AC

The specific surface area, total pore volume, average pore diameter, and pore size distribution of the prepared AC were examined by Brunauer–Emmett–Teller (BET) N_2 adsorption method. The presence of porosity, microanalysis, and surface morphology of the PTFAC carbonized at 400 °C were studied by using the SEM and TEM. The FTIR technique was used to study the effect of chemical treatment on biomasses and to recognize various functional groups present on the surface of the materials. This technique was used to examine the surface functional groups of the ACs.

Zeta potential measurement was used for determining the surface charge of AC particles in the solution. In other words, the possible difference between the dispersion medium and the stationary layer of fluid attached to the particle was measured by zeta potential. The pH of the solution with distilled water was measured as follows: the mixture was prepared in a ratio of 10 mL water to 1.0 g of carbon; this mixture was stirred, and the pH was measured several times until a constant value was reached. The pH of zero-point charge (pHzpc) for AC was determined by the batch equilibrium method. During the measurement, the solid/liquid ratio was 1:1000; that is, 0.05 g AC was added to 50 mL of water. This suspension was left for 48 h, and the pH was measured and found to be 3. In another experiment, 0.05 g of AC was mixed with 50 mL of 0.1 M NaCl, and the pH of solution was fixed between 2 and 12 by 0.1 M hydrochloric acid or 0.1 M sodium hydroxide. The solution was stirred for 48 h on a shaker bath at an agitation speed of 120 rpm and at room temperature (25 °C). Thereafter, the final pH values were recorded, and the difference between the original and final pH values ($\Delta pH = pH_{initial} - pH_{final}$) was calculated and compared with the initial ones.

2.5. Adsorption Processing

The adsorption of dyes onto PTFAC was assessed using batch adsorption experiments. The procedure was performed as follows: First, a series of dye solutions with initial concentrations of 5.0–50 mg L^{-1} was prepared. Second, dye solutions with different pH values (2.0–12) were prepared. Finally, 20 mL of dye solution was added into 50 mL volumetric flasks with different amounts of PTFAC adsorbent doses (0.05–0.25 g). Then, the mixtures were stirred in a shaker at an agitation speed of 250 rpm at different times (5.0–120 min) at room temperature. The solutions were filtered using the Whatman Qualitative Filter Paper Grade 2. The dye concentration was determined by using the UV-visible spectrophotometer at wavelength 554 nm and 496 nm for RhB and CR dyes, respectively. All the experiments were performed in triplicate, and the mean values were recorded. The amount of all dyes adsorbed by the PTFAC and the removal percentage was calculated by Equations (1) and (2), respectively:

$$Removal\ \% = \frac{C_o - C_e}{C_o} \times 100 \tag{1}$$

$$q_e = \frac{(C_o - C_e)}{m} \times V \tag{2}$$

where q_e is the quantity of adsorbed dye on 0.1 g of PTFAC (mg g^{-1}), C_o is the original dye concentration (mg L^{-1}), C_e is the residual concentration of the solution at equilibrium (mg L^{-1}), V is the volume of dye solution (L), and m is the amount of the adsorbent PTFAC (g).

2.6. Desorption Studies and Regeneration

To understand the sustainability of regenerating the AC and make the process more economical, desorption trials were performed to determine the efficiency of the AC. The traits of an excellent adsorbent should include not only high adsorption capacity but also reduction of the overall cost of the adsorbent by possessing high desorption capacity. Desorption studies of PTFAC were carried out using different concentrations of acetic acid CH_3COOH, sodium hydroxide NaOH, and different pH values of distilled water. The R-PTFAC is a sample after adsorption used to study the regeneration and recycle. Thereafter, 0.1 g R-PFAC was added to 50 mL desorption solutions (0.1, 0.5, 1.0 M NaOH; 0.1, 0.5, 1.0 M CH_3COOH) and distilled water and the pH was adjusted (2.5, 4.5, 6.5, 8.5, and 10.5). Then, it was stirred in a shaker at 250 rpm for 3 h at room temperature. After desorption, R-PTFAC was washed with distilled water until it became neutral and was reused for the removal of RhB and CR dyes from solution. The adsorption–desorption studies were performed for five sequential cycles. The dye concentration in the filtrate was examined by the UV–vis spectrophotometer. The effectiveness of the desorbed dye from AC was evaluated by Equation (2).

2.7. Application of Real Water Samples

The real water samples were collected from the water sewage treatment plant and tap water at King Abdulaziz University, Jeddah City. The well water and seawater samples were collected from Rabigh City. The samples were applied by the adsorption of dyes onto the PTFAC using batch experiment. Under the same condition of the adsorption process, the removal percentage was calculated by Equation (2).

3. Results and Discussion

3.1. Physical Properties, Characterization, and Morphology of Synthesized PTFAC

Surface area, total pore volume, and mean pore diameter of the prepared AC, as determined by adsorption/desorption isotherm of nitrogen, are shown in Figure 2a. As a result, the prepared AC has a 648.90 m^2 g^{-1} (BET) specific surface area and 2.83 cm^3 g^{-1} total pore volume. It is noteworthy to mention that the specific surface area for the raw material before activation was 9.0 m^2 g^{-1} [33]. As per the results of adsorption/desorption isotherms presented in Figure 2a,b, the AC carbon follows type V isotherm of mesoporous materials with a 0.79 nm mean pore diameter, and the process is proceeded with multilayer adsorption. The data found are in line with the surface area value reported in the literature based on the modification method such as chemicals, physical, or pyrolysis activation [17].

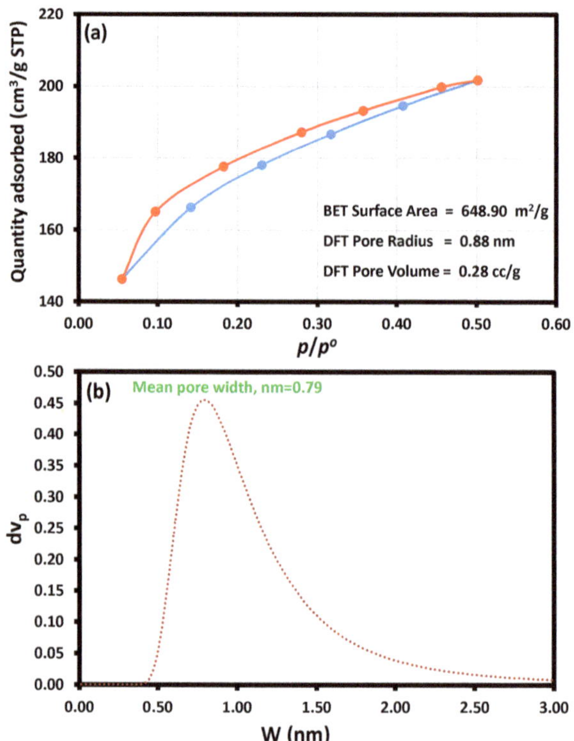

Figure 2. (**a**) Adsorption/desorption isotherm and (**b**) pore size distribution of activated carbon from palm tree fiber (PTFAC).

On the other hand, the zeta potential and zero-point charge of the prepared AC are presented in Figure 3. The prepared PTFAC shows a zeta potential value of −28.3 mV and has high surface electronegativity. The pH values for PTFAC in a solution and pHzpc were 3.56 and 3, respectively. This means that the surface of PTFAC behaves as positively charged under a pH value of 3 and negatively charged beyond a pH value of 3 [34]. In addition, similar values of pHzpc have been reported for white pine sawdust (pHzpc 3.65) [35] and Virgin Granular AC activation by HNO_3 (pHzpc 3.0) [36], which are close to the value observed in this study.

Figure 4 shows the SEM image of the palm fiber carbonized at 400 °C and TEM image of the AC after grinding. Figure 4a show that after thermal treatment, the fiber is converted into a skeleton of carbon grains with a size range between 1 and 4 μm. The TEM image of PTF shows a rough surface and porosity (Figure 4a). Meanwhile, the SEM image of raw biomass PTF is reported in ref. [33]. The PTF has a rough surface with no evidence of porosity and is covered by arrays of protrusions, which can be Si particles. The porosity of carbon is due to the decomposition of lignin, cellulose, and hemicellulose during carbonization, resulting in the formation of micropores and mesopores [37,38]. The cell structure of the PTFAC can be seen clearly by TEM. After grinding, the particles were converted into rounded and edged grains with a size range of 20–50 nm (Figure 4b).

Furthermore, the TEM image of raw biomass PTF indicates that the cell wall structure of the fiber is almost round [36].

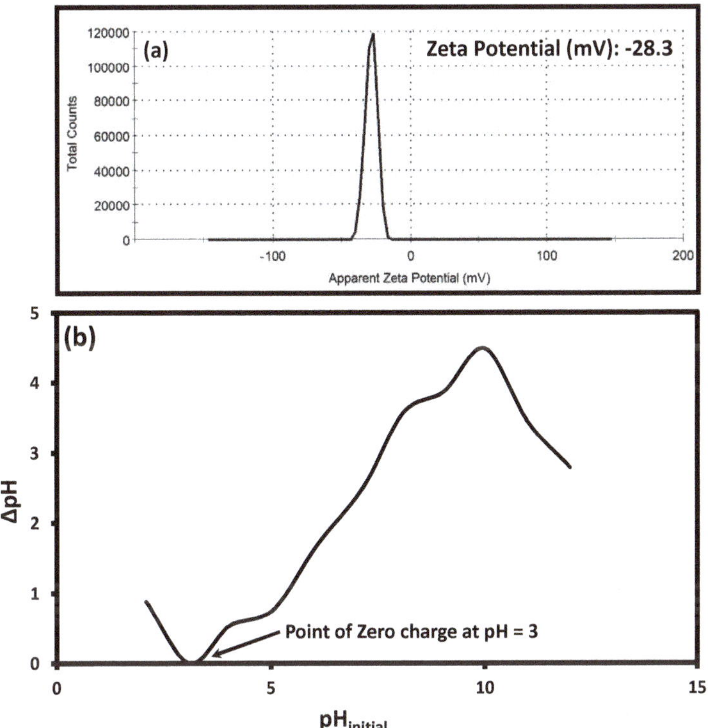

Figure 3. (**a**) Zeta potential and (**b**) Point of zero charge of PTFAC.

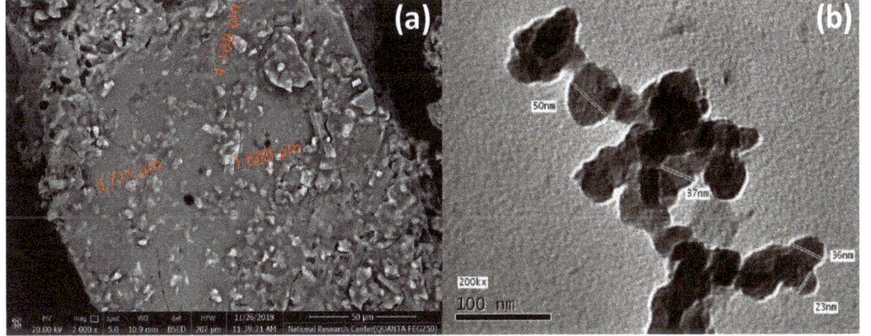

Figure 4. (**a**) SEM image of palm tree fiber carbonized at 400 °C and (**b**) TEM image of PTFAC.

The examination of the prepared AC before and after the adsorption of dyes by FTIR can give better insight into the expected attraction forces between the PTFAC and dyes. Figure 5 shows the FTIR spectra of the PTF and after the adsorption of CR and RhB dyes by PTFAC. As shown in Figure 5, the following function groups are detected in the spectra: the band observed at around 3335 cm^{-1} in the fiber and PTFAC spectra was attributed to the O–H stretching of the hydroxyl group of alcohol present in cellulose and lignin, and the band at 1625 cm^{-1} was attributed to O–H bending of adsorbed water [39,40]. The band at 2900 cm^{-1} was ascribed to the unsaturated C–H stretching vibration of the alkene or aromatic groups [39]. The band observed at around 2925 cm^{-1} in the spectrum of fiber was attributed to C–H vibration of organic matters in the fiber, which had nearly vanished in the PTFAC, indicating that most of the elements such as hydrogen and oxygen are removed after activation [16,41]. The bands at around 1605–1620 cm^{-1} were attributed to the C=C stretching vibration in the aromatic rings [42]. The bands between 1238.81 cm^{-1} and 1032.31 cm^{-1} were attributed to the C–O stretching in carboxylic groups present in cellulose, typically originating from the oxidized carbon [18,20]. The FTIR spectrum of PTFAC demonstrates significant change by lowering peaks intensities when contrasted to the PTF spectrum. The bands between 1220 and 1118 cm^{-1} can also ascribed to phosphorous species, such as hydrogen-bonded P=O, stretching vibrations O–C in P–O–C of aromatic, P=OOH, and polyphosphate chain P–O–P [29,31]. In addition, the band around 1032.31 cm^{-1} assigned to the silicon atom initially attached to the oxygen atom Si–O [43]. The bands observed at around 470 cm^{-1}, the spectra of carbons, were assigned to the siloxane Si–O–Si group [43]. When the surface of PTFAC was covered by dyes, most of its characteristic bands disappeared or mostly evanesced due to the adsorption of dye on its surface. For example, O–H, C=C, and C–O bands and all types of phosphorous species bands and Si–O–Si groups disappeared. This observation enhances the electrostatic attraction forces between the adsorbent and adsorbed dyes through noncovalent interactions, which may include dipole–dipole interaction, intermolecular forces, and hydrogen bonding interaction.

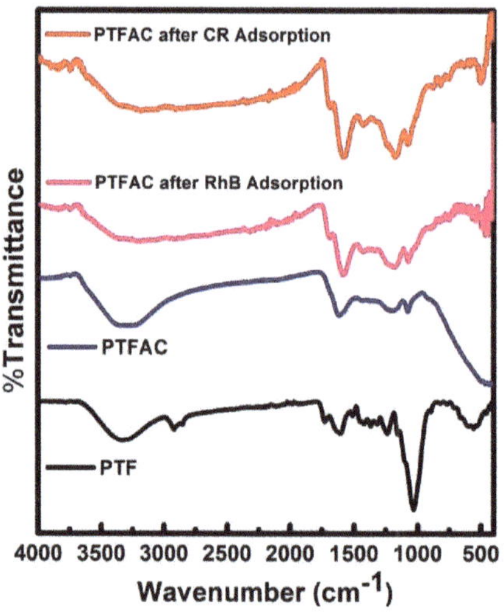

Figure 5. The comparison of Fourier transform infrared spectroscopy (FTIR) spectra of palm tree fiber (PTF) before and after the activation and after adsorption of Rhodamine B (RhB) and Congo red (CR) dyes by PTFAC.

3.2. Optimization of Preparation Conditions for PTFAC

To optimize the preparation conditions of PTFAC, adsorption trial tests were carried out using PTF activated with different chemical agents such as H_3PO_4 (85%), H_2SO_4 (98%), and KOH (2 M) with an impregnation ratio of 3:1 (chemical/biomass). In addition, adsorption trial tests were conducted on the AC prepared at different temperatures at various times. Figure 6 shows the removal percentage of CR and RhB dyes by PTF activated by H_2SO_4, H_3PO_4, and KOH. From the mentioned dyes, the fiber activated by H_2SO_4 showed the highest removal percentage; hence, the H_2SO_4 is considered a strong oxidizing and dehydrating agent. As a result of the strong acid and base, H_2SO_4 and KOH exert a high corrosive effect on the precursor that produces weight loss of the precursor, which is less than that of the raw material. Indeed, the activation of AC by H_3PO_4 improves the pore structure during the thermal treatment, which improves the porosity in terms of macropores and mesopores of the PTFAC [8,44]. Thus, the activation by H_3PO_4 is highly recommended for functionalization of the surface of the used bioadsorbent that has an environmentally friendly in nature and easy to recover and recycle properties [17,45].

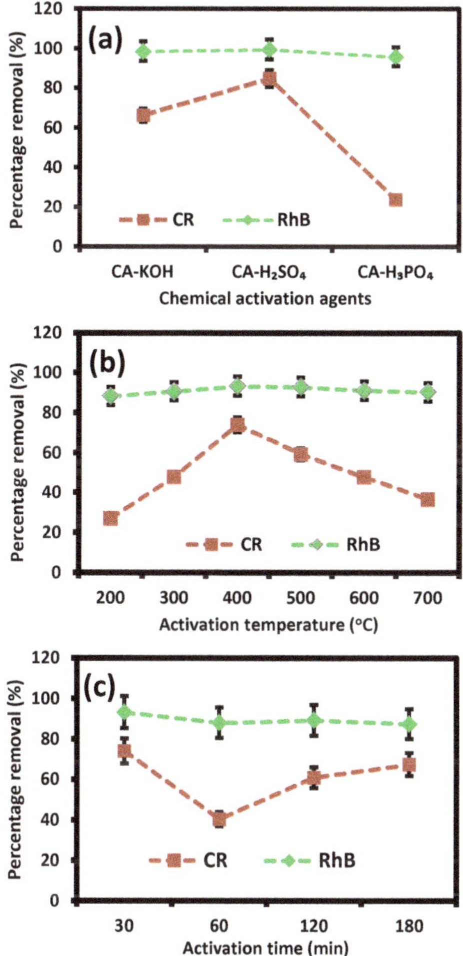

Figure 6. The removal percentage of CR and RhB dyes on the synthesized PTFAC with different (**a**) chemical activation reagents, (**b**) activation temperature of CA-PTF-H_3PO_4, and (**c**) activation times of CA-PTF-H_3PO_4 at 400 °C.

The removal percentage of dyes, adsorbed on PTFAC prepared at different temperatures and times, are shown in Figure 6b,c, respectively. Thermal treatment data revealed the chemical activation of AC with H_3PO_4 at 400 °C for 30 min in a facile and fast short analytical time process for AC activation. Thus, it can be concluded that the cellulose and hemicellulose in the PTF gradually decompose at 400 °C, resulting in a high specific surface area. On the other hand, a long activation time causes degradation and decreases the surface area and porosity of carbons, damaging the pores because of the excessive reaction with other activated agents. Hence, chemical activation with H_3PO_4 at 30 min is adopted in the subsequent study for dyes removal in batch mode.

3.3. Influence of Parameters on the Adsorption Process (pH, Duration, Initial Concentration, Adsorbent Dosage)

The batch experiment was done to study the influence parameters using $C_o = 25$ mg L^{-1}, PTFAC mass = 0.1 g, V = 20 mL, t = 25 °C). In the adsorption process, the charge on the adsorbent and the characteristics of adsorbate in solution are determined by the pH, which has a radical impact on the adsorption ability. Figure 7a shows the effect of pH between (2.0–12) on the removal percentage of CR and PhB dyes by PTFAC. Upon increasing the pH values, the removal percentage declined slightly from 98.3% to 95.2% in the case of CR and from 99.9% to 99.6% for RhB; apparently, the maximum adsorption is observed at pH 2.0. As the zero-point charge of the adsorbent is at pH 3, it means that the adsorption is high when pH is less than pHzpc and low when pH is more than pHzpc. However, the cation dye (RhB) and anion dye (CR) at different pH values show high percentage removal, which makes the PTFAC significant in the removal of wastewater, because dyes in wastewater are usually in a wide pH range. It is well known that the adsorption of dyes by the AC can proceed by van der Waals, electrostatic, and H-bonding interactions. The CR is a dipolar molecule; it exists in an anionic form with red color at basic pH and in a cationic form with blue color at acidic pH. It is also observed that the surface of AC is covered by phenolic and carboxylic groups due to oxidation from the atmosphere and water medium [46]. At lower pH, protonation of phenolic (PhOH) and carboxylic acid (COOH) groups occurs on the surface of PTFAC, which leads to an increase in electrostatic interactions with the cationic CR dye; therefore, the adsorption of CR on the AC surface is higher. However, at a higher pH, the carboxylic and phenolic groups on the surface of AC are expected to be completely ionized to carboxylate (COO−) and phenoxide (PhO−) ions. Thus, the electrostatic repulsion between the anionic–CR and anionic–AC surface decreases the adsorption capacity. In the case of RhB, the pH has no significant effect on the removal percentage. The RhB may exist in cationic, laconic, or zwitterionic forms depending on the pH of the solution medium. At lower pH, the RhB dye ions are cationic and present in a monomeric molecular form that could easily enter the pore structure of the carbon, resulting in higher adsorption. At higher pH, the RhB dye ions are zwitterionic and present in dimer molecular form. The electrostatic interaction between the carboxylate ion and xanthene groups in zwitterionic results in the formation of dimers of bigger molecular forms that cannot easily enter the pores of carbon, resulting in lower adsorption. The adsorption at all pH values in both the dyes shows a very insignificant variation, indicating no ion exchange has occurred. So, the physical forces such as van der Waals, hydrogen bonding and dipole–dipole interactions might participate in this developed adsorption [30]. Similar studies on the effect of pH at acidic media in removing RhB and CR dyes by other adsorbents reported that RhB and CR dyes have maximum adsorption at pH 3.0 [47].

The duration plays an important role in the adsorption process of RhB and CR dyes by PTFAC. Figure 7b shows that the removal percentage increases with the increase in the duration, with high values of percentage between the range (5.0–120 min). The dyes recorded a high absorption rate; their values were between 95.1% at 5 min and 99.5% at 120 min for CR dye, whereas they were 99.7% at 5.0 min and 99.9% at 120 min for RhB dye. It indicated that the PTFAC has a high efficiency in removing both adsorbed dyes. This is because of the very rapid adsorption that was found during the first 5.0 min, and even then, the amount of the adsorbate increased with time and reached a constant value of the maximum adsorption at 120 min. However, the availability of vacant active sites on the surface of the PTFAC is the result of the rapid increase in adsorption during the initial stage. Eventually, the process slows down due to the diffusion of dyes into the pores of the PTFAC, because the external sites were fully occupied [8]. A similar study reported an increase in the quantity of adsorption dyes with increased contact time. Using an initial concentration of 1.0 g L^{-1}, it was found that Raphia hookerie fruit epicarp could adsorb

312 mg/g RhB dye at acidic pH, and equilibrium was attained at 40 min for an initial dye concentration of 400 mg L^{-1} [47].

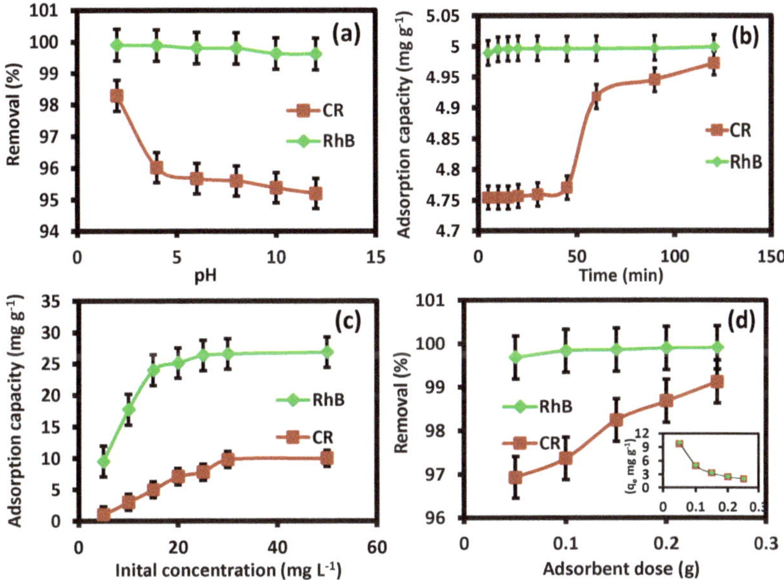

Figure 7. Influence of (**a**) pH, (**b**) contact time, (**c**) initial concentration, and (**d**) adsorbent dosage on elimination the RhB and CR dyes from solution by PTFAC (test conditions: T = 298 K, V = 20 mL).

The effect of the initial concentration parameter is significant in determining the effectiveness of the adsorbent. The influence of initial dye concentration 5.0 to 50 mg L^{-1} on the adsorption capacity of PTFAC is shown in Figure 7c. The adsorption capacities (q_e) were increased with the increase in the initial dye's concentration from 5.0 to 50 mg L^{-1}. The adsorption capacities were between 0.99 and 9.99 mg g^{-1} and 9.51 and 26.83 mg g^{-1} for CR and RhB days, respectively. After 25 mg L^{-1} initial concentration, no significant change was observed in the adsorption capacity; thus, the adsorption process reached the equilibrium. Initially, the adsorption process starts from the boundary layer; then, it diffuses into the adsorbent surface, and finally diffuses into the porous structure of the adsorbent [8]. In a similar study, an increase in the quantity of the dye adsorbed was observed on increasing the initial concentration of 1.0 g L^{-1} of R. hookerie fruit epicarp for adsorption capacity with 43.5 mg g^{-1} of RhB dye for an initial dye concentration of 50 mg L^{-1}, while it can have a high adsorption of 312 mg g^{-1} of RhB dye for an initial dye concentration of 400 mg L^{-1} [47].

The influence of adsorbent dosage (0.05 to 0.25 g) is shown in Figure 7d. The figure shows that the removal amount of RhB and CR dyes by PTFAC increased with the increase in the adsorbent dosage. Then, the equilibrium was attained after 0.1 g of PTFAC. There was not much increase in the removal percentage of the dye. The percentage adsorbed of CR dye increased from 97% (9.69 mg g^{-1}) for the adsorbent dosage of 0.05 g to 99% (1.99 mg g^{-1}) for the adsorbent dosage of 0.25 g. The adsorption capacity of RhB dye increased from 99.6% (9.96 mg g^{-1}) to 99.9% (1.99 mg g^{-1}) with an increase in the adsorbent dosage of 0.05 to 0.25 g, respectively. The high removal efficiency with a high amount of adsorbent (0.05–25 g) is due to the availability of binding sites for adsorption. The adsorption capacity (q_e) of the adsorbent was found to decrease with an increase in the adsorbent dosage, which could be due to the interaction of adsorbent particles such as aggregation or agglomeration, resulting in the decrease of the total surface area [8]. A similar study reported the increase in the percentage of adsorbed dye on increasing the adsorbent dosage [48]. Another study reported the RhB dye concentration to be 100 mg L^{-1}; at acidic pH, the percentage adsorption of RhB dye increased from 85.19% to 88.88% as the adsorbent dosage increased from 1.0 to 5.0 mg L^{-1} [47].

3.4. Adsorption Isotherms

Adsorption isotherm studies are very important as they describe how the adsorbed molecules are distributed between the PTFAC and the CR and RhB dyes in the solutions at equilibrium. They offer essential designing of the adsorption process by evaluating the adsorption effectiveness of the adsorbent. In this work, adsorption equilibrium data of CR and RhB dyes on PTFAC were analyzed by Langmuir and Freundlich adsorption isotherm models. The Langmuir isotherm model is the most extensively used system, which is based on a perfect hypothesis that adsorption occurs on the adsorbent at exact homogeneous sites and can be useful for monolayer adsorption [49]. The linear equation of the Langmuir isotherm model is presented below Equation (3).

$$\frac{C_e}{q_e} = \frac{1}{q\,K_l} + \frac{1}{q_{max}}C_e \tag{3}$$

where q_{max} is the maximum adsorption capacity and K_l (L/mg) is the Langmuir constant. q_{max} and K_l can be calculated by drawing C_e/q_e (specific adsorption) against C_e (Figure 8a). An additional important factor of Langmuir isotherm is the dimensionless separation factor (R_l), which provides an understanding on the favorability of the adsorption process. This has been given in Equation (4).

$$R_l = \frac{1}{1 + K_l\,C_o} \tag{4}$$

In the adsorption process, the separation factor R_l specifies whether the adsorption isotherm is unfavorable ($R_l > 1$), satisfactory ($0 < R_l < 1$), irreversible ($R_l = 0.0$), or linear ($R_l = 1.0$) [6,15]. The Freundlich isotherm model measures the adsorption capacity and intensity of the adsorbate on heterogeneous surfaces, but it does not predict the maximum adsorption [50]. The linear form of the Freundlich isotherm is expressed by Equation (5).

$$\log\,(q_e) = \log\,(K_f) + \frac{1}{n}\log\,(C_e) \tag{5}$$

As K_f and $1/n$ are the Freundlich constants, they help calculate the capacity and strength of adsorption, respectively. K_f and n can be obtained from the intercept and slope by drawing $\log q_e$ versus $\log C_e$, respectively (Figure 8b). The n value gives the degree of nonlinearity between the concentration of the solution and adsorbent. When $n = 1.0$, the adsorption is linear, whereas $n < 1$ suggests that the adsorption process is chemical. However, if $n > 1$, it means that the physical adsorption is satisfactory [15].

The Langmuir and Freundlich isotherms selected to explicate the interaction of CR and RhB dyes with PTFAC are shown in Figure 8. Figure 8a shows the Langmuir isotherm; as the linear correlation coefficient value (R^2) represents the accuracy of the model, it is found near the unity. The R^2 values for the CR were 0.996, and for RhB, these were 0.980. The Freundlich isotherm in Figure 8b shows R^2 values for CR and RhB to be 0.8575 and 0.8541, respectively. Thus, this suggests that the adsorption values of CR and RhB dyes are the best fit to the Langmuir adsorption model [49]. Based on the calculated data, the coefficients of Langmuir and Freundlich isotherms models are illustrated in Table 2. The Langmuir model shows a maximum adsorption capacity for CR dye at 10.4 mg g^{-1} and for RhB dye at 26.5 mg g^{-1} by PTFAC. The dimensionless separation factor (R_l) value found for CR dye was 0.04 and that for RhB dye was 0.034, indicating a favorable adsorption [6,15]. In addition, the highest value of the constant K_l for CR was 81.91 L mg^{-1}, and for RhB, it was 1264.5 L mg^{-1}, which indicates a strong bonding of dyes onto PTFAC [30]. In this study, the value of n for RhB adsorption was above unity, indicating favorable physical adsorption, but the value of n for CR was below unity, indicating chemical adsorption.

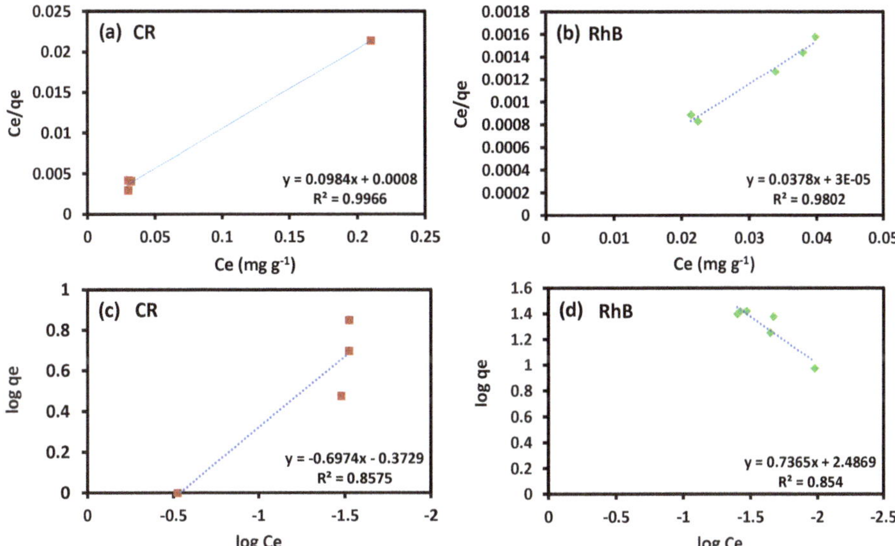

Figure 8. Adsorption isotherms: Langmuir model of (**a**) CR dye; (**b**) RhB dye and Freundlich model for the adsorption of (**c**) CR dye; (**d**) RhB dye by PTFAC (reaction conditions: T= 298 K, pH = 2.0, C_o = 25 mg L^{-1}, mass = 0.15 g, V = 20 mL, shaking time = 120 min).

Table 2. Adsorption isotherm parameters calculated for RhB and CR dyes adsorbed by PTFAC*.

Adsorption Isotherm	Parameters	CR	RhB
	q_{max} (mg g^{-1})/C_o (25 mg L$^{-1)}$	10.4	26.5
	K_l (L mg^{-1})	81.91	1264.5
Langmuir	R_l	0.04	0.034
	R^2	0.996	0.980
	χ^2	0.013	1×10^{-3}
	K_f (mg g^{-1}) (mg L^{-1})	0.24	3.07×10^2
	$1/n$	-1.43	1.357
Freundlich	R^2	0.857	0.854
	χ^2	4.131	3×10^{-3}

* Reaction conditions: T= 298 K, pH = 2.0, C_o = 25 mg L^{-1}, mass = 0.15 g/ 20 mL, shaking time = 120 min.

In addition, the chi-square test was considered to calculate the best fitted model and validation of adsorption kinetics [51], using Equation (6).

$$\chi^2 = \sum \frac{\left(q_{e,exp} - q_{e,cal} \right)^2}{q_{e,cal}} \tag{6}$$

where $q_{e,exp}$ and $q_{e,cal}$ are the experimental and calculated data of adsorption capacity at equilibrium (mg g^{-1}) for each model. This error equation is used to express the distributed error among the calculated and experimental values of the isotherm model. When the error is minor, it proves that the isotherm model is well fitted. According to the χ^2 analysis, it was found that the Langmuir isotherm model showed a lower χ^2 value for CR dye (0.013) and RhB dye (0.001), as depicted in Table 2, suggesting that the adsorption of both dyes was monolayer type. The previous study reported that the acid-activated red mud can adsorb the CR dye, and the Langmuir isotherm model is the best fit because the chi-square was low [52].

3.5. Adsorption Kinetics

Kinetic investigations offer an important indication to realize the adsorption dynamics or adsorption mechanism using pseudo-first-order and pseudo-second-order models. The mathematical term of the pseudo-first-order model [53] is expressed in Equation (7).

$$\log(q_e - q_t) = \log q_e - \frac{k_1}{2.303} t \tag{7}$$

where q_t (mg g^{-1}) and q_e (mg g^{-1}) are the quantities of dyes adsorbed at time t and at equilibrium, respectively. k_1 is the rate constant of the pseudo-first-order model (min^{-1}). The plot of log ($q_e - q_t$) versus t for the adsorption of CR and RhB dyes on PTFAC results in linear graphs with a negative slope (Figure 9a); k_1 and q_e are determined from the slope and intercept, respectively. In this case, the calculated q_e (CR 0.408, RhB 0.006 mg g^{-1}) values of both dyes were considerably different from the q_{exp} values. This suggests that the data do not fit well with the first-order model.

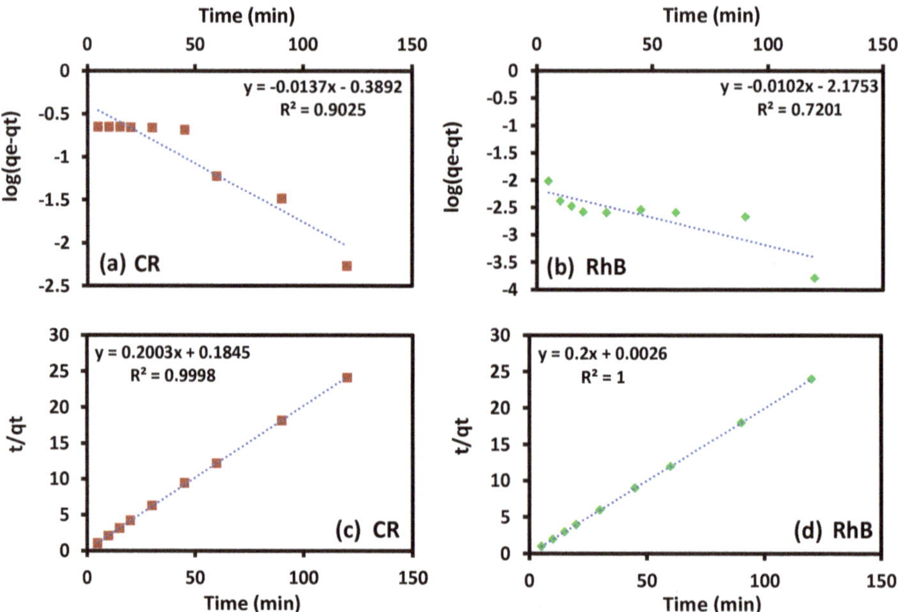

Figure 9. Adsorption kinetics: pseudo-first-order model of (**a**) CR dye; (**b**) RhB dye and pseudo-second-order model for the adsorption of (**c**) CR dye; (**d**) RhB dye by PTFAC (reaction conditions: T= 298 K, pH = 2.0, C_o = 25 mg L^{-1}, mass = 0.15 g, V = 20 mL).

The pseudo-second-order model describes that the adsorption process is a chemical process. It describes the proportion between the rate constant and the engaged active sites. The linear equation of the pseudo-second-order rate model is given in Equation (8).

$$\frac{t}{q_t} = \frac{1}{K_2 \, q_e^2} + \frac{t}{q_e} \tag{8}$$

where k_2 is the pseudo-second-order model adsorption rate constant. The drawing of t/q_t against t provides linear graphs. The values of q_e and k_2 can be determined from the slope and intercept of the PTFAC graph (Figure 9b).

The pseudo-first-order model in Table 3 shows that the calculated q_e value is 0.408 mg/g for CR dye and 0.006 mg g^{-1} for RhB dye. The values vary significantly lower than the experimental q_e values, which are 4.972 mg g^{-1} for CR dye and 4.999 mg g^{-1} for RhB dye. However, with the pseudo-second-order model, the calculated q_e values are 4.992 mg g^{-1} for CR dye and 4.998 mg g^{-1} for RhB dye, which are similar to the experimental q_e values. In addition, the obtained R^2 values are very close to ones for the pseudo-second-order kinetics adsorption process, which were 0.999 and 1.0 for CR dye and RhB dye, respectively. The χ^2 test was used to validate the kinetics models [47,54]. As χ^2 values are generally found to be low, the value for CR dye was 7.74×10^{-5}, and for RhB dye, it was 2.01×10^{-8}. This indicated that the adsorption kinetics data obey the pseudo-second-order model. This is similar to a report on the adsorption of RhB dye onto the area-modified R. hookerie epicarp [47].

Table 3. Adsorption kinetics parameters calculated for RhB and CR dyes adsorbed on PTFAC*.

Kinetic Models	Parameters	CR	RhB
Pseudo first order	$q_{e,cal}$ (mg g^{-1}) k_1 (1/min) R^2 χ^2	0.408 -0.031 0.90 51.05	0.006 -0.023 0.72 3732.3
Pseudo second order	$q_{e,cal}$ (mg g^{-1}) k_2 (g mg^{-1} min^{-1}) R^2 χ^2	4.992 1.35×10^{-3} 0.999 7.744×10^{-5}	4.998 9.75×10^{-3} 1 2.011×10^{-8}
$q_{e,exp}$		4.972	4.999

* Reaction conditions: T = 298 K, 0.15 g/20 mL, C_o = 25 mg L^{-1}.

The adsorption of CR and RhB dyes on PTFAC is an intricate process that involves physical forces acting through pores contained and chemical bonding derived from the active groups on the PTFAC adsorbent [43,54]. Activation by phosphoric acid also enhances the acidic-containing functional groups on PTFAC, resulting in better dye removal from the aqueous solution by the used adsorbent [33]. FTIR analysis revealed the existence of hydrogen bonds between CR, RhB, and the surface of PTFAC. The H bonding interaction occurred between hydroxyl groups at the PTFAC surface and nitrogen-containing functional groups of CR and RhB in acidic medium in addition to the specific attraction between negatively charged COO$^-$ and SO$_3{}^-$ groups of CR dye and the positively charged PTFAC surface [47,55]. Protonation of –OH and –COOH groups present at the PTFAC surface in addition to the van der Waals force also enhances the uptake of the used analytes [47,55].

3.6. Desorption and Regeneration Studies

Desorption studies help to explain the adsorption mechanism in the recovery of the adsorbent and to investigate the possibility of regeneration of used PTFAC loaded with RhB and CR dyes. Table 4 shows that the desorption using distilled water by adjusting pH to 2.5, 4.5, 6.5, 8.5, and 10.5 has a high desorption efficiency and is very close (in percentage) to the CH$_3$;COOH and NaOH (0.1, 0.5, and 1.0 M). Thus, the PTFAC loaded with RhB dye showed a high stability in the adsorption process. The desorption percentage at pH values of 2.5, 4.5, 6.5, 8.5, and 10.5 was 99.88%, 99.84%, 99.84%, 99.81%, and 99.77%, respectively, whereas in the desorption of CR loaded on PTFAC, the percentage decreased with increasing the pH from 2.5 to 10.5, which was 98.68%, 97.36%, 96.1%, 93.42%, and 93.42%, respectively. Hence, the optimum desorption percentage of CR and RhB was observed at pH 2.5 using distilled water as a green desorption reagent. This indicates that the adsorption of the dyes is mainly due to the Van der Waals, electrostatic, and H-bonding interactions, and chemisorption mechanism [48]. In addition, the regeneration study of the adsorbent might yield an efficient economical treatment process. Figure 10 shows the regeneration cycle of PTFAC with CR and RhB dyes. It has no significant change of percentage removal in CR dye. The removal percentage was between 98.68 and 96.1%, which indicated that the use of PTFAC could be increased to more than five cycles. Thus, PTFAC showed high resistance to adsorption efficiency after five regeneration cycles, whereas there was a decrease in percentage removal in RhB dye with the increase of the number of cycles from the first cycle (99.88%) to the fifth cycle (54.49%). That was due to the effect of the repeated use of the PTFAC on the adsorption process. These dyes loaded on the PTFAC surface can change superficial structures and result in the loss or blockage of the adsorption sites in the AC. The magnetic lignosulfonate FCS studied two regeneration cycles for CR, and removal was 83% after five cycles [56]. It can be concluded that PTFAC is an efficient high-potential adsorbent for the removal of cationic and anionic dyes and can be recycled several times.

Table 4. Desorption percentage of CR and RhB dyes by PTFAC with different desorption reagents.

Desorption Reagent	Desorption, %	
	CR	**RhB**
$CH_3;COOH$ (0.1 M)	90.78	99.69
$CH_3;COOH$ (0.5 M)	92.10	99.73
$CH_3;COOH$ (1 M)	89.47	99.65
NaOH (0.1 M)	93.42	99.77
NaOH (0.5 M)	92.10	99.81
NaOH (1 M)	92.10	99.84
H_2O (pH 2.5)	98.68	99.88
H_2O (pH 4.5)	97.36	99.84
H_2O (pH 6.5)	96.05	99.84
H_2O (pH 8.5)	93.42	99.81
H_2O (pH 10.5)	93.42	99.77

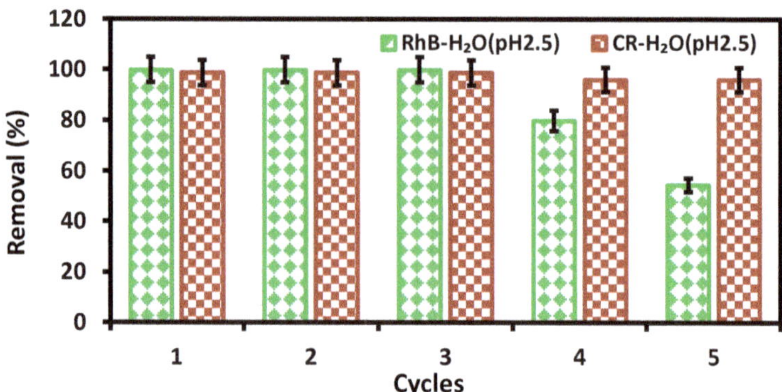

Figure 10. Regeneration cycle of PTFAC after adsorption by RhB and CR dyes.

3.7. Environmental Water Treatment

This study investigated the efficiency of PTFAC as an environmental application by determining the highest percentage removal of CR and RhB dyes. The environmental water samples were filtrated through No. 2 Whatman Qualitative filter paper; then, the spike method was used by obtaining a 15 mg L^{-1} dyes concentration. Table 5 shows the percentage removal of CR and RhB dyes from contaminated water. Thus, the percentage removal values for RhB dye were 99.8% for tap water, 99.3% for well water, 99.4% for seawater, and 98.3% for sewage water from the treatment plant. The percentage removal values for CR dye were 99.8% for tap water, 94.4% for well water, 93.3% for seawater, and 92.3% for sewage water from the treatment plant. This indicates the PTFAC has high stability to remediate water from CR and RhB dye. The removal percentages from the highest to the lowest are tap water > well water > seawater > sewage water from the treatment plant. These results confirmed that PTFAC could be used to treat RhB and CR wastewater in the environmental application without adjusting the pH of the wastewater. Thus, the PTFAC has high removal efficiency in the remediation of contaminated water from anion and cation dyes.

Table 5. Adsorptive percent removal of CR and RhB dyes from real water samples by PTFAC.

Dyes	Tap Water	Well Water	Seawater	Sewage Water Treatment Plant
% Removal of RhB	99.8	99.3	99.4	98.3
% Removal of CR	99.8	94.4	93.3	92.3

4. Conclusions

In the current study, palm fiber was successfully used as solid waste for the production of AC activated by H_3PO_4 at 400 °C after 30 min. The established method is a facile and fast method that can be used as an efficient adsorbent for the elimination of anionic and cationic dyes from wastewater. The texture and surface characteristics of the prepared PTFAC showed 648.90 m^2g^{-1} surface area, 2.83 cm^3g^{-1} total pore volume, and 0.79 nm mean pore diameter with a zeta potential value of -28.3 mV and point of zero charge (pHzpc) equal to 3. The FTIR study indicated the presence of various functional groups. The results of the adsorption process of the prepared PTFAC showed that pH, adsorbent dosage, contact duration, and initial dye concentration have a dominant effect on the adsorption of CR and RhB dyes. The Langmuir isotherm showed a maximum monolayer CR and RhB adsorption capacity of 10.4 mg g^{-1} and 26.5 mg g^{-1}, respectively, of an initial concentration 25 mg L^{-1} at pH 2.0, 25 °C, 120 min, and 0.1 g/20 mL adsorbent dose of AC. The kinetic studies showed that the adsorption process fits in with the pseudo-second-order kinetics. The desorption and regeneration studies of adsorbent were found to be effective until the fifth regeneration cycle by using distilled water an eco-friendly desorption reagent at pH 2.5, suggesting that the adsorbent could be regenerated and reused. Hence, the investigation concludes that the PTFAC prepared from PTF is a cost-effective and promising adsorbent toward dyes from environments.

Author Contributions: B.G.A. supervision, project administration, writing—original draft, Writing—review & editing; S.A., methodology, formal analysis, writing—original draft; E.A.B., supervision, validation; M.F.Z., investigation, formal analysis. All authors have read and agreed to the published version of the manuscript.

Funding: This work was funded under Deanship of Scientific Research (DSR), King Abdulaziz University, Jeddah, grant number G-569-247-39.

Institutional Review Board Statement: Not applicable.

Informed Consent Statement: Not applicable.

Data Availability Statement: Not applicable.

Acknowledgments: The authors express their sincere thanks and acknowledge the financial support by Deanship of Scientific Research (DSR), King Abdulaziz University, Jeddah for facilitating the research work.

Conflicts of Interest: The authors declare no conflict of interest.

References

1. Tian, C.; Feng, C.; Wei, M.; Wu, Y. Enhanced adsorption of anionic toxic contaminant Congo Red by activated carbon with electropositive amine modification. *Chemosphere* **2018**, *208*, 476–483. [CrossRef] [PubMed]
2. Moussavi, G.; Mahmoudi, M. Removal of azo and anthraquinone reactive dyes from industrial wastewaters using MgO nanoparticles. *J. Hazard. Mater.* **2009**, *168*, 2–3. [CrossRef]
3. Chen, Y.; Hermens, J.L.; Jonker, M.T.; Arnot, J.A.; Armitage, J.M.; Brown, T.; Nichols, J.W.; Fay, K.A.; Droge, S.T. Which Molecular Features Affect the Intrinsic Hepatic Clearance Rate of Ionizable Organic Chemicals in Fish? *Env. Sci. Technol.* **2016**, *50*, 12722–12731. [CrossRef] [PubMed]
4. Saxena, S.; Raja, A.S.M. Natural Dyes: Sources, Chemistry, Application and Sustainability Issues. In *Roadmap to Sustainable Textiles and Clothing*; Textile Science and Clothing Technology (Springer Singap.); Muthu, S., Ed.; Spring: Berlin/Heidelberg, Germany, 2014; pp. 37–80. [CrossRef]
5. Freundlich. Uber die Adsorption in Lösungen. *J. Phys. Chem.* **1906**, *57*, 385.
6. Aly, Z.; Graulet, A.; Scales, N.; Hanley, T. Removal of Aluminium from Aqueous Solutions Using PAN-Based Adsorbents: Characterisation, Kinetics, Equilibrium and Thermodynamic Studies. *Environ. Sci. Pollut. Res.* **2014**, *21*, 3972–3986. [CrossRef]
7. Suhas, V.K.; Gupta, P.J.M.; Carrott, R.; Singh, M.; Chaudhary, S.K. Cellulose: A review as natural, modified and activated carbon adsorbent. *Bioresour. Technol.* **2016**, *216*, 1066–1076. [CrossRef]
8. Danish, M.; Khanday, W.A.; Hashim, R.; Sulaiman, N.S.B.; Akhtar, M.N.; Nizami, M. Application of optimized large surface area date stone (Phoenix dactylifera) activated carbon for rhodamin B removal from aqueous solution: Box-Behnken design approach. *Ecotoxicol. Environ. Saf.* **2017**, *139*, 280–290. [CrossRef]
9. Zhu, L.; Wang, Y.; He, T.; You, L.; Shen, X. Assessment of Potential Capability of Water Bamboo Leaves on the Adsorption Removal Efficiency of Cationic Dye from Aqueous Solutions. *Polym. Environ.* **2016**, *24*, 148–158. [CrossRef]
10. Yao, T.; Guo, S.; Zeng, C.; Wang, C.; Zhang, L. Investigation on efficient adsorption of cationic dyes on porous magnetic polyacrylamide microspheres. *J. Hazard. Mater.* **2015**, *292*, 90–97. [CrossRef] [PubMed]
11. Bousba, S.; Meniai, A.H. Adsorption of 2-chlorophenol onto sewage sludge based adsorbent: Equilibrium and kinetic study. *Chem. Eng. Trans.* **2013**, *35*, 859–864.
12. Tian, H.; Peng, J.; Lv, T.; Sun, C.; He, H. Preparation and performance study of $MgFe_2O_4$/metal-organic framework composite for rapid removal of organic dyes from water. *J. Solid State Chem.* **2018**, *257*, 40–48. [CrossRef]
13. Fu, K.; Yue, Q.; Gao, B.; Wang, Y.; Li, Q. Activated carbon from tomato stem by chemical activation with FeCl2, COLSUA Colloids and Surfaces. *Phys. Eng. Asp.* **2017**, *529*, 842–849. [CrossRef]
14. Vohra, M.S. Adsorption-Based Removal of Gas-Phase Benzene Using Granular Activated Carbon (GAC) Produced from Date Palm Pits. *Arabian J. Sci Eng.* **2015**, *40*, 3007–3017. [CrossRef]
15. Somaia, G.M.; Sahar, M.A. Preparation of environmentally friendly activated carbon for removal of pesticide from aqueous media. *Int. J. Indust. Chem.* **2017**, *8*, 121–132.
16. Silva, T.L.; Cazetta, A.L.; Souza, P.S.C.; Zhang, T.; Asefa, T.; Almeida, V.C. Mesoporous activated carbon fibers synthesized from denim fabric waste: Efficient adsorbents for removal of textile dye from aqueous solutions. *J. Clean. Prod.* **2018**, *171*, 482–490. [CrossRef]
17. Mosbah, M.b.; Mechi, L.; Khiari, R.; Moussaoui, Y. Current State of Porous Carbon for Wastewater Treatment. *Processes* **2020**, *8*, 1651. [CrossRef]
18. Abdel-Ghani, N.T.; El-Chaghaby, G.A.; Rawash, E.S.A.; Lima, E.C. Adsorption of Coomassie Brilliant Blue R-250 dye onto novel activated carbon prepared from Nigella sativa L. waste: Equilibrium, kinetics and thermodynamics running. *J. Chil. Chem. Soc.* **2017**, *62*, 3505–3511. [CrossRef]
19. Choong, C.E.; Kim, M.; Yoon, S.; Lee, G.; Park, C.M. Mesoporous La/Mg/Si-incorporated palm shell activated carbon for the highly efficient removal of aluminum and fluoride from water. *J. Taiwan Inst. Chem. Eng.* **2018**, *93*, 306–314. [CrossRef]
20. Kaveeshwar, A.R.; Ponnusamy, S.K.; Revellame, E.D.; Gang, D.D.; Zappi, M.E.; Subramaniam, R. Pecan shell based activated carbon for removal of iron(II) from fracking wastewater: Adsorption kinetics, isotherm and thermodynamic studies. *Process. Saf. Environ. Prot.* **2018**, *114*, 107–122. [CrossRef]
21. Bhomick, P.C.; Supong, A.; Baruah, M.; Pongener, C.; Sinha, D. Pine Cone biomass as an efficient precursor for the synthesis of activated biocarbon for adsorption of anionic dye from aqueous solution: Isotherm, kinetic, thermodynamic and regeneration studies. *Sustainable Chem. Pharm.* **2018**, *10*, 41–49. [CrossRef]
22. Getu, A.; Sahu, Q. Removal of Reactive Dye using Activated Carbon from Agricul-tural Waste. *J. Eng. Geol. Hydrogeol.* **2014**, *2*, 23. [CrossRef]
23. Wongcharee, S.; Aravinthan, V.; Erdei, L. Mesoporous activated carbon-zeolite composite prepared from waste macadamia nut shell and synthetic faujasite. *Chin. J. Chem. Eng.* **2019**, *27*, 226–236. [CrossRef]
24. Pongener, C.; Bhomick, S.P.; Upasana Bora, R.L.; Goswamee, A.; Supong, D.S. Sand-supported bio-adsorbent column of activated carbon for removal of coliform bacteria and Escherichia coli from water. *Int. J. Environ. Sci. Technol.* **2017**, *14*, 1897–1904. [CrossRef]
25. Gumus, R.H.; Okpeku, I. Production of Activated Carbon and Characterization from Snail Shell Waste (Helix pomatia). *Adv. Chem. Eng. Sci.* **2015**, *5*, 51–61. [CrossRef]

26. Goswami, M.; Phukan, P. Enhanced adsorption of cationic dyes using sulfonic acid modified activated carbon. *J. Environ. Chem. Eng.* **2017**, *5*, 3508–3517. [CrossRef]

27. Shamsuddin, M.S.; Yusoff, N.R.N.; Sulaiman, M.A. Synthesis and Characterization of Activated Carbon Produced from Kenaf Core Fiber Using H_3PO_4 Activation. *Procedia Chem.* **2016**, *19*, 558–565. [CrossRef]

28. Chowdhury, S.; Pan, S.; Balasubramanian, R.; Das, P. Date Palm Based Activated Carbon for the Efficient Removal of Organic Dyes from Aqueous Environment. *Sustain. Agric. Rev.* **2019**, *34*, 247–263.

29. Kumar, A.; Jena, H.M. Preparation and characterization of high surface area activated carbon from Fox nut (Euryale ferox) shell by chemical activation with H_3PO_4. *Results Phys.* **2016**, *6*, 651–658. [CrossRef]

30. Thitame, P.V.; Shukla, S.R. Adsorptive removal of reactive dyes from aqueous solution using activated carbon synthesized from waste biomass materials. *Int. J. Environ. Sci Technol.* **2016**, *13*, 561–570. [CrossRef]

31. Gao, X.L.; Wu, Z.; Li, Q.; Xu, W.; Tian, R.W. Preparation and characterization of high surface area activated carbon from pine wood sawdust by fast activation with H3PO4 in a spouted bed, J Mater Cycles Waste Manag Journal of Material Cycles and Waste Management. *J. Mater. Cycles Waste Manag.* **2018**, *20*, 925–936. [CrossRef]

32. Al-Oqla, F.M.; Alothman, O.Y.; Jawaid, M.; Sapuan, S.M.; Es-Saheb, M.H. Processing and Properties of Date Palm Fibers and Its Composites. *Biomass Bioenergy* **2014**, 1–25. [CrossRef]

33. Alhogbi, B.G.; Salam, M.A.; Ibrahim, O. Environmental Remediation of Toxic Lead Ions from Aqueous Solution Using Palm Tree Waste Fibers Biosorbent. *Desal. Water Treat.* **2019**, *145*, 179–188. [CrossRef]

34. Chahinez, H.O.; Abdelkader, O.; Leila, Y.; Tran, H.N. One-stage preparation of palm petiole-derived biochar: Characterization and application for adsorption of crystal violet dye in water. *Env. Technol. Innov.* **2020**, *19*, 100872. [CrossRef]

35. Salazar-Rabago, J.; Leyva-Ramos, R. Novel Biosorbent with High Adsorption Capacity Prepared by Chemical Modification of White Pine (Pinus Durangensis) Sawdust. Adsorption of Pb (II) from Aqueous Solutions. *J. Environ. Manag.* **2016**, *169*, 303–312. [CrossRef] [PubMed]

36. Wang, S.; Zhu, Z.; Coomes, A.; Haghseresht, F.; Lu, G. The Physical and Surface Chemical Characteristics of Activated Carbons and the Adsorption of Methylene Blue from Wastewate. *J. Colloid Interface Sci.* **2005**, *284*, 440–446. [CrossRef] [PubMed]

37. Jabar, J.M.; Odusote, Y.A.; Alabi, K.A.; Ahmed, I.B. Kinetics and mechanisms of congo-red dye removal from aqueous solution using activated Moringa oleifera seed coat as adsorbent. *Appl. Water Sci.* **2020**, *10*, 1–11. [CrossRef]

38. Deng, J.; Xiong, T.; Wang, H.; Zheng, A.; Wang, Y. Effects of Cellulose, Hemicellulose, and Lignin on the Structure and Morphology of Porous Carbons. *Sustain. Chem. Eng.* **2016**, *4*, 3750–3756. [CrossRef]

39. Ojedokun, A.T.; Bello, O.S. Kinetic modeling of liquid-phase adsorption of Congo red dye using guava leaf-based activated carbon. *Appl. Water Sci.* **2017**, *7*, 1965–1977. [CrossRef]

40. Yakout, S.M.; El-Deen, G.S. Characterization of activated carbon prepared by phosphoric acid activation of olive stones. *Arabian J. Chem.* **2016**, *9*, S1155–S1162. [CrossRef]

41. Nasser, R.A.; Salem, S.M.Z.; Hiziroglu, H.A.; Al-Mefarrej, A.S.; Mohareb, M.; Alam, I.M.A. Chemical Analysis of Different Parts of Date Palm (Phoenix Dactylifera L.) Using Ultimate, Proximate and Thermo-Gravimetric Techniques for Energy Production. *Energies* **2016**, *9*, 374. [CrossRef]

42. Daoud, M.; Benturki, O.; Kecira, Z.; Daoud, M.; Girods, P.; Donnot, A. Removal of Reactive dye (BEZAKTIV Red S-MAX) from Aqueous Solution by Adsorption onto Activated Carbons Prepared from Date Palm Rachis and Jujube Stones. *J. Mol. Liq.* **2017**, *243*, 799–809. [CrossRef]

43. Buhani1, S.; Aditiya, I.; Al Kausar, R.; Sumadi; Rinawati. Production of a Spirulina sp. Algae Hybrid with a Silica Matrix as an Effective Adsorbent to Absorb Crystal Violet and Methylene Blue in a Solution. *Sustain. Environ. Res.* **2019**, *29*, 27.

44. Marsh, H.; Rodríguez-Reinoso, F. Activated Carbon, Activation Processes (Chemical). *Elsevire (Lond. UK)* **2006**, *6*, 322–365.

45. Yorgun, S.; Yildiz, D. Preparation and Characterization of Activated Carbons from Paulownia Wood by Chemical Activation with H_3PO_4. *J. Taiwan Inst. Chem. Eng.* **2015**, *53*, 122–131. [CrossRef]

46. Pego, M.F.F.; Veiga, T.R.L.A.; Bianchi, M.; Carvalho, L. Surface modification of activated carbon by corona treatment. *J. Ann. Braz. Acad. Sci.* **2019**, *91*, 1678–2690. [CrossRef] [PubMed]

47. Inyinbor, A.A.; Adekola, F.A.; Olatunji, G.A. Kinetics and isothermal modeling of liquid phase adsorption of rhodamine B onto urea modified Raphia hookerie epicarp. *App. Water Sci.* **2017**, *7*, 3257–3266. [CrossRef]

48. Sivarajan, A.; Shanmugapriya, V. Determination of isotherm parameters for the adsorption of Rhodamine B dye onto activated carbon prepared from Ziziphus jujuba seeds. *Asian J. Res. Chem.* **2017**, *10*, 362. [CrossRef]

49. Langmuir, I. The Adsorption of Gases on Plane Surfaces of Glass, Mica and Platinum. *Am. Chem. Soc.* **1968**, *40*, 1361–1402. [CrossRef]

50. Urano, K.; Koichi, Y.; Nakazawa, Y. Equilibria for adsorption of organic compounds on activated carbons in aqueous solutions I. Modified Freundlich isotherm equation and adsorption potentials of organic compounds. *J. Colloid Interface Sci.* **1981**, *81*, 477–485. [CrossRef]

51. Zamri, T.K.; Abdul Munaim, M.S.; Abdul Wahid, Z. Regression analysis for the adsorption isotherms of natural dyes onto bamboo yarn. *Int. Res. J. Eng. Technol.* **2017**, *4*, 1699–1703.

52. Tor, A.; Cengeloglu, Y. Removal of Congo Red from Aqueous Solution by Adsorption onto Acid Activated Red Mud. *J. Hazard. Mater.* **2006**, *138*, 409–415. [CrossRef]

53. Pathak, U.; Jhunjhunwala, A.; Roy, A.; Das, P.; Kumar, T.; Mandal, T. Efficacy of spent tea waste as chemically impregnated adsorbent involving ortho-phosphoric and sulphuric acid for abatement of aqueous phenol—Isotherm, kinetics and artificial neural network modelling. *Environ. Sci. Pollut. Res.* **2020**, *27*, 20629–20647. [CrossRef] [PubMed]
54. Auta, M.; Hameed, B. Coalesced Chitosan Activated Carbon Composite for Batch and Fixed-Bed Adsorption of Cationic and Anionic Dyes. *Colloids Surf. B Biointerfaces* **2013**, *105*, 199–206. [CrossRef] [PubMed]
55. Ahmad, R.; Kumar, R. Adsorptive Removal of Congo Red Dye from Aqueous Solution Using Bael Shell Carbon. *Appl. Surf. Sci.* **2010**, *257*, 1628–1633. [CrossRef]
56. Hu, L.; Guang, C.; Liu, Y.; Su, Z.; Gong, S.; Yao, Y.; Wang, Y. Adsorption behavior of dyes from an aqueous solution onto composite magnetic lignin adsorbent. *Chemosphere* **2020**, *246*, 125757. [CrossRef] [PubMed]

Article

Sequential Abatement of Fe^{II} and Cr^{VI} Water Pollution by Use of Walnut Shell-Based Adsorbents

Marius Gheju [1,*] and **Ionel Balcu** [2]

[1] Faculty of Industrial Chemistry and Environmental Engineering, Politehnica University Timisoara, Bd. V. Parvan Nr. 6, 300223 Timisoara, Romania

[2] National Institute for Research and Development in Electrochemistry and Condensed Matter, Str. Dr. Aurel Paunescu Podeanu Nr. 144, 300587 Timisoara, Romania; incemc@incemc.ro

* Correspondence: marius.gheju@upt.ro; Tel.:+0040-256-404185; Fax: 0040-256-403060

Abstract: In this study walnut shells, an inexpensive and readily available waste, were used as carbonaceous precursor for preparation of an innovative adsorbent (walnut-shell powder (WSP)) which was successfully tested for the removal of Fe^{II} from synthetic acid mine drainage (AMD). Then, the exhausted iron-contaminated adsorbent (WSP-Fe^{II}) was recovered and treated with sodium borohydride for the reduction of adsorbed Fe^{II} to Fe^0. The resulting material (WSP-Fe^0) was subsequently tested for the removal of Cr^{VI} from aqueous solutions. Treatability batch experiments were employed for both Fe^{II} and Cr^{VI}-contaminated solutions, and the influence of some important experimental parameters was studied. In addition, the experimental data was interpreted by applying three kinetic models and the mechanism of heavy metal removal was discussed. The overall data presented in this study indicated that fresh WSP and WSP-Fe^0 can be considered as promising materials for the removal of Fe^{II} and Cr^{VI}, respectively. Furthermore, the present work clearly showed that water treatment residuals may be converted in upgraded materials, which can be successfully applied in subsequent water treatment processes. This is an example of sustainable and environmentally-friendly solution that may reduce the adverse effects associated with wastes and delay expensive disposal methods such as landfilling or incineration.

Keywords: sustainable water treatment; water treatment residuals; innovative adsorbent; heavy metals; acid mine drainage; hexavalent chromium

Citation: Gheju, M.; Balcu, I. Sequential Abatement of Fe^{II} and Cr^{VI} Water Pollution by Use of Walnut Shell-Based Adsorbents. *Processes* **2021**, *9*, 218. https://doi.org/10.3390/pr9020218

Received: 20 December 2020
Accepted: 22 January 2021
Published: 25 January 2021

Publisher's Note: MDPI stays neutral with regard to jurisdictional claims in published maps and institutional affiliations.

1. Introduction

In last decades, water pollution with heavy metals has become an increasingly important worldwide threat. Numerous heavy metals have been introduced into natural water environments, especially as a result of human industrial activities, but also due to agricultural, transport and waste disposal practices. Among the most important industrial activities that contribute to contamination of aquatic systems by metallic ions (Cr, Ni, Zn, Cu, Zn, Pb, Fe, Cd etc.) are: electroplating, surface finishing of metals, production and recycling of electronics, metallurgy, mining, leather tanning, paper and pulp production, fertilizer and pesticide production, batteries production [1–4]. Most heavy metals cause toxic effects to living species, not only at excessive exposures, but also at low concentrations, because they do not have any biological role in living cells, do not degrade into harmless end products, and are bioaccumulative in nature. In addition, even heavy metals which are necessary in small amounts as micronutrients for the normal development of biological systems (e.g., Cu, Fe, Zn, Cr etc.) exert harmful effects to biological organisms at high concentrations [4–6]. Therefore, the removal of heavy metals from contaminated waters, prior to their discharge into natural effluents, is a necessary step in order to reduce their adverse effects. The World Health Organization guidelines of some heavy metals in drinking water are summarized in Table S1.

Various methods have been applied for the removal of heavy metals from aqueous solutions; these include chemical precipitation, membrane technologies, ion exchange, electrochemical treatment, floatation, coagulation-flocculation, adsorption, evaporation, photocatalysis [1,3,7–11]. Among all these methods, adsorption is generally considered to be one of the best for the removal of heavy metals, due to its simplicity, ease of operation, selectivity and high efficiency. Activated carbon is the most widely used adsorbent in water and wastewater treatment, owing to its high surface area and high degree of surface porosity and reactivity. The heavy metal adsorption capacities reported in the literature for commercial activated carbon are in the range of 0.29–146 mg/g [12]. Unfortunately, in spite of its excellent adsorption capacity, this material is also expensive, which makes the costs of adsorption processes high, and, therefore, less economically viable, especially in low income developing countries [4,13,14]. Therefore, over the last decades, there has been an intense activity directed at the development of inexpensive and easily available alternatives to commercially available activated carbon (e.g., clay minerals, industrial/agricultural wastes/byproducts etc.) which should decrease the cost of the adsorption process, while still being efficient in improving the quality of the treated effluents [3,4,14–19]. The heavy metal adsorption capacities reported in [12] for natural materials, agricultural and industrial wastes are in the range of 0.003–83.3, 0.001–158 and 0.0002–133.35 mg/g, respectively; thus, it is obvious that low-cost adsorbents may exert excellent metal removal capabilities, comparable to commercial activated carbon [12].

On the other hand, water treatment technologies should be not only efficient and affordable, but also environmental-friendly. Therefore, great attention has been paid in the last years to handling and disposal of water treatment residues (WTRs), which are challenging tasks for today environmental scientists. One of the traditional and most common methods of WTR management is by landfilling. However, today, this is no longer considered a viable solution because of: (1) soil and groundwater contamination, (3) high operating costs, and (3) difficulties in finding and operating new landfill sites, under the circumstances of more and more strict environmental regulation. Hence, recovery, recycling and reuse should be the preferred solution for the sustainable management of WTR [20]. The use of groundwater treatment residuals (Fe and Mn (hydr)oxides) and floculation-coagulation residuals (Fe and Al (hydr)oxides) as adsorbents, to remove pollutants (e.g., heavy metals, metalloids, pesticides etc.) from aqueous solutions, has been reported in the last years by numerous studies [21–24]. Instead, to our knowledge, few researchers have addressed the issue of reusability of cheap exhausted adsorbents, resulted from the removal of a particular heavy metal, in treatment processes of aqueous effluents polluted with a different type of heavy metal [25]. In this previous study, bentonite was used for sequential adsorption of heavy metals from aqueous solutions, proving not only that reusability of exhausted adsorbents is possible, but also that some adsorbed metals may have a beneficial role in the subsequent adsorption process. However, this research also revealed that using bentonite as adsorbent suffers from multiple drawbacks, including low adsorption capacity and leaching of structural iron at strong acidic pH [25].

The fruit and vegetable processing industry operates globally, producing huge amounts of products and being a well-known generator of large volumes of agricultural wastes [26,27]. The world production of walnuts has been relatively stable over the last years, at about 2 million tons (in-shell basis), with China and the USA accounting for nearly three-quarters of the total production [28]. Since the kernel represents approximately 45% of the walnut mass (depending on the size and the variety of the walnut), it is obvious that the remaining shells are an abundantly available waste that could be used as cheap adsorbent materials. Consequently, the present study has two main objectives. Firstly, to investigate the use of walnut shells, in powdered form (WSP), as cheap adsorbent in the remediation process of synthetic acid mine drainage (AMD) containing Fe^{II}. To the best of our knowledge, removal of Fe^{II} with agricultural waste derived adsorbents are few [29,30], while walnut shells in their natural form (i.e., not activated by any chemical or physical method) were not researched yet. The second objective of this paper was to recover the water treatment residue

(WSP-Fe^{II}) resulted from the AMD remediation process, for further reuse in the removal of Cr^{VI} from aqueous solutions. Since during the last two decades Fe^0 was acknowledged as an efficient reactive material for remediation of heavy metal contaminated effluents [31], the water treatment residue (WSP-Fe^{II}) resulted from AMD remediation was treated with sodium borohydride, in order to reduce the adsorbed Fe^{II} to Fe^0, and the resulted material (WSP-Fe^0) was then used for the removal of Cr^{VI}. The effect of several important experimental parameters (pH, heavy metal concentration, temperature, and ionic strength) on efficiency of both treatment processes was investigated. Furthermore, the kinetic parameters of the remediation processes were determined and the mechanisms of Fe^{II} and Cr^{VI} removal were discussed.

2. Results and Discussion

2.1. Adsorbent Characterization

2.1.1. Fourier-Transform Infrared Spectroscopy (FTIR) Analysis

The FTIR spectra of the adsorbents were recorded within the range of 500–4000 cm^{-1}. The spectra of native WSP and WSP-Fe^{II} are given in Figures S1 and S2 (Supplementary Material). The analysis of Figure S1 reveals a wide band near 3440 cm^{-1}, indicating the presence of hydrogen-bonded hydroxyl groups on the WSP surface [32]. This can be correlated with the intense band around 1040 cm^{-1}, characteristic for the valence vibration of C-O bond in primary alcohols [33,34]. Peaks observed around 2920 and 1380 cm^{-1} can be assigned to the stretching vibration of C-H bonds in methyl and methylene groups [35]. The flat peak located at about 2100–2200 cm^{-1} corresponds to C≡C groups stretching vibration [34]. The bands around 1750 and 1720 cm^{-1} are indicative for the C=O group stretching vibration in carboxyl and carbonyl groups [35,36]. Peaks around 1370 and 1330 cm^{-1} may be attributed to the O-H in-plane deformation characteristic for alcohols and phenols [33,37]. Absorption bands near 1650, 1600, 1510, 1260, 800, 670 and 600 cm^{-1} can be related to complex vibrations related to aromatic compounds (e.g., lignin) [32–35,37]. By comparing the spectra of WSP (Figure S1) and WSP-Fe^{II} (Figure S2), a strong decrease in intensity of peaks at 3440, 1600, 1260, 1040 and 800 cm^{-1} could be observed after the adsorption of Fe^{II}; if we assume that FTIR spectra were recorded by following the same procedure (i.e., pellets were prepared by mixing and pressing exactly the same amounts of sample and KBr), this may suggest participation of some of the above mentioned functional groups (hydroxyl, carboxyl and carbonyl) in metal binding. Similar changes in intensity of the bands was observed also after the reaction of Cr^{VI} solution with WSP and WSP-Fe^0 (Figures S1, S3–S5, Supplementary Material).

2.1.2. Scanning Electron Microscopy (SEM) Analysis

The SEM analysis enables the observation of the surface morphology of the studied adsorbents materials. Visual examination of the SEM micrographs (Figures S6–S10, Supplementary Material) showed that external surface of prepared materials has an irregular rugged morphology, containing macropores with sizes of 1–2 μm, homogeneously distributed all over the surface. No noticeable differences can be observed in SEM micrographs before and after the adsorption process; no accumulation of contaminant on the exhausted adsorbent can be discerned too, presumably due to the low amount of retained metal at surface of the adsorbents.

2.1.3. Energy Dispersive X-ray Spectroscopy Analysis (EDX) Analysis

The EDX spectra of the adsorbents before and after Cr^{VI} adsorption are shown in Figures S11–S15 (Supplementary Material). The absence of alkali and alkaline earth metals (Ca^{2+} and K$^+$) in the WSP-Fe^{II} sample (Figure S10) indicated that the adsorption process may have involved an ion-exchange mechanism; furthermore, the EDX spectra of WSP-Fe^{II} revealed additional Fe signals in comparison to WSP, indicating retention of Fe at the surface of adsorbent. The EDX analysis also confirmed the retaining of Cr^{VI} onto the surface of both WSP (control experiments) and WSP-Fe^0; nevertheless, a visual comparison of EDX spectra of Cr^{VI}-loaded WSP-Fe^0 and WSP (Figures S14 and S15) clearly reveals that more Cr was bound on WSP-Fe^0. In addition, the suppression of one Fe band in the EDX spectra of WSP-Fe^0 after reaction with Cr^{VI} may suggest that the Fe^0 sites were involved in removal of Cr^{VI} anions.

2.2. AMD Treatability Experiments

2.2.1. Effect of pH

Earlier studies have shown that pH of the aqueous solution is a highly important factor in adsorption processes, capable to control the mechanism, and therefore, to enhance or decrease the amount of metal retained at the adsorbent surface [38]. The effect of pH on the removal of Fe^{II} was studied by varying the pH of the metal ion solution from 1.0 to 4.1. These pH values were selected because they are within the range of levels reported for pH in AMD environments [39,40]. In addition, pH values below 4.5 also ensure that removal of Fe^{II} occurs solely due to adsorption. Figures 1 and 2 clearly show that both efficiency of Fe^{II} removal and adsorption capacity of WSP increased with increasing pH from 1.0. to 4.1. While only limited AMD remediation was observed at pH 1.0 and 2.1, an important enhancement of the adsorption process was achieved as pH was increased to 2.5, and then further gradually raised up to 4.1. This is in accord with results of previous works using alternative vegetal adsorbents like thermochemically-activated walnut shells and orange peels, which indicated increased removal efficiency with increasing pH of the solution [29,30].

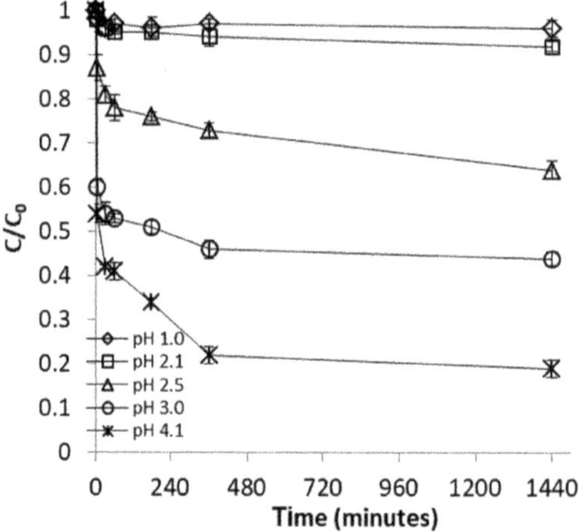

Figure 1. Effect of pH on Fe^{II} removal by walnut-shell powder WSP. The lines are not fitting models; they simply connect points to facilitate visualization.

Carboxyl and hydroxyl (fenolic) groups are among the most important functional oxidized groups (active centers) existent at surface of natural carbon-based agricultural residues, which are able to take part in specific adsorption processes with metal cations, according to [20,41]:

$$WSP\text{-}C_6H_5\text{-}OH + Me^{n+} \Leftrightarrow WSP\text{-}C_6H_5\text{-}OM^{(n-1)+} + H^+ \qquad (1)$$

$$WSP\text{-}COOH + Me^{n+} \Leftrightarrow WSP\text{-}COOM^{(n-1)+} + H^+ \qquad (2)$$

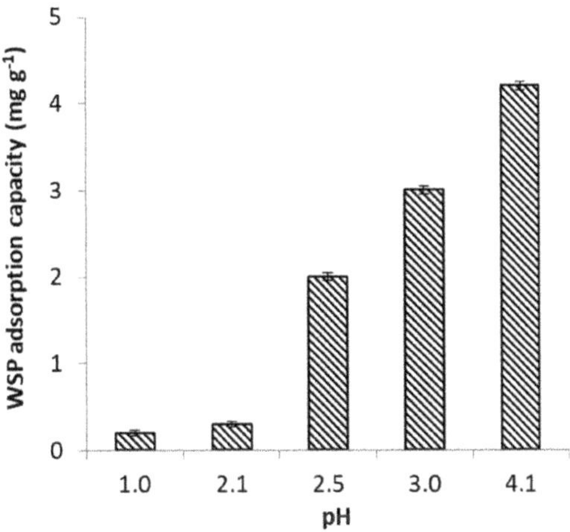

Figure 2. Effect of pH on Fe^{II} adsorption capacity of WSP.

However, carboxyl and hydroxyl groups are also involved in acido-basic equilibria which may be described as following:

$$WSP\text{-}C_6H_5\text{-}OH + H^+ \Leftrightarrow WSP\text{-}C_6H_5\text{-}OH_2^+ \tag{3}$$

$$WSP\text{-}C_6H_5\text{-}OH + HO^- \Leftrightarrow WSP\text{-}C_6H_5\text{-}O^- + H_2O \tag{4}$$

$$WSP\text{-}COOH + H^+ \Leftrightarrow WSP\text{-}COOH_2^+ \tag{5}$$

$$WSP\text{-}COOH + HO^- \Leftrightarrow WSP\text{-}COO^- + H_2O \tag{6}$$

In the present work, the pH_{pzc} of the WSP was found to be 6.4; accordingly, the net charge of WSP surface was positive over the entire studied pH range. Nevertheless, it is clear from the above equations that an increase in solution pH (i.e., more HO^- anions available for Equations (4) and (6) causes an increase in the number of negative charges existent at WSP surface, even though the net charge still remains positive at pH < 6.4. Hence, on the one hand, the efficiency of adsorption will increase at higher pH due to enhanced electrostatic attraction between cationic Fe^{II} species and negatively charged centers at WSP surface. On the other hand, the competition with hydronium cations for anionic exchanging sites at WSP surface also decreased as the pH was raised in the range 1.0–4.1, contributing thus to the increased sorption of Fe^{II} cations.

2.2.2. Effect of Fe^{II} Initial Concentration

The influence of Fe^{II} concentration was studied within the concentration range of 25–100 mg L^{-1}. These concentrations were selected because they are within the range of Fe^{II} levels reported in AMD environments [39,40]. As revealed in Figure 3, the efficiency of Fe^{II} uptake by WSP was found to decrease proportionally with the increase of Fe^{II} concentration. This is attributable to the fact that available sorption sites become progressively insufficient for the increasingly number of Fe^{II} ions at higher concentrations; hence, a more rapid saturation of the adsorption centers will occur and, as a result, the percentage removal of Fe^{II} ions decreases. Figure 3 also shows that adsorption of Fe^{II} proceeds in two steps: a rapid decrease of metal concentration within the first stage (first 60 min), when the amount of available sites was still much higher than the amount of Fe^{II} ions to be adsorbed, followed by a strong decrease in the adsorption rates in the second phase, due to continuous diminishing of the number of negatively charged functional groups throughout the adsorption experiment. This phenomenon was previously reported by several studies employing agro-based waste adsorbents in the process of heavy metal removal from aqueous solutions [16]. In contrast, Figure 4 indicates that adsorption capacity of WSP firstly increased with increasing the initial Fe^{II} concentration, and then reached a saturation value. The maximal adsorption capacity of WSP was found to be about 5.8 mg g^{-1}, achieved at the concentration of 100 mg L^{-1} Fe^{II}. The higher amount of Fe^{II} retained per unit mass of adsorbent (mg g^{-1}) at higher initial concentration is the result of increased Fe^{II} concentration gradient at solution-adsorbent interface (i.e., increased probability of collision between metal ions and adsorbent surface), which led to enhanced mass transfer driving forces to overcome all mass transfer resistances [42]. Our results are in agreement with findings reported by several earlier workers for Fe^{II} adsorption on agro-wastes, such as thermochemically-activated walnut shells and orange peels [29,30].

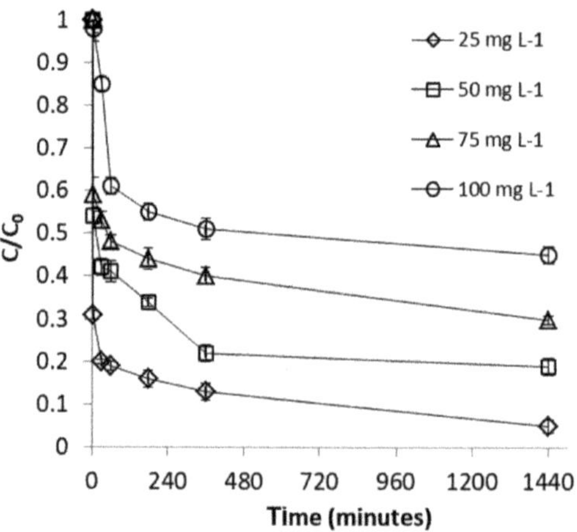

Figure 3. Effect of initial concentration on Fe^{II} removal by WSP. The lines are not fitting models; they simply connect points to facilitate visualization.

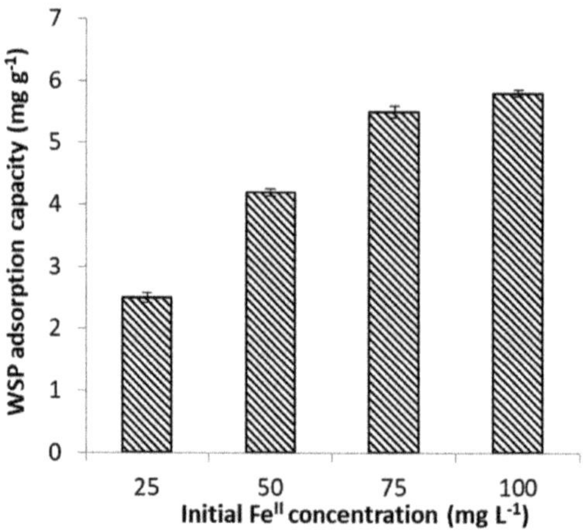

Figure 4. Effect of initial concentration on Fe^{II} adsorption capacity of WSP.

2.2.3. Effect of Temperature

The effect of temperature was investigated over the range of 6–33 °C. The results presented in Figures 5 and 6 show that uptake of Fe^{II} on WSP was positively affected by the increase of temperature; nevertheless, it is important to point out that an improvement in Fe^{II} adsorption was observed only when temperature was increased from 6 to 22 °C;

a subsequent rise of temperature to 33 °C led to no discernible enhancement the Fe^{II} uptake. The observed temperature dependence is indicative of an endothermic adsorption process. The enhancement of adsorption efficacy with increasing temperature may be attributed to better interactions between Fe^{II} and WSP as a result increased rates of intraparticle diffusion of Fe^{II} ions into the pores of WSP, or to creation of new adsorption sites at higher temperatures [43]. The positive effect of temperature on the adsorption efficacy was reported also in early works investigating removal of Fe^{II} from aqueous solutions by adsorption on thermochemically-activated walnut shells and orange peels [29,30].

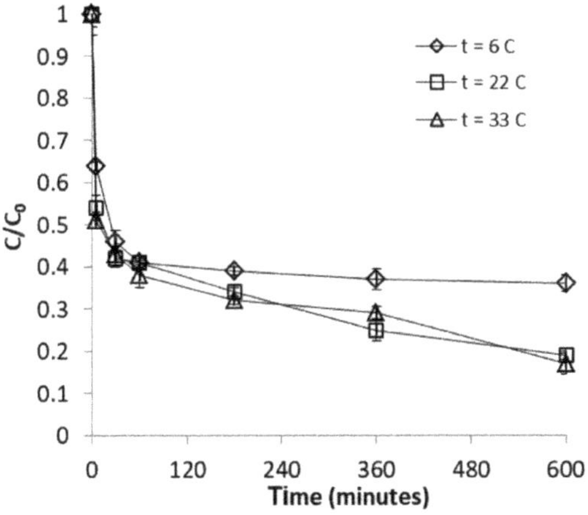

Figure 5. Effect of temperature on Fe^{II} removal by WSP. The lines are not fitting models; they simply connect points to facilitate visualization.

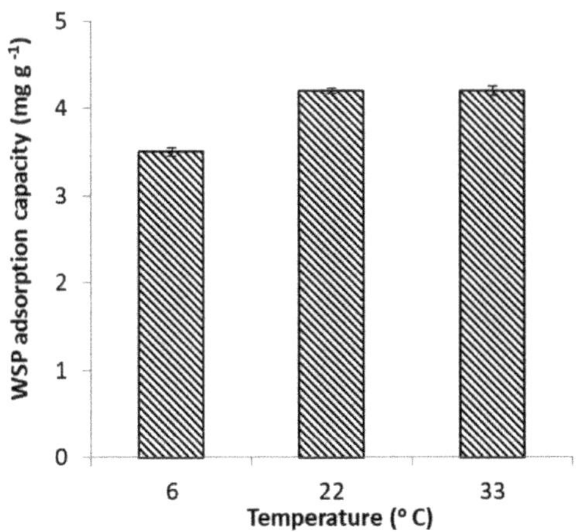

Figure 6. Effect of temperature on Fe^{II} adsorption capacity of WSP.

2.2.4. Effect of Ionic Strength

To investigate the influence of ionic strength, adsorption of Fe^{II} on WSP was conducted in the co-presence of NaCl concentrations of 0, 0.01, 0.03 and 0.05 M as background electrolyte. NaCl was used as indifferent electrolyte, in accord to previous studies investigating the effect of ionic strength [44]. From Figures 7 and 8 it results that the process of Fe^{II} adsorption was progressively hindered in the presence of increasingly concentrations of competing Na^+ cations. The trend of the change of metal adsorption with ionic strength can be used for differentiating between the two main adsorption processes that may be involved in binding of anions onto minerals: physical (non-specific) adsorption, and chemical (specific) adsorption. In our case, the observed effect can be interpreted as indicating non-specific weak interactions (physisorption) being involved in adsorption mechanism of Fe^{II} [44].

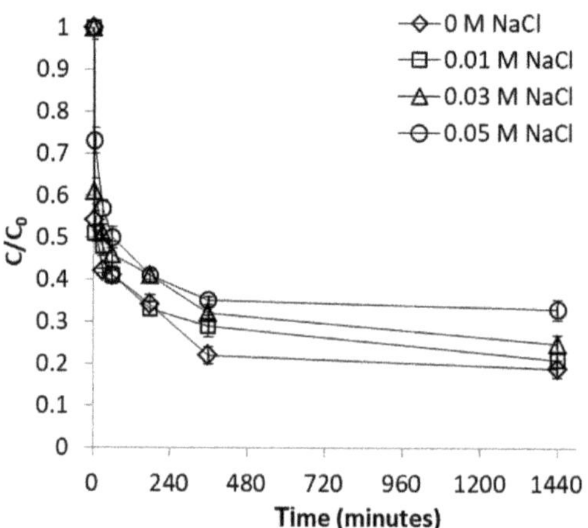

Figure 7. Effect of ionic strength on Fe^{II} removal by WSP. The lines are not fitting models; they simply connect points to facilitate visualization.

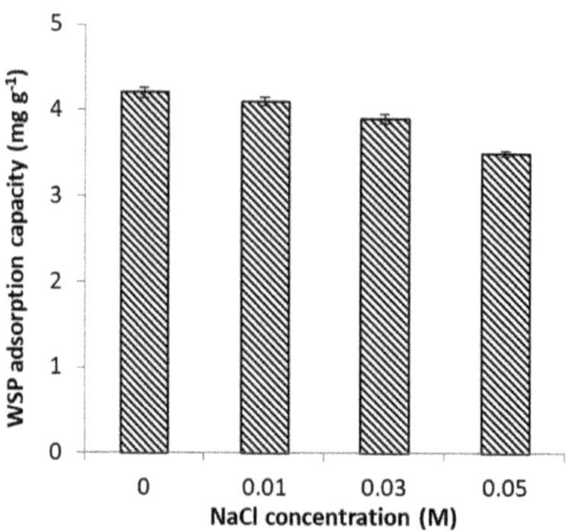

Figure 8. Effect of ionic strength on Fe^{II} adsorption capacity of WSP.

2.3. Cr^{VI} Treatability Experiments

2.3.1. Effect of pH

In this series of tests the impact of initial pH was studied within the pH range of 1.0–5.9. The results of the present experiments (Figure 9) indicated that Cr^{VI} removal with WSP-Fe^0 was significantly hindered by the increase of pH; moreover, at pH $\geq$ 5.1 Cr^{VI} removal was almost totally inhibited. Control experiments with WSP showed the same trend of decreasing efficacy of Cr^{VI} removal with increasing pH (Figure 9). Our results may be attributed, on the one hand, to a decrease in the number of positive charges existent at WSP surface with increasing pH, which hinders electrostatic attraction of anionic Cr^{VI} species. On the other hand, removal of Cr^{VI} at Fe^0 centers is known to be a complex process also inhibited by the increase of pH [45]. Similar maximum adsorption efficiency in the acidic range has been most often reported in the literature for retaining of Cr^{VI}

on adsorbents developed from different agricultural wastes (acid-activated rice husk, $ZnCl_2$-activated wood, acid-activated saw dust, $ZnCl_2$-microwave-activated sawdust, date pits, tea-waste) [32,46–48]. However, different influence of pH has also been observed; for instance, the effective pH range for *Chrysophyllum albidum* seed shells-based adsorbents was found to be 4.5–5 [49].

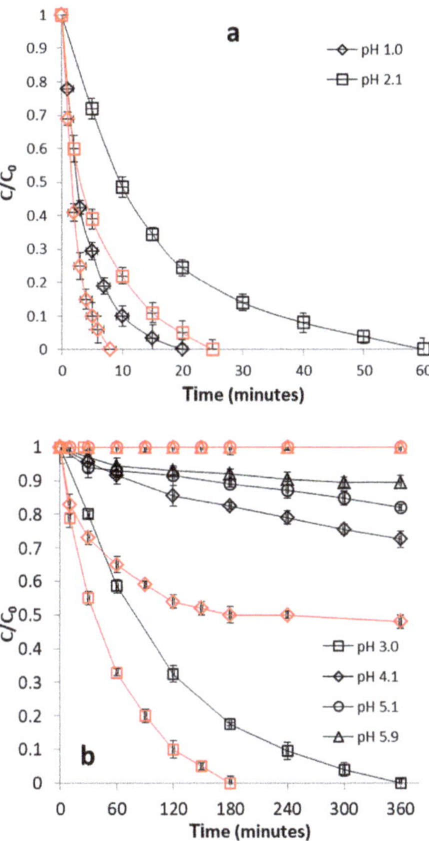

Figure 9. Effect of pH on Cr^{VI} removal by WSP-Fe^0 (red curves) and WSP (black curves). (**a**) pH 1.0–2.1; (**b**) pH 3.0–5.9. The lines are not fitting models; they simply connect points to facilitate visualization.

Two important observations can be made by analyzing the findings of pH influence experiments: (1) higher Cr^{VI} removal efficiencies for WSP-Fe^0 than for WSP were observed over the pH range of 1.0–4.1, and (2) no Cr^{VI} removal and low Cr^{VI} removal efficiency was noticed over the pH range of 5.1–5.9 for WSP-Fe^0 and WSP, respectively. The existence of Fe^0 centers at surface of WSP-Fe^0 may explain both the better Cr^{VI} removal at pH 1.0–4.1 and the lack of Cr^{VI} removal at pH 5.1–5.9, observed for WSP-Fe^0. It is well known that Cr^{VI} removal at Fe^0 surface (adsoption + possible reduction) is pH-dependent: the lower the pH, the higher the removal efficiency [45]. However, Cr^{VI} adsorption at WSP surface is also favored by an acidic pH. Therefore, it is apparent that adsorption at surface of Fe^0 centers was more severely hindered at pH 5.1–5.9 than adsorption at surface of WSP surface. This is a relevant evidence of the importance of Fe^0 centers in the process of Cr^{VI} removal with WSP-Fe^0.

2.3.2. Effect of Cr^{VI} Initial Concentration

The influence of initial concentration was examined by varying the initial metal concentration from 1 to 5 mg L^{-1}. These concentrations were selected because they are within the common levels both for subsurface Cr^{VI}-contaminated groundwater [50] and for wastewater effluents [51,52]. Figure 10 depicts the influence of initial Cr^{VI} concentration on removal efficiency. It can be easily seen from this figure that initial concentration of Cr^{VI} is another parameter which plays an important role in the process of Cr^{VI} removal with WSP-Fe^0: the higher the initial Cr^{VI} concentration, the lower the efficacy of the removal process. The same outcome was noticed also for the control experiments conducted with WSP (Figure 10): uptake of Cr^{VI} was inhibited at higher Cr^{VI} concentrations; nevertheless, Figure 10 clearly reveals that, for same Cr^{VI} initial concentration, better removal yields were always obtained for WSP-Fe^0 than for WSP, which is attributable to existence of Fe^0 at surface of WSP-Fe^0. The results of the influence of initial concentration are consistent with previous findings reporting removal of Cr^{VI} from aqueous effluents by use of other biosorbents (acid-activated rice husk, $ZnCl_2$-activated wood, acid-activated saw dust, $ZnCl_2$-microwave-activated sawdust) [32,47,48]. The negative effect of initial Cr^{VI} concentration is similar to the one observed in the process of Fe^{II} removal, and has an identical explanation: the more rapid saturation of the reactive centers existent at surface of WSP-Fe^0 (available for the interaction with Cr^{VI}) with increasing Cr^{VI} concentration.

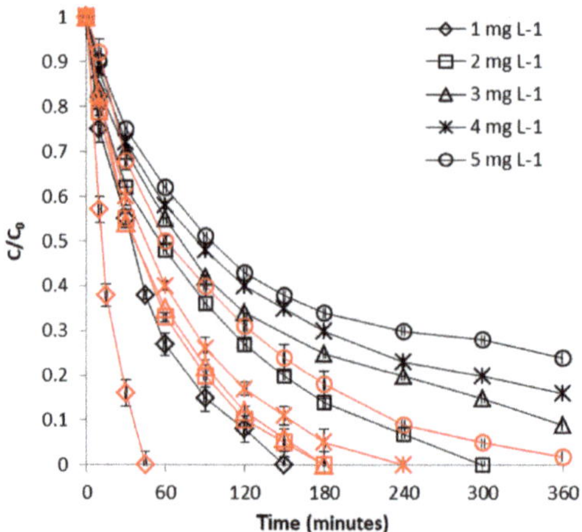

Figure 10. Effect of initial concentration on Cr^{VI} removal by WSP-Fe^0 (red curves) and WSP (black curves). The lines are not fitting models; they simply connect points to facilitate visualization.

2.3.3. Effect of Temperature

The dependence of the Cr^{VI} removal process with temperature was investigated over the range of 6–32 °C. It is evident from Figure 11 that removal of Cr^{VI} with WSP-Fe^0 was highly dependent on the temperature: an increased trend in removal efficiency was noticed with rise in temperature, indicating the endothermic nature of the process. The observed influence of increasing temperature can be most probably ascribed to increase in rate of diffusion of the Cr^{VI} ions across the boundary layer. Even though the same effect of temperature was observed also in control experiments with WSP (Figure 11), however, for same temperature, higher Cr^{VI} removal efficiencies were always obtained for WSP-Fe^0 than for WSP, attributable to existence of Fe^0 at surface of WSP-Fe^0. Our results are in line with previous findings indicating favorable binding of Cr^{VI} on different biosorbents (acid-activated rice husk, $ZnCl_2$-activated wood, acid-activated saw dust, $ZnCl_2$-microwave-activated sawdust) at higher temperature [32,47,48].

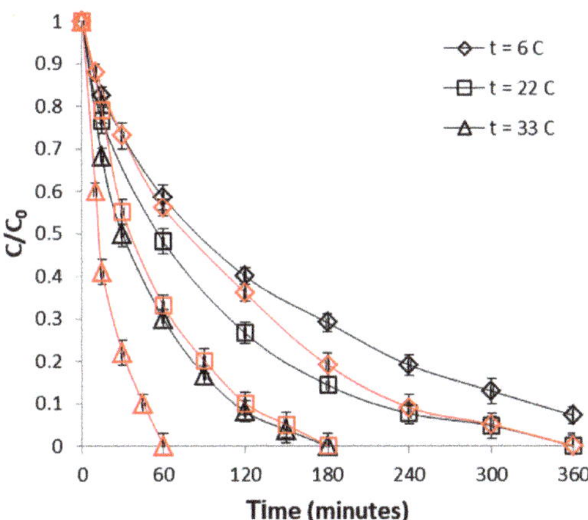

Figure 11. Effect of temperature on Cr^{VI} removal by WSP-Fe^0 (red curves) and WSP (black curves). The lines are not fitting models; they simply connect points to facilitate visualization.

2.3.4. Effect of Ionic Strength

To study the influence of this parameter, the ionic strength of Cr^{VI} solutions was adjusted using NaCl as background electrolyte, in the concentration range of 0–0.05 M. The extent of Cr^{VI} removal with WSP-Fe^0 as a function ionic strength is depicted in Figure 12. As revealed by this figure, the addition of NaCl (i.e., increase of ionic strength) led to a slight increase in Cr^{VI} removal efficiency. The highest improvement in removal efficacy was noticed as the ionic strength was increased from 0 to 0.01 M; a further increase in ionic strength to 0.03 and 0.05 M lead to removal yields higher than for 0 M, but lower than for 0.01 M. Thus, it can be concluded that optimal ionic strength for Cr^{VI} removal with WSP-Fe^0 was 0.01 M. On the other hand, control experiments conducted with WSP revealed that removal of Cr^{VI} was practically not influenced by the increase of ionic strength (Figure 12). Conversely, other authors reported a more or less significant adverse influence of ionic strength on the removal of Cr^{VI} with grape stalks, cork, olive stones, thermochemically-activated walnut shells or surfactant modified spent mushroom [46,53–55]. The two different effects exerted by the background ionic strength on removal of Cr^{VI} with WSP-Fe^0 and with WSP are indicative of two distinct removal mechanisms involved in the two cases. On the one hand, the absence of any visible influence of ionic strength on Cr^{VI} removal with WSP can be interpreted as indicating a specific adsorption mechanism [44]; on the other hand, the higher removal efficiencies obtained with WSP-Fe^0 at higher ionic strengths may be ascribed to existence of Fe^0 active sites at surface of WSP-Fe^0. This is in accord with findings of previous studies which demonstrated that Cl^- anion can accelerate Fe^0 corrosion by forming soluble complexes with Fe^{II}, which are carried away from the metal surface; the as formed Fe^{II} complexes have two important roles: (1) to delay the formation of oxide layers at surface of Fe^0, and (2), to act as secondary reducing agents for the reduction of Cr^{VI} [56,57].

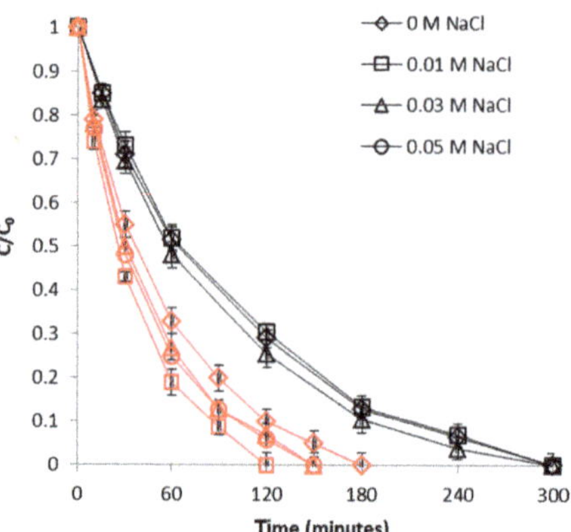

Figure 12. Effect of ionic strength on Cr^{VI} removal by WSP-Fe^0 (red curves) and WSP (black curves). The lines are not fitting models; they simply connect points to facilitate visualization.

2.4. Kinetic Modeling

2.4.1. Identification of the Kinetic Order

The statistical fits of Fe^{II} and Cr^{VI} removal experimental data to pseudo first and pseudo second-order equations, and the parameters of the two kinetic models are summarized in Table 1. With regard to Fe^{II} removal, as evidenced by the correlation coefficients, pseudo second-order kinetic model provided the best match for the experimental data. This conclusion is confirmed also by the fact that equilibrium adsorption capacity value (q_e) predicted by the pseudo second-order model fits the best the experimental value (q_e^{exp}). These evidences indicate that pseudo second-order kinetic model was the more appropriate to describe Fe^{II} adsorption. This is consistent with results from a previous study using thermochemically-activated orange peels [30]. The pseudo second-order kinetic model assumes that the rate-limiting step of the adsorption process is of chemisorption nature, involving sharing or exchange of electrons between adsorbent and adsorbate [58].

Table 1. Kinetic parameters of Fe^{II} and Cr^{VI} removal.

Test	Pseudo 1st Order			Pseudo 2nd Order			q_e^{exp} (mg g^{-1})
	k_1 (min^{-1})	q_e (mg g^{-1})	R^2	k_2 (g mg^{-1} min^{-1})	q_e (mg g^{-1})	R^2	
Fe^{II} + WSP	1.4×10^{-3}	1.20	0.7936	18.8×10^{-3}	3.93	0.9998	4.20
Cr^{VI} + WSP	6.9×10^{-3}	0.35	0.9891	1.9×10^{-2}	0.33	0.9874	0.40
Cr^{VI} + WSP-Fe^0	1.8×10^{-2}	0.46	0.9948	5.9×10^{-2}	0.56	0.9923	0.50

From the kinetic data of Cr^{VI} removal with WSP-Fe^0 it can be seen that regression coefficient of the second-order model is lower than of the pseudo first-order model, which implies that removal of Cr^{VI} with WSP-Fe^0 follows the pseudo first-order kinetics. In addition, the calculated q_e value obtained from the pseudo first-order model agrees better with the experimental q_e^{exp} value than the one obtained from the second-order model. Consequently, the retention of Cr^{VI} onto WSP-Fe^0 could be best described by the pseudo first-order kinetic model. Control experiments with WSP are in good agreement with WSP-Fe^0 experiments, revealing that adsorption onto WSP also fitted well to the pseudo first-order kinetic model. The pseudo first-order kinetic model was successfully applied in early works investigating Cr^{VI} removal by different biomaterials (thermochemically-modified *Terminalia arjuna* nuts, Fe^{III} impregnated biochar and tea-waste [59–61]), being indicative for existence of relatively weak electrostatic interactions between Cr^{VI} and adsorbent [62]. Nevertheless, several other studies, working with $ZnCl_2$-activated *S. guttata* shell waste or acid-activated pomegranate husk, indicated that pseudo-second order was the applicable kinetic model for Cr^{VI} removal [63,64].

2.4.2. Identification of the Rate Limiting Step

Removal of a contaminant via adsorption occurs through a mechanism comprising the following consecutive steps: (1) transport of contaminant in the bulk of the solution, (2) transport of contaminant through the liquid film surrounding the adsorbent particle, to its external surface (film diffusion), (3) transport of contaminant from the adsorbent surface into its pores (intraparticle diffusion), and (4) retention of the contaminant inside the pores. Generally, phase (1) and (4) are very rapid and do not represent the rate determining step [65]. The Weber and Morris model was applied in this study to determine whether film diffusion or intraparticle diffusion is the rate limiting step. If intra-particle diffusion would be the rate-limiting step, then Weber and Morris plots should pass through the origin and have a good linearity. Figures 13–15 clearly reveal that, for both Fe^{II} and Cr^{VI} removal, the q_t versus $t^{0.5}$ plots show multilinearity, indicating that at least two steps take place. This implies that intraparticle diffusion was not the only rate-controlling step, and that diffusion through the liquid film around the adsorbent toward particle surface is also involved in metal binding onto adsorbent. The first (sharper) region of the Weber and Morris plots corresponds to the phase of the adsorption which is predominantly controlled by film diffusion, while the second region describes the adsorption stage where intraparticle diffusion played the major role, being thus rate limiting [66,67]. Accordingly, the k_{dif} intraparticle diffusion rate constants were derived from the slope of the second linear portion, while the C values were computed from the intercept of the first linear portion. The k_{dif} intraparticle diffusion rate constant can be used for evaluation of the effect of intraparticle diffusion on the adsorption process: the higher the k_{dif}, the lower the resistance to diffusion inside the pores. The intercept C value provides information about the thickness of the boundary layer: the larger the intercept value, the greater the resistance of external mass transfer across the boundary layer [64,67].

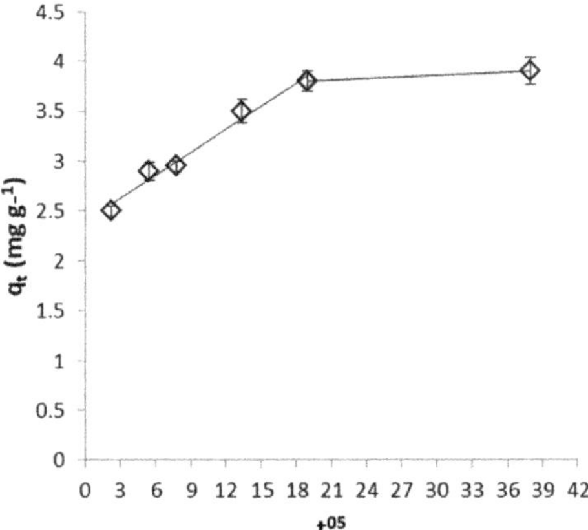

Figure 13. Weber and Morris plot for Fe^{II} removal by WSP.

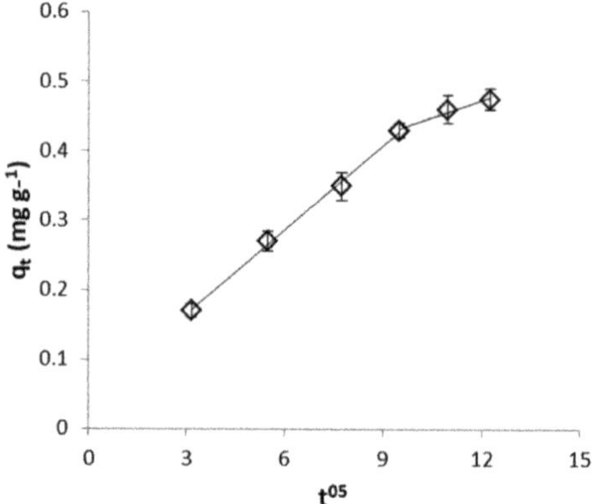

Figure 14. Weber and Morris plot for Cr^{VI} removal by WSP-Fe^0.

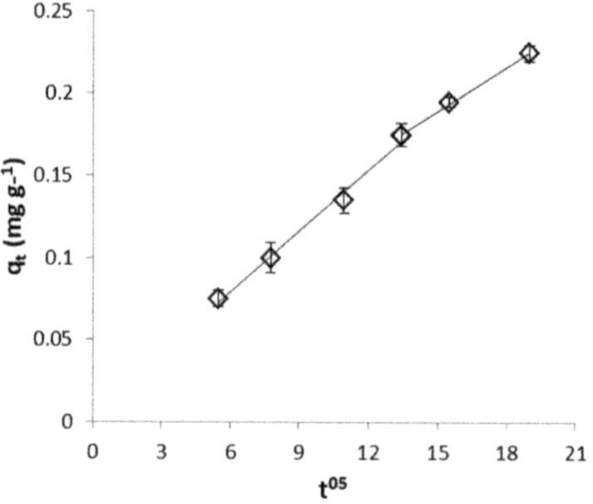

Figure 15. Weber and Morris plot for Cr^{VI} removal by WSP.

From the values of intraparticle diffusion model parameters (Table 2) it can be seen that higher diffusion rate was observed for removal of Cr^{VI} with WSP-Fe^0 than with WSP, while similar low C values were determined in both cases. On the other hand, Fe^{II} removal with WSP is characterized by much lower diffusion rate constant and much higher boundary layer effect than removal of Cr^{VI} with both WSP-Fe^0 and WSP. Similar Weber and Morris plots exhibiting multilinearity and not intersecting the origin were previously reported for the adsorption of acid-activated date palm seed, *Eichhornia crassipes* biomass and tea-waste [61,68,69].

Table 2. Weber and Morris diffusion model parameters.

	k_{diff} (mg g^{-1} min$^{-0.5}$)	C
Fe^{II} removal by WSP	5.3×10^{-3}	2.4
Cr^{VI} removal by WSP	0.9×10^{-2}	2×10^{-2}
Cr^{VI} removal by WSP-Fe^0	1.6×10^{-2}	3×10^{-2}

2.5. Mechanism of Metal Removal

The remediation of the AMD solution occurs through a pure adsorption process of Fe^{II} at surface of WSP, via mixed physical and chemical mechanisms. Similarly, Cr^{VI} removal with WSP-Fe^0 can also be ascribed to adsorption processes. However, in this case, the higher removal efficiencies observed for WSP-Fe^0 than for WSP are indicative of existence of differences in mechanism of metal removal. This is attributable to Fe^0 centers formed at surface of WSP as a result of reaction between WSP-Fe^{II} and sodium borohydride. Over the last three decades, Fe^0 has been demonstrated to represent a highly efficient reagent in remediation of water contaminated with a wide variety of pollutants, including Cr^{VI}. Removal of Cr^{VI} with Fe^0 occurs through a very complex mechanism, which may involve physicochemical processes such as adsorption, direct reduction, indirect reduction, co-precipitation/enmeshment in the mass of precipitates; generally, both adsorption and reduction processes of Cr^{VI} in Fe^0/H$_2$O system are favored by an acidic pH, being strongly hindered at pH levels close to neutral values [45]. After being adsorbed at surface of WSP-Fe^0, Cr^{VI} can be reduced to Cr^{III} by WSP functional groups, by Fe^0 (at very acidic pH, when it's not covered by oxides), by Fe^{II}-based corrosion products formed at surface of Fe^0 or by dissolved Fe^{II}. In addition, under acidic conditions, Cr^{VI} reduction may take place also homogeneously with dissolved Fe^{II}. Thus, we can suggest that removal of Cr^{VI} with WSP-Fe^0 occurred through a combined adsorption-reduction process. However, since very low concentrations of dissolved Cr^{III} (~0.2–0.4 mg L^{-1}) were detected only at pH 1.0 and 2.1, and Cr^{III} adsorption/precipitation is inhibited at acidic pH, it can be assumed that adsorption processes played the most important role in removal of Cr^{VI}. This is in good agreement with similar observations reported by other researchers, indicating that Cr^{VI} removal by natural biomaterials occurs via an adsorption-coupled reduction mechanism [70,71].

3. Materials and Methods

3.1. Preparation of the Adsorbent for AMD Treatability Experiments

Walnuts (*Juglans regia*) were obtained from a local market in Timisoara (Romania). After crushing the walnuts by hand, shells were separated, rinsed several times with distilled water to remove impurities, and dried in an oven at 80 °C for 24 h. Then, the dried shells were powdered using an electric grinder. The resultant WSP was washed with distilled water until no brown coloration of the water was noticed, and then dried again in oven at 80 °C for 24 h, to remove moisture. After cooling, the WSP was ground with a mortar and pestle, and subsequently sieved to particles size of 0.5–1.25 mm for further treatability experiments with synthetic AMD solutions.

3.2. Preparation of the Reactive Material for Cr^{VI} Treatability Experiments

After each AMD treatability experiment, the exhausted adsorbent (WSP-Fe^{II}) was recovered and dried at room temperature. By means of mass balance calculation, the concentration of adsorbed iron was determined to be about 3 mg Fe^{II}/g WSP. Then, the adsorbed Fe^{II} was reduced to Fe^0 via the liquid-phase reduction method, using sodium borohydride (NaBH$_4$) as reducing reagent [72]:

$$Fe(H_2O)_6{}^{2+} + 2BH_4{}^- \rightarrow Fe^0 + 2B(OH)_3 + 7H_2 \tag{7}$$

About 250 mL distilled water were added over 80 g of WSP-Fe^{II} and the obtained slurry was stirred at a rate of 200 rpm, in order to keep solid particles in suspension. Then, 0.6 g NaBH$_4$ was added in small portions while stirring, in a fume hood; NaBH$_4$ was used in excess to the stoichiometric needed amount, in order to account for any that may decompose during the course of the reaction with water. The usual brown color of the solid material immediately darkened to a black appearance, indicating the formation of Fe^0 centers at surface of WSP (Figure S16) [73]. After the addition of NaBH$_4$ was completed (~60 min), the resulted mixture was stirred for an additional 60 min. The resulted WSP-Fe^0 was separated from the solution, washed with distilled water, dried at 80 °C for 24 h in an oven, and kept in vacuum desiccator prior to being used in treatability experiments with Cr^{VI} solution.

3.3. AMD Treatability Experiments

Synthetic AMD stock solution (1000 mg L^{-1}) was prepared by dissolving the required amount of AR grade FeSO$_4$·7H$_2$O in distilled de-ionized water. Then, AMD working solutions with desired Fe^{II} concentrations were prepared by appropriate dilution of stock solution, knowing that Fe is often the main heavy metal present in acid mine drainage [74]. AMD treatability experiments were conducted in batch system, using an Ovan jar tester. 500 mL AMD solution was poured in 800 mL Berzelius flasks, followed by addition of 5 g WSP. The mixture was stirred (200 rpm) and, at timed intervals, samples were withdrawn, filtered using a 0.45 µm filter and analyzed for Fe^{II}. The pH of Fe^{II} solutions was adjusted before experiments to the required value by addition of small amounts of concentrated H$_2$SO$_4$. Detailed conditions of AMD treatability experiments are summarized in Table 3.

Table 3. Setup design of AMD treatability experiments·

	Investigation of the Influence of			
	pH	**Fe^{II} Concentration**	**Temperature**	**Ionic Strength**
pH	1.0–4.1	4.1	4.1	4.1
Fe^{II} concentration (mg L^{-1})	50	25–100	50	50
Temperature (°C)	22	22	6–33	22
Ionic strength (mole L^{-1} NaCl)	0	0	0	0–0.05

3.4. Cr^{VI} Treatability Experiments

Cr^{VI} stock solution (1000 mg L^{-1}) was prepared by dissolving the required amount of AR grade K$_2$Cr$_2$O$_7$ in distilled de-ionized water. The stock solution was then further diluted with de-ionized distilled water in order to prepare the working Cr^{VI} solutions. Batch Cr^{VI} treatability experiments were conducted by mixing 1 g of WSP-Fe^0 with a volume 500 of mL Cr^{VI} solution in 800 mL Berzelius flasks. The mixture was stirred using an Ovan jar tester (200 rpm) and, at predetermined times, supernatant aliquots were collected, filtered through a 0.45 µm filter and sent to Cr^{VI} analysis. The pH of Cr^{VI} solution was set by addition of small amounts of concentrated H$_2$SO$_4$ or 1 M NaOH solution. For comparison purposes, control Cr^{VI} treatability experiments with raw WSP were also conducted, by keeping unchanged all experimental conditions. Detailed conditions of Cr^{VI} treatability experiments are summarized in Table 4.

Table 4. Setup design of Cr^{VI} treatability experiments·

	Investigation of the Influence of			
	pH	**Cr^{VI} Concentration**	**Temperature**	**Ionic Strength**
pH	1.0–5.9	3.0	3.0	3.0
Cr^{VI} concentration (mg L^{-1})	2	1–5	2	2
Temperature (°C)	22	22	6–33	22
Ionic strength (mole L^{-1} NaCl)	0	0	0	0–0.05

3.5. Analytical Procedure

Cr^{VI} and Fe^{II} concentrations in the filtrate were analyzed by the 1,5-diphenylcarbazide method (at 540 nm) and, 1,10-ortophenantroline colorimetric method (at 510 nm), respectively, by using a 200 PLUS spectrophotometer (Specord, Germany). Cr^{total} was determined by treating the sample with $KMnO_4$ to oxidize any present Cr^{III}, followed by analysis as Cr^{VI}; then, Cr^{III} was determined from the difference between Cr^{total} and Cr^{VI} [75]. The pH of the samples was measured using a 7320 pH-meter (Inolab, Germany); three standard buffer solutions at pHs 4.0, 7.0 and 10.0 were employed for calibration. Point of zero charge (pH_{pzc}) of WSP surface was determined using the pH drift method [76]. The prepared adsorbents were characterized by Fourier transform infrared spectrometry (FTIR: VERTEX 70, Bruker, Germany) and scanning electron microscopy (SEM: Inspect S, FEI, The Netherlands) coupled with energy dispersive X-ray spectroscopy (EDX: GENESIS XM 2i, The Netherlands).

3.6. Kinetic Modeling of Experimental Data

The kinetics of contaminant removal was analyzed using the linearized forms of Lagergren pseudo first-order model and Ho's pseudo second-order model [58,77,78]:

$$\log(q_e - q_t) = \log q_e - \frac{k_1}{2303}t \tag{8}$$

$$\frac{t}{q_t} = \frac{1}{k_2 q_e^2} + \frac{t}{q_e} \tag{9}$$

where q_e is the equilibrium adsorption capacity (mg g^{-1}), q_t is the adsorption capacity at time t (mg g^{-1}), k_1 (min^{-1}) and k_2 (g mg^{-1} min^{-1}) are the pseudo first-order and, respectively, pseudo second-order adsorption rate coefficients. The product $k_2 q_e^2$ also represents the initial sorption rate. q_e and q_t were calculated as follows:

$$q_t = \frac{(C_0 - C_t)V}{M} \tag{10}$$

$$q_e = \frac{(C_0 - C_e)V}{M} \tag{11}$$

where M is the mass of adsorbent used in the kinetic experiments (g), C_e the equilibrium concentration of metal (mg L^{-1}), C_t the metal concentration at time t (mg L^{-1}), C_0 the initial concentration of metal (mg L^{-1}); V the volume of solution used in the kinetic experiments (L).

The slope and intercept of the plots of $log(q_e - q_t)$ vs. t allows computation of pseudo first-order k_1 and of equilibrium adsorption capacity q_e; similarly, the plot of t/q_t vs. t enables the pseudo second-order rate constant k_2 and q_e to be determined from intercept and slope. In order to further assess the nature of the rate-limiting step of the process (film diffusion or intraparticle diffusion), experimental data was fitted also to the Weber and Morris intraparticle diffusion model [64,79]:

$$q_t = k_{diff} \cdot t^{0.5} + C \tag{12}$$

where q_t (mg g^{-1}) is the adsorption capacity at time t, k_{diff} (mg g^{-1} min$^{-0.5}$) is the intraparticle diffusion rate constant and C is a constant linked to the apparent thickness of the film boundary layer.

Kinetic modeling was conducted using experimental data acquired at pH 4.1, 50 mg L^{-1}, 22 °C, and pH 4.1, 2 mg L^{-1}, 22 °C, for Fe^{II} and Cr^{VI}, respectively.

3.7. Statistical Analysis

All the data represent the mean of two independent experiments and relative error less than 2% were obtained. Statistical analysis was performed using Microsoft Excel 2016 statistical tool.

4. Conclusions

In last years, the use of agricultural wastes/byproducts as cost-effective alternative adsorbents for the treatment of water contaminated with a large variety of pollutants has attracted significant interest. However, much less interest has shown for finding environmentally-friendly solutions for the management of residual solids resulted from such water treatment processes. The present paper presents data on the use of WSP, a local agricultural waste, for the sequential removal of two heavy metals, namely Fe^{II} and Cr^{VI}, from aqueous effluents. Results presented herein clearly demonstrated that WSP can be considered as a promising adsorbent for the removal of Fe^{II} from AMD, while WSP-Fe^0, obtained by treating the Fe^{II}-contaminated solid residue (WSP-Fe^{II}) with sodium borohydride, is a suitable reactive reagent in the process of Cr^{VI} removal from contaminated waters. The better capacity of WSP-Fe^0 to remove Cr^{VI}, compared to fresh WSP, was ascribed to existence of Fe^0 centers at surface of WSP-Fe^0. Adsorption kinetics of Cr^{VI} and Fe^{II} was successfully fitted by the pseudo first- and pseudo second-order model, respectively. While binding of Fe^{II} on WSP occurred via physical and chemical mixed adsorption, removal of Cr^{VI} with WSP-Fe^0 took place through a more complex mechanism, involving both adsorption and reduction processes. This study provides compelling evidence that residues resulted from a water adsorption treatment process can be successfully converted into reactive materials for a subsequent water treatment technology. The major challenge of this strategy is to identify water treatment processes with fully compatible pollutants.

Supplementary Materials: The following are available online at https://www.mdpi.com/2227-9717/9/2/218/s1.

Author Contributions: Conceptualization, M.G. and I.B.; methodology, M.G. and I.B.; software, M.G.; validation, M.G. and I.B.; formal analysis, M.G. and I.B.; investigation, M.G. and I.B.; resources, M.G. and I.B.; data curation, M.G. and I.B.; writing—original draft preparation, M.G.; writing—review and editing, M.G. and I.B.; visualization, M.G. and I.B.; supervision, M.G. All authors have read and agreed to the published version of the manuscript.

Funding: This research received no external funding.

Institutional Review Board Statement: Not applicable.

Informed Consent Statement: Not applicable.

Data Availability Statement: Data is contained within the article.

Acknowledgments: We sincerely thank the three anonymous reviewers from Processes whose insightful comments and suggestions provided on earlier version of this manuscript helped improve and clarify this study.

Conflicts of Interest: The authors declare no conflict of interest.

References

1. Vardhan, K.H.; Kumar, P.S.; Panda, R.C. A review on heavy metal pollution, toxicity and remedial measures: Current trends and future perspectives. *J. Mol. Liq.* **2019**, *290*, 111197. [CrossRef]
2. Bozic, D.; Stankovic, V.; Gorgievski, M.; Bogdanovic, G.; Kovacevic, R. Adsorption of heavy metal ions by sawdust of deciduous trees. *J. Hazard. Mater.* **2009**, *171*, 684–692. [CrossRef] [PubMed]
3. Barakat, M.A. New trends in removing heavy metals from industrial wastewater. *Arab. J. Chem.* **2011**, *4*, 361–377. [CrossRef]
4. Gautam, R.K.; Mudhoo, A.; Lofrano, G.; Chattopadhyaya, M.C. Biomass-derived biosorbents for metal ions sequestration: Adsorbent modification and activation methods and adsorbent regeneration. *J. Environ. Chem. Eng.* **2014**, *2*, 239–259. [CrossRef]
5. Akhbarizadeh, R.; Shayestefar, M.R.; Darezereshki, E. Competitive removal of metals from wastewater by maghemite nanoparticles: A comparison between simulated wastewater and AMD. *Mine Water Environ.* **2014**, *33*, 89–96. [CrossRef]
6. Gheju, M.; Pode, R.; Manea, F. Comparative heavy metal chemical extraction from anaerobically digested biosolids. *Hydrometallurgy* **2011**, *108*, 115–121. [CrossRef]
7. Fu, F.; Wang, Q. Removal of heavy metal ions from wastewaters: A review. *J. Environ. Manag.* **2011**, *92*, 407–418. [CrossRef]
8. Bolisetty, S.; Peydayesh, M.; Mezzenga, R. Sustainable technologies for water purification from heavy metals: Review and analysis. *Chem. Soc. Rev.* **2019**, *48*, 463–487. [CrossRef]
9. Hashim, M.A.; Mukhopadhyay, S.; Sahu, J.N.; Sengupta, B. Remediation technologies for heavy metal contaminated groundwater. *J. Environ. Manag.* **2011**, *92*, 2355–2388. [CrossRef]

10. Selvi, A.; Rajasekar, A.; Theerthagiri, J.; Ananthaselvam, A.; Sathishkumar, K.; Madhavan, J.; Rahman, P.K.S.M. Integrated remediation processes toward heavy metal removal/recovery from various environments–A review. *Front. Environ. Sci.* **2019**, *7*, 66. [CrossRef]

11. Hu, M.; Zhang, S.; Pan, B.; Zhang, W.; Lv, L.; Zhang, Q. Heavy metal removal from water/wastewater by nanosized metal oxides: A review. *J. Hazard. Mater.* **2012**, *211–212*, 317–331. [CrossRef] [PubMed]

12. Kurniawan, T.A.; Chan, G.Y.S.; Lo, W.H.; Babel, S. Comparisons of low-cost adsorbents for treating wastewaters laden with heavy metals. *Sci. Total Environ.* **2006**, *366*, 409–426. [CrossRef] [PubMed]

13. Heidarinejad, Z.; Dehghani, M.H.; Heidari, M.; Javedan, G.; Ali, I.; Sillanpää, M. Methods for preparation and activation of activated carbon: A review. *Environ. Chem. Lett.* **2019**, *18*, 393–415. [CrossRef]

14. Dias, J.M.; Alvim-Ferraz, M.C.M.; Almeida, M.F.; Rivera-Utrilla, J.; Sanchez-Polo, M. Waste materials for activated carbon preparation and its use in aqueous-phase treatment: A review. *J. Environ. Manag.* **2007**, *85*, 833–846. [CrossRef]

15. Sirry, S.M.; Aldakhil, F.; Alharbi, O.M.L.; Ali, I. Chemically treated date stones for uranium (VI) uptake and extraction in aqueous solutions. *J. Mol. Liq.* **2019**, *273*, 192–202. [CrossRef]

16. Mosoarca, G.; Vancea, C.; Popa, S.; Gheju, M.; Boran, S. Syringa vulgaris leaves powder a novel low-cost adsorbent for methylene blue removal: Isotherms, kinetics, thermodynamic and optimization by Taguchi method. *Sci. Rep.* **2020**, *10*, 17676. [CrossRef]

17. Abdel-Ghani, N.T.; El-Chaghaby, G.A. Biosorption for metal ions removal from aqueous solutions: A review of recent studies. *Int. J. Lat. Res. Sci. Technol.* **2014**, *3*, 24–42.

18. Mohan, D.; Pittman, C.U., Jr. Activated carbons and low cost adsorbents for remediation of tri- and hexavalent chromium from water. *J. Hazard. Mater.* **2006**, *B137*, 762–811. [CrossRef]

19. Joseph, L.; Jun, B.M.; Flora, J.R.V.; Park, C.M.; Yoon, Y. Removal of heavy metals from water sources in the developing world using low-cost materials: A review. *Chemosphere* **2019**, *229*, 142–159. [CrossRef]

20. Ahmad, T.; Ahmad, K.; Alam, M. Sustainable management of water treatment sludge through 3'R' concept. *J. Clean. Prod.* **2016**, *124*, 1–13. [CrossRef]

21. Ippolito, J.A.; Barbarick, K.A.; Elliott, H.A. Drinking water treatment residuals: A review of recent uses. *J. Environ. Qual.* **2011**, *40*, 1–12. [CrossRef]

22. Turner, T.; Wheeler, R.; Stone, A.; Oliver, I. Potential alternative reuse pathways for water treatment residuals: Remaining barriers and questions–A Review. *Water Air Soil Pollut.* **2019**, *230*, 227. [CrossRef]

23. Babatunde, A.; Zhao, Y. Constructive approaches toward water treatment works sludge management: An international review of beneficial reuses. *Crit. Rev. Environ. Sci. Technol.* **2007**, *37*, 129–164. [CrossRef]

24. Wolowiec, M.; Komorowska-Kaufman, M.; Pruss, A.; Rzepam, G.; Bajda, T. Removal of heavy metals and metalloids from water using drinking water treatment residuals as adsorbents: A review. *Minerals* **2019**, *9*, 487. [CrossRef]

25. Gheju, M.; Balcu, I. Mitigation of Cr(VI) aqueous pollution by reuse of iron-contaminated water treatment residues. *ChemEngineering* **2017**, *1*, 9. [CrossRef]

26. Sagar, N.A.; Pareek, S.; Sharma, S.; Yahia, E.M.; Lobo, M.G. Fruit and vegetable waste: Bioactive compounds, their extraction, and possible utilization. *Compr. Rev. Food Sci. Food Saf.* **2018**, *17*, 512–531. [CrossRef]

27. Gowe, C. Review on potential use of fruit and vegetables by-products as a valuable source of natural food additives. *Food Sci. Qual. Manag.* **2015**, *45*, 47–61.

28. USDA. *Tree nuts: World Market and Trade*; United States Department of Agriculture Foreign Agricultural Service: Washington, DC, USA, 2017.

29. Ghasemi, M.; Ghoreyshi, A.A.; Younesi, H.; Khoshhal, S. Synthesis of a high characteristics activated carbon from walnut shell for the removal of Cr(VI) and Fe(II) from aqueous solution: Single and binary solutes adsorption. *Iran. J. Chem. Eng.* **2015**, *12*, 28–51.

30. Adebayo, G.B.; Mohammed, A.A.; Sokoya, S.O. Biosorption of Fe(II) and Cd(II) ions from aqueous solution using a low cost adsorbent from orange peels. *J. Appl. Sci. Environ. Manag.* **2016**, *20*, 702–714. [CrossRef]

31. Noubactep, C. Metallic iron for environmental remediation: A review of reviews. *Water Res.* **2015**, *85*, 114–123. [CrossRef]

32. Bansal, M.; Singh, D.; Garg, V.K. A comparative study for the removal of hexavalent chromium from aqueous solution by agriculture wastes' carbons. *J. Hazard. Mater.* **2009**, *171*, 83–92. [CrossRef] [PubMed]

33. Savy, D.; Piccolo, A. Physical-chemical characteristics of lignins separated from biomasses for second-generation ethanol. *Biomass Bioenergy* **2014**, *62*, 58–67. [CrossRef]

34. Coates, J. Interpretation of infrared spectra, a practical approach. In *Encyclopedia of Analytical Chemistry*; Meyers, R.A., Ed.; John Wiley & Sons: Chichester, UK, 2000; pp. 10815–10837.

35. Domínguez-Robles, J.; Sánchez, R.; Espinosa, E.; Savy, D.; Mazzei, P.; Piccolo, A.; Rodríguez, A. Isolation and characterization of *Gramineae* and *Fabaceae* soda lignins. *Int. J. Mol. Sci.* **2017**, *18*, 327. [CrossRef] [PubMed]

36. Aravindhan, R.; Madhan, B.; Rao, J.R.; Nair, B.U.; Ramasam, T. Bioaccumulation of chromium from tannery wastewater: An approach for chrome recovery and reuse. *Environ. Sci. Technol.* **2004**, *38*, 300–306. [CrossRef] [PubMed]

37. Bykov, I. Characterization of Natural and Technical Lignins Using FTIR Spectroscopy. Master's Thesis, Lulea University of Technology, Lulea, Sweden, 2008.

38. Bhattacharyya, K.G.; Sen Gupta, S. Adsorption of chromium(VI) from water by clays. *Ind. Eng. Chem. Res.* **2006**, *45*, 7232–7240. [CrossRef]

39. Espana, J.S.; Pamo, E.L.; Pastor, E.S. The oxidation of ferrous iron in acidic mine effluents from the Iberian Pyrite Belt (Odiel Basin, Huelva, Spain): Field and laboratory rates. *J. Geochem. Explor.* **2007**, *92*, 120–132. [CrossRef]
40. Espana, J.S. Acid mine drainage in the Iberian pyrite belt: An overview with special emphasis on generation mechanisms, aqueous composition and associated mineral phases. *Macla* **2008**, *10*, 34–43.
41. Yu, L.J.; Shukla, S.S.; Doris, K.L.; Shukla, A.; Margrave, J.L. Adsorption of chromium from aqueous solutions by maple sawdust. *J. Hazard. Mater* **2003**, *B100*, 53–63. [CrossRef]
42. Baral, S.S.; Das, S.N.; Rath, P. Hexavalent chromium removal from aqueous solution by adsorption on treated sawdust. *Biochem. Eng. J.* **2006**, *31*, 216–222. [CrossRef]
43. Das, D.D.; Mahapatra, R.; Pradhan, J.; Das, S.N.; Thakur, R.S. Removal of Cr(VI) from aqueous solution using activated cow dung carbon. *J. Coll. Interf. Sci.* **2000**, *232*, 235–240. [CrossRef]
44. McBride, M.B. A critique of diffuse double layer models applied to colloid and surface chemistry. *Clays Clay Miner.* **1997**, *45*, 598–608. [CrossRef]
45. Gheju, M. Decontamination of hexavalent chromium-polluted waters: Significance of metallic iron technology. In *Enhancing Cleanup of Environmental Pollutants. Volume 2. Non biological Approaches*; Anjum, N., Gill, S., Tuteja, N., Eds.; Springer International Publishing: Cham, Switzerland, 2017; pp. 209–254.
46. Albadarin, A.B.; Mangwandi, C.; Walker, G.M.; Allen, S.J.; Ahmad, M.N.M.; Khraisheh, M. Influence of solution chemistry on Cr(VI) reduction and complexation onto date-pits/tea-waste biomaterials. *J. Environ. Manag.* **2013**, *114*, 190–201. [CrossRef] [PubMed]
47. Acharya, J.; Sahu, J.N.; Sahoo, B.K.; Mohanty, C.R.; Meikap, B.C. Removal of chromium(VI) from wastewater by activated carbon developed from Tamarind wood activated with zinc chloride. *Chem. Eng.J.* **2009**, *150*, 25–39. [CrossRef]
48. Chen, C.; Zhao, P.; Li, Z.; Tong, Z. Adsorption behavior of chromium(VI) on activated carbon from eucalyptus sawdust prepared by microwave-assisted activation with $ZnCl_2$. *Desal. Water Treat.* **2016**, *57*, 12572–12584. [CrossRef]
49. Amuda, O.S.; Adelowo, F.E.; Ologunde, M.O. Kinetics and equilibrium studies of adsorption of chromium(VI) ion from industrial wastewater using *Chrysophyllum albidum* (Sapotaceae) seed shells. *Colloids Surf. B Biointerfaces* **2009**, *68*, 184–192. [CrossRef]
50. Flury, B.; Eggenberger, U.; Mader, U. First results of operating and monitoring an innovative design of a permeable reactive barrier for the remediaton of chromate contaminated groundwater. *J. Appl. Geochem.* **2009**, *24*, 687–697. [CrossRef]
51. Gao, R.M. Simultaneous determination of hexavalent and total chromium in water and plating baths by spectrophotometry. *Talanta* **1993**, *40*, 637–640. [CrossRef]
52. Ye, J.; Yin, H.; Mai, B.; Peng, H.; Qin, H.; He, B.; Zhang, N. Biosorption of chromium from aqueous solution and electroplating wastewater using mixture of *Candida lipolytica* and dewatered sewage sludge. *Biores. Technol.* **2010**, *101*, 3893–3902. [CrossRef]
53. Altun, T.; Pehlivan, E. Removal of Cr(VI) from aqueous solutions by modified walnut shells. *Food Chem.* **2012**, *132*, 693–700. [CrossRef]
54. Fiol, N.; Villaescusa, I.; Martinez, M.; Miralles, N.; Poch, J.; Serarols, J. Biosorption of Cr(VI) using low cost sorbents. *Environ. Chem. Lett.* **2003**, *1*, 135–139. [CrossRef]
55. Jing, X.; Cao, Y.; Zhang, X.; Wang, D.; Wu, X.; Xu, H. Biosorption of Cr(VI) from simulated wastewater using a cationic surfactant modified spent mushroom. *Desalination* **2011**, *269*, 120–127. [CrossRef]
56. Tepong-Tsindé, R.; Phukan, M.; Nassi, A.; Noubactep, C.; Ruppert, H. Validating the Efficiency of the MB Discoloration Method for the Characterization of Fe0/H2O systems using accelerated corrosion by chloride ions. *Chem. Eng. J.* **2015**, *279*, 353–362. [CrossRef]
57. Gheju, M.; Balcu, I.; Enache, A.; Flueras, A. A kinetic approach on hexavalent chromium removal with metallic iron. *J. Environ. Manag.* **2017**, *203*, 937–941. [CrossRef] [PubMed]
58. Mohan, D.; Singh, K.P.; Singh, V.K. Removal of hexavalent chromium from aqueous solution using low-cost activated carbons derived from agricultural waste materials and activated carbon fabric cloth. *Ind. Eng. Chem. Res.* **2005**, *44*, 1027–1042. [CrossRef]
59. Wang, H.; Tian, Z.; Jiang, L.; Luo, W.; Wei, Z.; Li, S.; Cui, J.; Wei, W. Highly efficient adsorption of Cr(VI) from aqueous solution by Fe^{3+} impregnated biochar. *J. Disp. Sci. Technol.* **2017**, *38*, 815–825. [CrossRef]
60. Mohanty, K.; Jha, M.; Meikap, B.C.; Biswas, M.N. Removal ofchromium (VI) from dilute aqueous solutions by activated carbon developed from *Terminalia arjuna* nuts activated with zinc chloride. *Chem. Eng. Sci.* **2005**, *60*, 3049–3059. [CrossRef]
61. Malkoc, E.; Nuhoglu, Y. Potential of tea factory waste for chromium(VI) removal from aqueous solutions: Thermodynamic and kinetic studies. *Sep. Purif. Technol.* **2007**, *54*, 291–298. [CrossRef]
62. Maitlo, H.A.; Kim, K.H.; Kumar, V.; Kim, S.; Park, J.W. Nanomaterials-based treatment options for chromium in aqueous environments. *Environ. Int.* **2019**, *130*, 104748. [CrossRef]
63. Rangabhashiyam, S.; Selvaraju, N. Adsorptive remediation of hexavalent chromium from synthetic wastewater by a natural and $ZnCl_2$ activated Sterculia guttata shell. *J. Molec. Liq.* **2015**, *207*, 39–49. [CrossRef]
64. El Nemr, A. Potential of pomegranate husk carbon for Cr(VI) removal from wastewater: Kinetic and isotherm studies. *J. Hazard. Mater.* **2009**, *161*, 132–141. [CrossRef]
65. Zhu, Q.; Moggridge, G.D.; D'Agostino, C. Adsorption of pyridine from aqueous solutions by polymeric adsorbents MN 200 and MN 500. Part 2: Kinetics and diffusion analysis. *Chem. Eng. J.* **2016**, *306*, 1223–1233. [CrossRef]
66. Golban, A.; Lupa, L.; Cocheci, L.; Pode, R. Synthesis of MgFe layered double hydroxide from iron-containing acidic residual solution and its adsorption performance. *Crystals* **2019**, *9*, 514. [CrossRef]

67. Khambhaty, Y.; Mody, K.; Basha, S.; Jha, B. Kinetics, equilibrium and thermodynamic studies on biosorption of hexavalent chromium by dead fungal biomass of marine *Aspergillus niger*. *Chem. Eng. J.* **2009**, *145*, 489–495. [CrossRef]
68. El Nemr, A.; Khaled, A.; Abdelwahab, O.; El-Sikaily, A. Treatment of wastewater containing toxic chromium using new activated carbon developed from date palm seed. *J. Hazard. Mater.* **2009**, *152*, 263–275. [CrossRef] [PubMed]
69. Mohanty, K.; Jha, M.; Meikap, B.C.; Biswas, M.N. Biosorption of Cr(VI) from aqueous solutions by *Eichhornia crassipes*. *Chem. Eng. J.* **2006**, *117*, 71–77. [CrossRef]
70. Park, D.; Lim, S.R.; Yun, Y.S.; Park, J.M. Reliable evidences that the removal mechanism of hexavalent chromium by natural biomaterials is adsorption-coupled reduction. *Chemosphere* **2007**, *70*, 298–305. [CrossRef]
71. Lin, Y.C.; Wang, S.L. Cr K-edge X-ray absorption and FTIR spectroscopic study on the reaction mechanisms of Cr(III) and Cr(VI) with lignin. *Desal. Water Treat.* **2016**, *57*, 21598–21609. [CrossRef]
72. Uzum, C.; Shahwan, T.; Eroglu, A.E.; Lieberwirth, I.; Scott, T.B.; Hallam, K.R. Application of zero-valent iron nanoparticles for the removal of aqueous Co^{2+} ions under various experimental conditions. *Chem. Eng. J.* **2008**, *144*, 213–220. [CrossRef]
73. Ali, S.W.; Mirza, M.L.; Bhatti, T.M. Removal of Cr(VI) using iron nanoparticles supported on porous cation-exchange resin. *Hydrometallurgy* **2015**, *157*, 82–89. [CrossRef]
74. Rakotonimaro, T.V.; Neculita, C.M.; Bussiere, B.; Zagury, G.J. Comparative column testing of three reactive mixtures for the bio-chemical treatment of iron-rich acid mine drainage. *Miner. Eng.* **2017**, *111*, 79–89. [CrossRef]
75. APHA, AWWA, WEF. *Standard Methods for the Examination of Water and Wastewater*, 19th ed.; United Book Press: Baltimore, MD, USA, 1995.
76. Zach-Maor, A.; Semiat, R.; Shemer, H. Synthesis, performance, and modeling of immobilized nano-sized magnetite layer for phosphate removal. *J. Colloid Interface Sci.* **2011**, *357*, 440–446. [CrossRef] [PubMed]
77. Cocheci, L.; Pode, R.; Popovici, E.; Dvininov, E.; Iovi, A. Sorption removal of chromate in single batch systems by uncalcined and calcined Mg/Zn-Al-type hydrotalcites. *Environ. Eng. Manag. J.* **2009**, *8*, 865–870. [CrossRef]
78. Ho, Y.S. Review of second-order models for adsorption systems. *J. Hazard. Mater.* **2006**, *B136*, 681–689. [CrossRef] [PubMed]
79. Albadarin, A.B.; Mangwandi, C.; Al-Muhtase, A.H.; Walker, G.M.; Allen, S.J.; Ahmad, M.N.M. Kinetic and thermodynamics of chromium ions adsorption onto low-cost dolomite adsorbent. *Chem. Eng. J.* **2012**, *179*, 193–202. [CrossRef]

Review

New Perspectives on Iron-Based Nanostructures

Seyedeh-Masoumeh Taghizadeh [1,2], **Aydin Berenjian** [3,*], **Marziyeh Zare** [4]
and Alireza Ebrahiminezhad [5,*]

[1]   Biotechnology Research Center, Shiraz University of Medical Sciences, Shiraz 71468-64685, Iran;
       taghizm@sums.ac.ir
[2]   Department of Pharmaceutical Biotechnology, School of Pharmacy and Pharmaceutical Sciences Research
       Center, Shiraz University of Medical Sciences, Shiraz 71468-64685, Iran
[3]   School of Engineering, Faculty of Science and Engineering, The University of Waikato,
       Hamilton 3240, New Zealand
[4]   Department of Molecular Medicine, School of Advanced Medical Sciences and Technologies,
       Shiraz University of Medical Sciences, Shiraz 71336-54361, Iran; taghizm@yahoo.com
[5]   Department of Medical Nanotechnology, School of Advanced Medical Sciences and Technologies,
       Shiraz University of Medical Sciences, Shiraz 71336-54361, Iran
*   Correspondence: aydin.berenjian@waikato.ac.nz (A.B.); a_ebrahimi@sums.ac.ir (A.E.)

Received: 5 August 2020; Accepted: 24 August 2020; Published: 10 September 2020

Abstract: Among all minerals, iron is one of the elements identified early by human beings to take advantage of and be used. The role of iron in human life is so great that it made an era in the ages of humanity. Pure iron has a shiny grayish-silver color, but after combining with oxygen and water it can make a colorful set of materials with divergent properties. This diversity sometimes appears ambiguous but provides variety of applications. In fact, iron can come in different forms: zero-valent iron (pure iron), iron oxides, iron hydroxides, and iron oxide hydroxides. By taking these divergent materials into the nano realm, new properties are exhibited, providing us with even more applications. This review deals with iron as a magic element in the nano realm and provides comprehensive data about its structure, properties, synthesis techniques, and applications of various forms of iron-based nanostructures in the science, medicine, and technology sectors.

Keywords: Fe nanomaterials; Fe nanoparticles; ferric; ferrous; iron ores; synthesis methods

---

1. Introduction

Iron, in terms of mass, is the most abundant element on Earth. This element is an important constituent of the earth's outer and inner cores, and the fourth most important element in the earth's crust. Among all minerals, iron is one of the early elements identified by the human beings to be taken advantage of and used. It made an era in the ages of humanity; namely, the Iron Age is identified as the final epoch of the three-age division of the prehistoric ages. This came after the Bronze Age, when the human beings reached superior strength and hardness in tools and weapons by iron. In fact, the age of iron did not come to an end, but developed until the 21st century life. The United States of America alone produced ~49 million metric tons of iron ore in 2018. This number is larger for the world's largest iron ore mine producers, with Australia and Brazil, countries which produce 900 million and 490 million metric tons, respectively [1].

Pure metallic iron is rarely found on the lithosphere because, in the presence of oxygen and moisture, it will cause it to readily oxidize. In order to obtain metallic iron, oxygen must be removed from natural ores by reduction. From a chemical point of view, iron, with the chemical symbol "Fe" (from Latin: *ferrum*), is an element with an atomic number of 26, and is located in the first round of the transition metals. It is also the head of ferromagnetic materials. The D orbital of iron ($[Ar] 3d^6 4s^2$)

is not totally full (Figure 1) and can have different oxidation states. The most important states of iron oxidation are ferrous (Fe^{2+}) and ferric (Fe^{3+}). Ferrous iron links to sulfur and nitrogen ligands as well as oxygen ligands. Ferric iron tends to react to hard ligands, such as oxygenated ligands. The most common coordination number for iron (II) and iron (III) is six, which has an octagon spatial arrangement. Furthermore, for ferrous/ferric iron, there is also a tetrahedral arrangement (a coordination number of four), and also a square pyramid or a trigonal bipyramid (especially in a complex coordination of five).

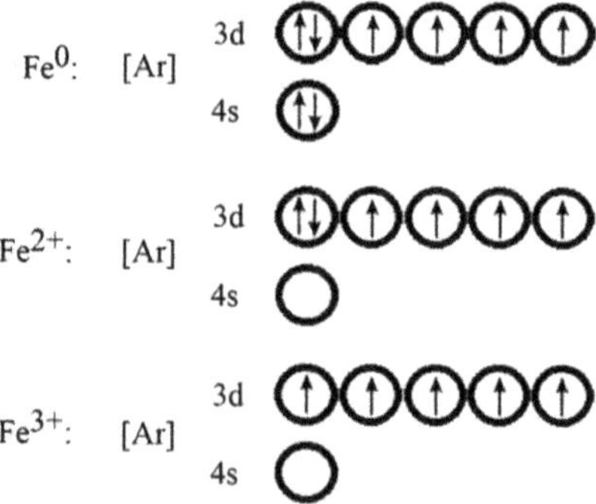

Figure 1. Orbital diagram for iron and iron ions.

Iron has a shiny grayish-silver color but after combination with oxygen and water can make various forms of iron oxide, iron oxide hydroxide, and iron hydroxide. This diversity sometimes appears ambiguous but provides a colorful array of materials with divergent properties and applications. By taking this color pen set to the nano realm, it provides us with even more colors. The current review is dealing with this magic element and provides a wide view of iron, iron oxide, iron oxide hydroxide, and iron hydroxide in the nano realm.

2. Zero-Valent Iron

The appearance of pure iron ranges from silver to black or gray and has a metallic sheen with ferromagnetic properties. Pure iron is not used in the industry because of its softness and lack of sturdiness. Iron is usually used along with nonmetal elements (carbon, sulfur, phosphorus, and silicon), and sometimes with other metals (chromium, nickel, vanadium, and molybdenum) in alloy form, where the resultant product would be steel or cast iron. The cheap price and high strength of iron alloys have commoditized their use in the automotive industry, the hull of large ships, and buildings. Nanoparticles of pure iron are known as zero-valent iron nanoparticles (ZVINPs) and are usually synthesized by chemical reduction of iron ions [2]. In this manner, reducing agents, such as sodium borohydride ore hydrazine, are used to reduce iron ions to iron atoms and subsequently iron atoms aggregate and form ZVINPs [3,4]. In addition to chemical reducing agents, some biochemical species are also able to reduce iron ions, and hence can be employed for the bio-reduction of iron ions to ZVINPs [5,6]. In comparison to chemical reduction reactions, in the bio-reduction reactions, no additional molecule is required to stabilize nanoparticles. Biologic molecules are sufficient for both the reduction and protection of the prepared particles [5,7–9]. However, when chemical reducing agents are employed, a second protective molecule is required to stabilize the particles (Figure 2).

Generally, ZVINPs are unstable and tend to oxidize and hydrate iron into oxides and iron hydroxides (4). In fact, ZVINPs usually exhibit the dual characteristics of iron (Fe^{0}) and iron oxides or hydroxides. This feature is due to oxidation of exposed iron atoms on the surface of particles and the formation of an iron oxide/hydroxide shell around the zero-valent core (see Section 4.1.3) [2,7–9]. In order to prevent the oxidation reactions and maintain the properties of ZVINPs, coating with noble metals can be used. For instance, ZVINPs with gold or silver coating are effective in the treatment of cancer, and can suppress the growth of cancer cells while possessing adequate biocompatibility with nonmalignant cells [10,11]. Minerals are the other applicable material to protect zero-valent iron nanoparticles. For instance, montmorillonite, which is a clay mineral, has been used as an effective protective agent and support [2]. In some investigations chelating agents, such as diethylenetriaminepentaacetic acid (DTPA), were used as a stabilizer for ZVINPs [7]. Biologic compounds were also shown to be effective for protection from oxidation. These compounds can make a thick coating around the particles and protect them from oxidation. It has been shown that by using biologic coatings any detectable iron oxides can be formed on the particles [5,12].

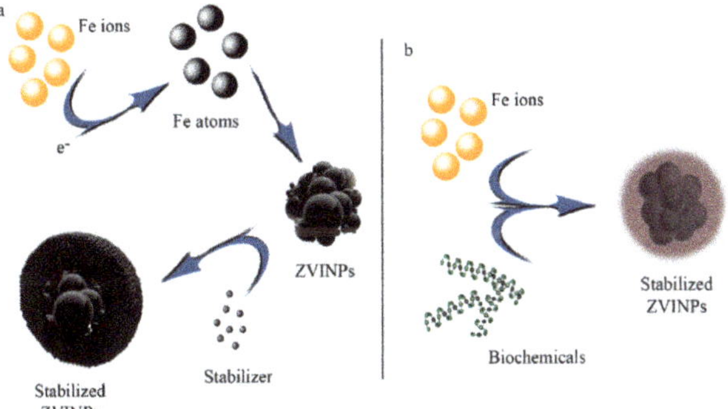

Figure 2. Chemical (**a**) and biological (**b**) reduction of iron ions to zero-valent iron nanoparticles (ZVINPs) and their stabilization.

ZVINPs gained diverse applications in environmental sciences and remediation proposes. These nanoparticles are used to remove arsenic from contaminated aquatic environments [13,14]. Extensive lab studies have shown the potency of these nanoparticles in the adsorption and degradation of halogenated hydrocarbons, such as trichloroethylene (TCE) [15]. These nanoparticles are also effective against pesticides, transition metals, organic dyes, and radioactive contaminations [7,16–18]. ZVINPs can also act as an antimicrobial agent which have a better performance in the absence of oxygen. The antimicrobial activity of these nanoparticles against *Escherichia coli* has been studied under anaerobic and aerobic conditions. In the absence of oxygen, when *E. coli* is exposed to 9 mg/L of iron nanoparticles for 10 min, bacterial cells were effectively disabled. In aerobic conditions, when the bacteria were exposed to 90 mg/L of iron nanoparticles for 90 min, the antimicrobial effects were negligible, since, in the presence of oxygen, ZVINPs are oxidized [19].

3. Iron Oxides

In the presence of oxygen and humidity, iron easily combines with oxygen and transforms to iron oxides. Iron oxides are widespread in nature and can be found in soil, rocks, lakes, rivers, seafloors, and even in living creatures [20]. Generally iron oxides receive much attention from scientists and researchers because of the variety in their physical and chemical properties. Iron reacts with oxygen in various states and therefore can be found in diverse chemical formulas with different properties as described in the following.

3.1. Ferrous Oxide (FeO)

Fritz Wüst, a German metallurgist (1860–1938), named iron (II) oxide (ferrous oxide, Fe=O, Figure S1) as wüstite. It is a gray-black ore with a greenish shade. The FeO formula is not exactly correct because wüstite is a typical non-stoichiometric mineral. In this mineral the amount of iron is always less than the stoichiometric ratio, and there is no temperature at which a one-to-one ratio can be seen between iron and oxygen. In fact, there is deficient iron in the structure with compositions ranging from $Fe_{0.84}O$ to $Fe_{0.95}O$ $(Fe_{1-x}O)$ [21]. It has an isometric (cubic) crystal system and has anti-ferromagnetic properties [22].

The most common methods for preparing wüstite nanoparticles are co-precipitation (Scheme 1), polyol method (Scheme 2), pyrolysis (thermal decomposition) of organometallic compounds (Scheme 3), micro emulsion, hydrothermal (Scheme 4), and sol-gel synthesis (Scheme 5) [23]. Among these methods, thermal decomposition of iron (II) or iron (III) complexes is the most employed technique [22,24,25]. wüstite nanoparticles can also be fabricated by physical approaches, such as ball milling, using a mixture powder of hematite and iron [26]. The main challenge in all the above-mentioned methods for preparing wüstite nanoparticles is to achieve nanoparticles with controlled size, shape, and composition [23].

Wüstite nanoparticles are semi-stable, and are therefore used in the creation of mixed phase nanoparticles [27]. Moreover, in order to create stability in wüstite nanoparticles, they are covered with organic ligands. In another approach, wüstite nanoparticles can be doped with other divalent metals (Dm), such as cobalt and magnesium, to fabricate $Dm_{(x)}Fe_{(1-x)}O$. These structures can be synthesized by the thermal decomposition of DmFe organometallic complexes or by solution-combustion methods [28,29]. The surface of the doped wüstite nanoparticles is oxidized to $DmFe_2O_4$ and form a shell to protect the core from oxygen transmission. The thickness of the oxidized shell can be tuned through a controllable oxidation process and has significant effects on the core/shell ratio and exchange bias properties. For instance, it has been shown that a shell of $CoFe_2O_4$ with 4 nm thickness is ideal for resistance to oxygen transmission [29].

From application point of view, there are reports for the antimicrobial properties of wüstite nanostructures. It has been shown that the nanochains of carbon coated wüstite nanoparticles can kill bacterial cells within a short incubation time at minimal dosages [30]. Wüstite nanoparticles can be applied as radiation filters. Glasses with thin films of FeO and Fe_2O_3 nanoparticles are used as solar radiation filters. These filters are applicable on the building windows to prevent heat transmittance while visible light is passing through [31].

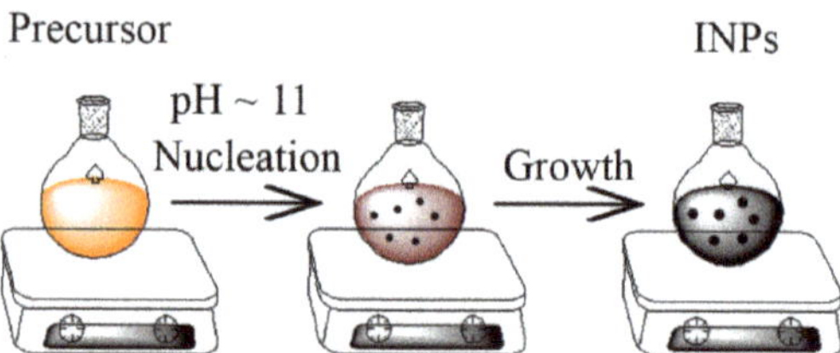

Scheme 1. Schematic illustration of a common precipitation reaction, INPs stands for iron nanoparticles.

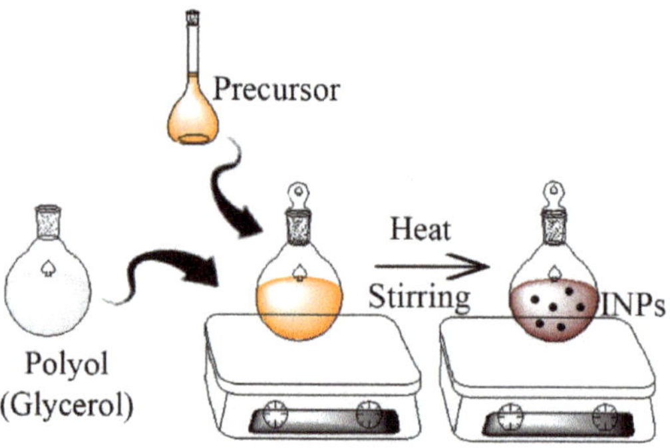

Scheme 2. Schematic illustration of a polyol reaction.

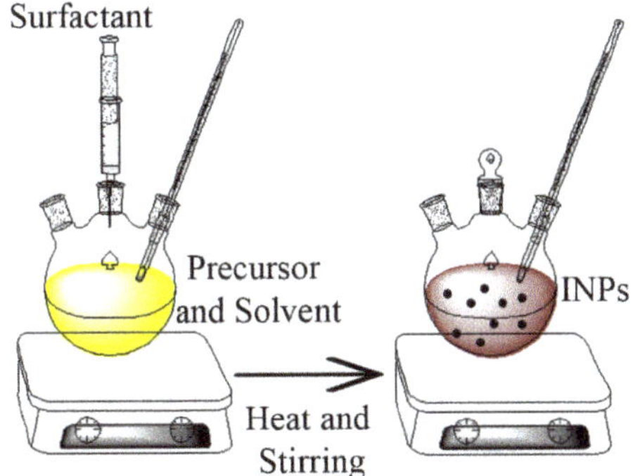

Scheme 3. Schematic illustration of a pyrolysis reaction.

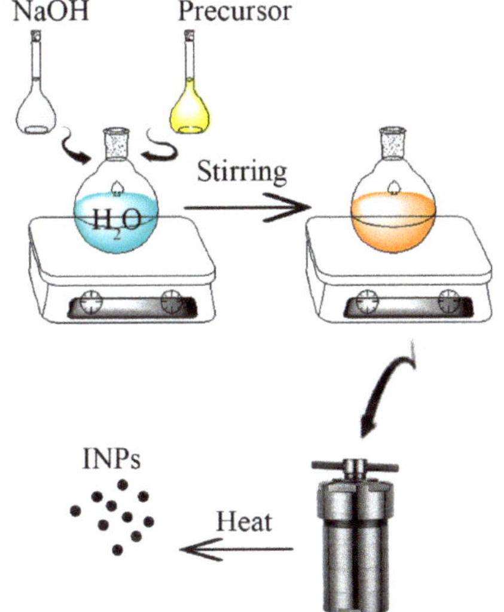

Scheme 4. Schematic illustration of hydrothermal synthesis reaction.

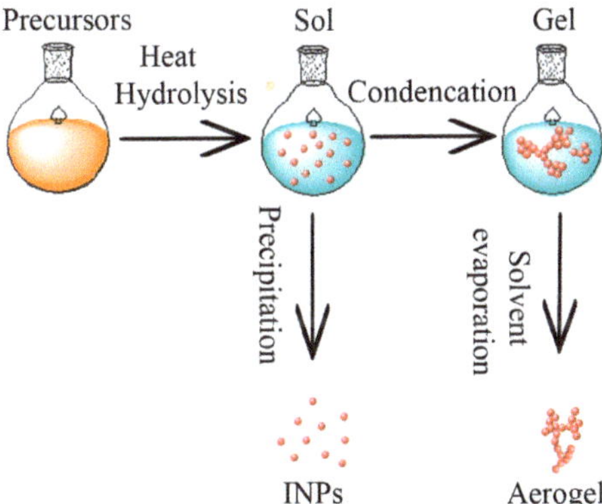

Scheme 5. Schematic illustration of a sol-gel reaction.

3.2. Ferric Oxide (Fe_2O_3)

As its name indicates, ferric oxide is the oxide of iron (III) (O=Fe–O–Fe=O, Figure S1). This form of iron oxide is well-known as rust and can be found in red and brown. Ferric oxide is the result of dehydration reaction of FeO(OH) and Fe(OH)$_3$ at temperatures above 200 °C (Equations (1) and (2)). It can be found in its hydrated (see Section 3.2.2) or anhydrous forms. In the following we made a detailed reviews on the various phases of anhydrous ferric oxides.

$$2FeO(OH) \rightarrow Fe_2O_3 + H_2O \tag{1}$$

$$2Fe(OH)_3 \rightarrow Fe_2O_3 + 3H_2O \tag{2}$$

3.2.1. Anhydrous Ferric Oxides

Anhydrous forms of ferric oxide can be found in different polymorphs, such as alpha (α-Fe$_2$O$_3$), beta (β-Fe$_2$O$_3$), gamma (γ-Fe$_2$O$_3$), and epsilon (ε-Fe$_2$O$_3$). These minerals are the same in chemical structure but have differences in their crystal systems [32].

Hematite (α-Fe$_2$O$_3$)

The word hematite is taken from the Greek word hematicos, which means bloody. The chemical reaction to get hematite from magnetite is shown in Equation (3). Hematite is a widespread iron oxide and can be found in rocks and soil. It is the oldest known ore of iron that has ever formed on the Earth.

$$2Fe_3O_4 + 1/2O_2 \rightarrow 3(\alpha\text{-Fe}_2O_3) \tag{3}$$

Hematite has a rhombohedral and hexagonal crystal system, and its crystal habit can be seen as either spindle, bar, oval, square, or spherical. It has a red metallic appearance. Hematite has different forms, but all forms have a red sheen. Hematite iron deposits are most significant in the steel industry and in the preparation of cast iron. This mineral is one of the precious rocks, and the fine cuts of its black crystals are used in making jewelry. Moreover, its powder is used in pigments, such as red pigment 22 and antirust [20].

The most common ways to produce hematite nanoparticles are chemical precipitation, the polyol process, pyrolysis of organometallic compounds, micro emulsion, hydrothermal synthesis, and sol-gel process [23,33,34]. These nanoparticles can also be obtained by calcination of mixtures of iron oxide nanoparticles at high temperatures (about 500 °C) [35]. Biosynthesis (plant mediated synthesis) is another route, in fact, a green route, for fabrication of hematite nanoparticles [36]. These particles can be fabricated by using various plants, such as root extract of *Arisaema amurense* [37], Tragacanth gel [38], *Citrus reticulum* peels extract [39], the extract of green tea leaves (*Camellia sinensis*) [40], *Ailanthus excelsa* [36], and guava (*Psidium guajava*) [41].

Hematite nanoparticles are the most stable iron nanostructure in the ambient conditions and have size and shape-dependent magnetic properties. The magnetic saturation is in direct relation to particle size, and by increasing the size, an increase in the magnetic saturation value can be observed. On the other hand, it has been shown that spherical nanoparticles have the highest magnetic saturation, while cubic and ellipsoidal nanoparticles are ranked in the next positions, respectively [42]. These nanoparticles can exhibit ferromagnetism or superparamagnetic behavior at room temperatures. However, in contrast to magnetite (Fe$_3$O$_4$) nanoparticles, saturation magnetization values of hematite nanoparticles are very low [43]. These nanostructures have gained applications in the development of electrochemical devices, as cathode in lithium batteries, the fabrication of photo electrochemical systems to produce hydrogen from water using solar radiation, as photo reactive nanomaterial for the enhanced performance of photoelectrochemical cells, and as reusable catalysts [23,40,44]. These nanoparticles can be employed for the enhancement of the thermal conductivity of the base fluids in nanofluidic systems [41]. Hematite nanoparticles have a high capacity for metal ions' removal from aqueous solutions. These particles can be employed for metal cations removal from acid mine drainage. It has been shown that hematite nanoparticles can completely remove aluminum, magnesium, manganese, and iron ions, while zinc and nickel ions remove over 80% and sodium and calcium ions remove up to 72% [35]. Hematite nanostructures are also able to decontaminate aqueous systems from different organic pollutants and can be used as an effective material in fertilizers fabrication [39]. The surface modification of hematite nanoparticles with the organic functional groups makes these particles more biocompatible and suitable to treat the chlorosis in plants, such as *Glycine max* [45].

Beta-Fe$_2$O$_3$ (β-Fe$_2$O$_3$)

β-Fe$_2$O$_3$ is a scarce form of iron (III) oxides. This polymorph is a metastable phase and easily transformed into the more stable alpha (α-) phase of ferric oxide. The transformation occurs at high temperatures, above 500 °C, but the mechanism of transformation and the changes which occur in crystal structure are, as yet, unclear. In fact, the exact crystal structure of the β-Fe$_2$O$_3$ is not defined, and this is due to difficulties in obtaining a β phase in pure monophasic crystals [46]. β-Fe$_2$O$_3$ was discovered in 1958 and believed to have a cubic body centered crystal structure [47]. However, techniques for the fabrication of β-Fe$_2$O$_3$ nanostructures have been developed and these structures can be obtained via simple chemical reactions. Solid state reaction is one of the most common techniques for the fabrication of β-Fe$_2$O$_3$ nanoparticles, where NaFe(SO$_4$)$_2$ is used as the precursor. In brief, a solution mixture of Fe$_2$(SO$_4$)$_3$ and Na$_2$SO$_4$ is evaporated in an oven at ambient atmosphere to obtain NaFe(SO$_4$)$_2$. The resulting powder is then fully grounded and mixed with NaCl, in a 1:2 ratio, and heated up to 400–500 °C in a muffle furnace. After cooling, the product should be dispersed in distilled water and washed to remove NaFe(SO$_4$)$_2$ and Na$_3$Fe(SO$_4$)$_3$ [46,47]. In addition, colloidal β-Fe$_2$O$_3$ nanostructures can be fabricated by hydrolysis of ferric ions in boiling water. The shape of prepared nanostructures can be controlled by the employing cobalt ions (Co^{2+}). It has been shown that spherical and rod nanoparticles can be obtained in this order [48].

Water splitting and hydrogen production is one of the most promising approaches to overcoming the global energy crisis. Water splitting can be performed via photoelectrochemical reactions which employ ferric oxide nanoparticles as the photoanode material. Due to availability, photochemical stability, and high efficiency for light absorption, in the past decades, hematite (α-Fe$_2$O$_3$) has been envisioned as a strong semiconductor photocatalyst. However, there are some obstacles with hematite. For instance, its low conductivity reduces the photoelectrochemical activity and there are also some other defects, such as large over potential and short hole transport distance. Recent studies introduced β-Fe$_2$O$_3$ as a more promising phase for light assisted water splitting. β-Fe$_2$O$_3$ has a smaller band gap than hematite and therefore is more efficacious in light absorption and photoelectrochemical reactions [47].

Maghemite (γ-Fe$_2$O$_3$)

Maghemite is the second most abundant form of ferric oxide. It has a cubic and tetragonal crystal system and its crystal habit is cube, plate, or spindle in shape. The color of maghemite is brown or reddish brown. Its combination is metastable and at high temperatures (about 500 °C), it turns into the alpha phase. Maghemite has ferrimagnetic properties and its nanoparticles show superparamagnetic properties [49,50]. The most common techniques for maghemite nanoparticles production are co-precipitation, polyol process, micro emulsion, hydrothermal synthesis, sol-gel process, and the pyrolysis of organometallic compounds [23]. These particles can also be obtained by the oxidation of magnetite (Fe$_3$O$_4$) nanoparticles via thermic treatments [50–52]. The conversion of magnetite to maghemite can also be seen on the surface of magnetite nanoparticles that forms a core-shell structure of magnetite-maghemite [53]. To improve the properties of the prepared nanostructure, the surface of the maghemite, nanoparticles can be functionalized using divergent biocompatible compounds, such as amino acids, fatty acids, silica, and hydroxyapatite [52]. Surface modifications were also done by multiwall carbon nanotubes (MWCNTs) and ethylenediaminetetraacetic acid (EDTA) [52].

Even polymeric materials have been used [54–56]. Doping is another approach to improve the properties and functionalities of maghemite nanoparticles. For instance, these nanoparticles can be doped with rare-earth elements, such as yttrium [57]. On the other hand, maghemite nanoparticles themselves can be used for the decoration of other nanostructures, such as carbon nanotubes. Maghemite nanoparticles decorated on carbon nanotubes have been used for the fabrication of oxygen evolution reaction (OER) electrocatalysts [58]. Also, maghemite nanoparticles are used in magnetic recording media, biocompatible magnetic fluid production, and electro chromic devices [23].

The other well studied application of maghemite nanoparticles is in the environmental sciences. These nanoparticles are efficient to remove heavy metals, such as uranium [56], hexavalent chromium ions [59,60], copper(II), and lead(II) ions [52], from aqueous systems. Maghemite nanoparticles are so promising in this field that they can clean up contaminated water to drinking water standards [60].

Like hematite nanoparticles, maghemite nanoparticles can also be used in fertilizer formulations. In contrast to chelated irons, employment of maghemite nanoparticles in plant fertilizers resulted in more improved agronomic properties such as growth rate of leaves and chlorophyll content [57]. Also, maghemite nanoparticles can improve plants' abiotic stress tolerance. Delivery of maghemite nanoparticles by irrigation in a nutrient solution to plants resulted in a significant reduction of hydrogen peroxide and malondialdehyde formation. These effects were discussed as nanozyme properties of maghemite nanoparticles [57].

Maghemite nanoparticles have been approved by the FDA to be employed in medical and pharmaceutical sciences [61,62]. These nanoparticles can be used for theranostic applications [63]. Nanostructures of maghemite can illustrate the high values of magnetic saturation (near or equal to magnetite nanoparticles) that makes them ideal for biomedical magnetic applications, such as magnetic labeling, magnetically guided deliveries, and hyperthermia therapies [51,63–65]. For instance, it has been shown that by using 10 nm maghemite nanoparticles an increase in the temperature of up to ~50 °C is possible [51]. Maghemite nanoparticles are also a promising candidate for the fabrication of dual functional nanostructures. Fluorescent magnetic nanoparticles are one of them. In an experiment rhodamine isothiocyanate (RITC) is covalently encapsulated within maghemite nanoparticles and used for cell labeling. These nanostructures were found to be pH sensitive and introduced as a pH sensor for animal tissues and cell compartments [66]. In another experiment rhodamine was used for labeling a maghemite–albumin nanohybrid to combine targeting by the iron-acquisition pathway and photothermal therapies [67].

Maghemite nanoparticles can be employed in magnetofection techniques. Nanoparticles with efficient coatings, such as dendrimers, can couple fragments of DNA and take them into the cell under magnetic stimulus [68].

Drugs can also be delivered by these nanoparticles. In an attempt to fabricate a potential drug carrier, maghemite nanoparticles were functionalized with divergent molecules, such as human serum albumin (HSA), aminobenzoic acid, poly (ethylene imine) (PEI), and poly (ethylene oxide)-block-poly(glutamic acid) (PEO–PGA) [69–71]. Maghemite nanoparticles were also employed for the fabrication of multi stimuli-responsive smart nanogels. By combining thermo-responsive polymers, such as poly (vinyl alcohol)-*b*-poly(*N*-vinylcaprolactam) copolymers with maghemite nanoparticles, an innovative drug delivery system was obtained that is glucose-, pH-, thermo-, and magnetic-responsive. In addition, the magnetic properties of the maghemite nanoparticles make nanogel effective for magnetically-induced heating and magnetic resonance imaging (MRI) performance [72].

Maghemite nanoparticles can be used as a high sensitivity magnetic biosensor. In an experiment, it has been shown that, based on the giant magneto-impedance (GMI) effect, MAT-LyLu cells that grow in the presence of 0.1 mg/mL nanoparticles for 24 h make detectable modifications in the magnetic field at 1 MHz frequency [65].

Beside all the mentioned applications, maghemite nanoparticles have gained applications in tissue engineering and biologic scaffolds preparation. The inclusion of these nanoparticles in the scaffolds' polymer network resulted in the significant improvement in physical and chemical characteristics of the prepared scaffolds. It has been shown that gradual increase of nanoparticle concentration in the polymer network significantly increases the polymer's hydrophilicity, water swelling, Young modulus (mechanical stiffness), effective viscosity, and negative value of electrical potential [73,74]. Maghemite nanoparticles also improve biologic properties of the scaffolds. Polymers imbedded with maghemite nanoparticles pose increased biocompatibility and cell adhesion index [73]. It has been shown that addition of maghemite nanoparticles to the scaffolds can increase initial cell adhesion by 1.4-fold, enabling earlier cellular proliferation confluence [74]. In some cases, doping with maghemite nanoparticles provides scaffolds with specific advantages for particular tissues. For instance, the application of these nanoparticles in scaffolds for bone repair and regeneration resulted in higher mineral induction [74]. In regard to skin repair, nanoparticle-doped scaffolds provide a higher cell proliferation rate for human skin fibroblast cells [75]. Maghemite nanoparticles were also used for tissue engineering a heart valve to obtain the required properties of an aortic heart valve. Fabricated materials using this technology are desirable scaffolds for human aortic smooth muscle cells migration and proliferation. In addition, these scaffolds provide a mutual adaption in terms of clotting time and hemolysis percent [76].

ε-Fe$_2$O$_3$

ε-Fe$_2$O$_3$ has an orthorhombic crystal system and shows properties between the alpha and gamma phases [32,77]. The ε-polymorph of Fe$_2$O$_3$ is unique due to its significant ferromagnetic resonance and coupled magnetoelectric features that cannot be found in any other phases [78]. This compound is metastable and at temperatures from 500 °C to 750 °C it turns into hematite (alpha phase). The preparation of the pure epsilon phase is also very challenging due to the contamination with alpha and gamma phases. There are some techniques to achieve a high proportion of epsilon phase [79]. However, the optimal reaction condition to achieve single-phase ε-Fe$_2$O$_3$ remains as a challenge for the future [78]. Recent studies have shown that this compound was used in the glazing of ceramics in ancient China [20]. Rod-shaped ε-Fe$_2$O$_3$ with a large coercive field were used in a mesoscopic ferrite bar magnet which can be used as a probe in magnetic force microscopes [80].

Zeta-Fe$_2$O$_3$ (ζ-Fe$_2$O$_3$)

As mentioned in detail previously, there are four well-known polymorphs for ferric oxide (α-Fe$_2$O$_3$, β-Fe$_2$O$_3$, γ-Fe$_2$O$_3$, and ε-Fe$_2$O$_3$), which are similar in chemical structure but differ in crystal structures and, hence, physical properties However, in 2015, via synchrotron X-ray diffraction experiments, a new phase was introduced for ferric oxide. It evolved from cubic β-Fe$_2$O$_3$ during a high pressure (above 30 GPa) treatment and was termed zeta-Fe$_2$O$_3$ (ζ-Fe$_2$O$_3$). After pressure release the structure was retained and was stable at an ambient atmosphere. The structure was also stable at high pressures up to 70 GPa, which is above the pressures that induced phase transfer in other ferric oxide phases. This structure shows antiferromagnetic properties with ~69° K Neel transition temperature. The discoverers claimed that this novel phase may have unique physicochemical properties that would lend it particular applications [81].

3.2.2. Hydrous Ferric Oxide (Ferrihydrite)

Hydrous ferric oxide or ferrihydrite is one of the well-known iron compounds with a 5Fe$_2$O$_3$.9H$_2$O chemical formula. However, there are also other reported formulas such as Fe$_5$HO$_8$.4H$_2$O and Fe$_2$O$_3$.2FeO(OH).2.6H$_2$O [82,83]. The indeterminate chemical formula of ferrihydrite is due to the variability of the content of combined water. Ferrihydrite has a hexagonal crystal system (hcp) and a spherical habit crystal. Based on the constitutive crystallites, ferrihydrite can be considered between two diffraction end-crystalline structures which are described as two- or six-line ferrihydrite. This nomination is based on the number of scattering bands in their X-ray diffraction patterns.

It is interesting that ferrihydrite can just be found in the nature as nanocrystals in a dark brown or reddish brown color [20,83]. It is a widespread mineral on the Earth and can be found in soil and sediments of hot springs, marine environments, and fresh waters. In these environments ferrihydrite nanostructures precipitate by the oxidation of aqueous Fe(II) solutions that contain dissolved organic matter. The oxidation occurs at neutral pH and it is obvious that presence of organic compounds significantly affects the crystallinity of the resulted nanostructures [84,85]. Ferrihydrite can be mineralized as nanoparticles in the ferritin protein of animals, plants, and microorganisms, where it is utilized in intracellular iron storage [86]. It has been shown that crystalline and amorphous ferrihydrite nanoparticles in the size range of 3–6 nm can be assembled within ferritin [87]. Some microbial cells are also able to mineralize ferrihydrite nanostructures on the cell surface. The structure is mostly nanostructured polysaccharide-iron hydrogel and produced under microaerophilic conditions. Nanostructures which are produced by the bacterium *Klebsiella oxytoca* are one of the well-known and studied mineralize ferrihydrite nanostructures. This iron reducing bacterium (FeRB) is able to ferment ferric citrate under microaerophilic or anaerobic conditions and produces a Fe(III)-hydrogel [88–91]. Other bacterial cells, such as *Staphylococcus warneri*, *Xanthomonas campestris*, and *Ralstonia* sp., were also reported to be able to fabricate an Fe(III)-binding exopolysaccharide and develop polysaccharide-iron nanostructures [92–94]. Synthetic ferrihydrite nanoparticles can also be prepared chemically. The reaction can be performed at ambient atmosphere by adding an alkaline solution (NaOH) to the solution of ferric ions (e.g., ferric nitrate, $Fe(NO_3)_3.9H_2O$, or ferric chloride, $FeCl_3.6H_2$) under constant stirring [95,96].

Both synthetic and biologic ferrihydrite nanostructures exhibit weak magnetic (paramagnetic) behavior. Low magnetic response beside good contrasting properties and biostability make the ferrihydrite a valuable contrasting agent for MRI imaging [97]. However, there are some techniques to improve the magnetic moment of ferrihydrite nanoparticles. It has been shown that the heat treatment of ferrihydrite nanoparticles at low temperatures (about 160 °C) enhances the average magnetic moment of biosynthesized nanoparticles. It is claimed that an enhanced in magnetic moment can be due to the partial agglomeration of nanoparticles throughout heat treatment process [98].

Ferrihydrite nanoparticles have gained applications in environmental remediation activities. These particles can be employed to remove mineral contaminants such as Pb(II), Cd(II), Cu(II), Zn(II), phosphate, and arsenate from aqueous solutions [99,100]. In the case of organic contaminants, ferrihydrite nanoparticles are known to effectively improve the bioremediation activities. Microorganisms play the main role in this regard and ferrihydrite nanoparticles can enhance microbial degradation of a wide range of contaminants [101]. In recent attempts scientists tried to inject a colloidal ferrihydrite in the subsurface to create reactive zones and promote biodegradation reactions [101]. There are also some attempts to enhance particles' mobility in the soil column via polymer coatings as a surface engineering strategy or post flushing with particle-free electrolyte solutions [101,102]. The injection of ferrihydrite nanoparticles in the soil column is not just profitable for the soil microorganisms; the soil that is treated with ferrihydrite nanoparticles is even more beneficial for plants. It has been shown that soil enrichment with ferrihydrite nanoparticles enhances the growth and chlorophyll content in the maize seedlings [103].

3.3. Ferrous Ferric Oxide (Fe_3O_4)

Ferrous ferric oxide is well known as magnetite and its name is taken from the Greek word "magnet". Its IUPAC (International Union of Pure and Applied Chemistry) name is iron (II, III) oxide, and its common chemical name is ferrous-ferric oxide. It has two iron (III) and one iron (II) in combination with four oxygen. The structural formula of Fe_3O_4 is shown in Figure S1. Magnetite has a crystal cubic-hexoctahedral system with an octahedral habit crystal. It has a black appearance with a metallic opaque sheen. This combination is fragile and has caught the attention of researchers because of its outstanding properties and has therefore has been thoroughly studied. Magnetite is ferrimagnetic and can be employed as a permanent magnet [20,32].

It is interesting that magnetite shows superparamagnetic qualities in the nanometer range (1 to 100 nanometers). This means that nanoscale magnetite fails to maintain magnetization after removal of the external magnetic field. This unique property makes magnetite nanoparticles one of the most employed nanostructures in technology, science, and especially in medicine. Wide applications of magnetite nanoparticles have led to the development of various methods for their synthesis. Some of the common methods of magnetite nanoparticle synthesis are pyrolysis of organometallic compounds, polyol process, micro emulsion, hydrothermal synthesis, sol-gel, and coprecipitation [104]. Coprecipitation is the most commonly employed technique. In this method, magnetite nanoparticles are produced using bivalent and trivalent hydrated salt ions in the presence of a strong base under controlled atmosphere. The chemical reaction is as provided in Equation (4).

$$2\,Fe^{+3} + Fe^{+2} + 8\,OH^- \rightarrow Fe_3O_4 + 4\,H_2O \tag{4}$$

It is fair to mention that magnetite nanoparticles are the most studied and employed iron-based nanostructures. These particles have gained myriad divergent applications in engineering, such as data storage and transfer [105], the rising up and the aggregation of activated sludge [106], heavy metal absorbents [107,108], nanocatalysts [109], and various others. In addition to the various engineering applications of magnetite nanoparticles, plenty of studies were also performed for the medical application of magnetite nanoparticles in drug delivery, magnetic resonance imaging (MRI), hyperthermia therapy, and immunoassay [110,111]. In order to employ magnetic iron oxide nanoparticles in medical applications, they are usually coated with biocompatible organic or inorganic molecules, such as silica, amino acids, lipoamino acids, carbohydrates, proteins, and polymers [112–122]. Biocompatible magnetite nanoparticles have also been widely used in biological sciences for magnetic labeling and the magnetic separation of the cells and biomolecules (e.g., nucleic acids, enzymes, and antibodies) [123–136]. These nanoparticles gained some applications to increase the performance and addressing the defects of fermentation industries. Magnetite nanoparticles can increase the performance of microbial cells and make high performance cells. It showed that magnetite nanoparticles can increase the permeability of the cell membrane toward improvement in mass transfer and increasing the cell functions [123,137,138]. On the other hand, certain concentrations of magnetite nanoparticles can affect biofilm formation by microbial cells which is a major problem of the fermenters [139,140]. A recent study showed that by using magnetite nanoparticles in a fermentation period, valuable industrial strains can be protected from potential phage infections [141].

It is interesting that living creatures employ magnetite nanoparticles for navigation. Magnetotactic bacteria are Gram-negative bacteria that are able to build specialized magnetite nanostructures. Members of this bacterial family are able to fabricate magnetosome nanostructures that are biogenerated magnetite nanocrystals engulfed by a phospholipid bilayer. Magnetosomes are arranged in a line in the cytoplasm of bacteria and act as a compass needle to provide the ability of magnetotaxis for the cell [142]. Magnetosomes can be in diverse morphologies and it has been shown that environmental conditions have an immense effect on their morphology [143]. Biogenic magnetite nanoparticles are not restricted to the bacterial realm and there are obvious reports for the presence of these nanostructures in migratory and non-migratory animals. It has been shown that, like arrangement of magnetosome in the bacterial cytoplasm, magnetic nanoparticles are arranged in short or long chains in the ethmoid and lateral ethmoid bones of migratory and non-migratory fishes [144]. The chains of magnetic nanoparticles were also reported in the plants. Magnetic nanoparticles are located in the form of chains in the wall of the phloem sieve tubes (i.e., the vascular tissue of plants) [145]. Interestingly, magnetite nanoparticles have been identified in human tissues (e.g., brain, meninges, heart, liver, spleen, and cervical skin) [146]. In addition, human stem cells are able to degrade and synthesize magnetic nanoparticles [147].

4. Ferric Oxide-Hydroxide (Ferric Oxyhydroxide)

Ferric oxyhydroxide is composed of iron, oxygen, and hydrogen with the FeO(OH) chemical formula (O=Fe–OH, Figure S1). In aqueous matrixes, ferric ions can hydrolyze to some extent and produce ferric oxyhydroxide. As shown in Equation (5), this hydrolysis reaction produces protons and, to some extent, makes the environment acidic. Ferric oxyhydroxide is usually encountered as hydrated forms with the FeO(OH)·nH$_2$O general formula (see the Section 4.2) and anhydrous ferric oxyhydroxide. These two forms are discussed in the following sections.

$$Fe(III) + 2H_2O \rightarrow FeO(OH) + 3H^+ \tag{5}$$

4.1. Anhydrous Ferric Oxyhydroxides

Like ferric oxide, anhydrous ferric oxyhydroxide can be found in four different polymorphs, denoted as the α, β, γ, and δ phases. Different phases are distinguished by their distinct crystalline structure [20].

4.1.1. Goethite (α-FeOOH)

Goethite is the main component of iron rust and is also the main compound of bog iron ores. In recent years, this mineral has benn used as iron ore and known as brown iron ore. Goethite is an antiferromagnetic mineral [148]. It has an orthorhombic crystal system and its crystal habitat is usually either needle, star, or hexagonal [20,148]. Its color in the crystalline state is dark brown or black and yellow. When powder, it is semi-transparent to opaque.

Recent investigations revealed that nanostructures of goethite can be formed in nature. By cryogenic Fe MÖssbauer spectroscopy studies it was discovered that in the marine and lacustrine environments the ferric oxyhydroxide phase is the nanogoethite [149]. The laboratory synthesis of goethite nanoparticles is usually performed by using ferric nitrate (Fe(NO$_3$)$_3$.9H$_2$O) or ferric chloride hexahydrate (FeCl$_3$.6H$_2$O) as an iron precursor. In this order a basic solution, mostly sodium hydroxide (NaOH), is added to the iron solution with continuous stirring until a clear solution of yellowish-brown color is formed. This yellowish-brown precipitate is ferrihydrite and needs to age to goethite [96,150]. The aging process is done at high temperatures, about 60–100 °C [96,151,152]. The speed of the base addition is a critical point in the process and is inversely proportional to the resulting particles' specific surface area. It is shown that by varying the rate of base addition from 1 mL/min to 10 mL/min the obtained specific surface area is reduced from 101 ± 5 m^2/g to 64 ± 2 m^2/g [96]. The major problem with these techniques is that the produced nanoparticles are not uniform in shape and size. Recent experiments indicated that this problem can be solved by controlled synthesis reactions. These reactions are based on the fact that presence of additional molecules, such as carbohydrates, can control the crystals' growth pattern. It has been shown that such compounds can drive the reaction to induce isotropic growth of goethite nanocrystals [150].

High quality goethite is rare and usually polished to be used in making jewelry [20]. From an industrial standpoint, goethite is among the main pigments (brown ochre; a yellowish-brown pigment) which has been used since ancient times. Goethite nanoparticles have gained some environmental applications and are used to remove metallic cation pollutants, such as arsenic and chromium [152–154]. These nanoparticles are also used as an absorbent for the removal of fluoride from contaminated waters [83]. Surface modification studies were also performed to enhance the nanostructures' stability and mobility in porous media. Humic acid is reported to be efficient in this regard. However, the major drawback is aggregation in the presence of divalent cations, e.g., calcium and magnesium ions, which is due to complexation phenomena related to the interaction of divalent cations with humic acid [155].

Goethite nanoparticles are able to quench free radicals. This property can be used to improve the oxidative stability of commercial products to enhance the storage time. For instance, gasoline oxidation is a factor that significantly decreases the gasoline storage time. A study on the gasoline induction period indicated that goethite nanoparticles can inhibit gasoline oxidation and significantly elongate the induction period of the commercial gasoline [156].

4.1.2. Akaganeite (β-FeOOH)

The name of this compound has been taken from the Akagan mine in Japan. It is an antiferromagnetic mineral and has a monoclinic crystal system with hexagonal, bar-like, star-shaped, and prismatic crystals [157]. Its appearance has a tan-brown color with a metallic sheen. This mineral has been found in different parts of Earth as well as in the rocks from the Moon. This mineral can be found in corrosion of iron and hot springs that contain chloride [158]. Although akaganeite is known as one of the anhydrous forms of FeOOH, in chloride-containing environments. In such conditions, chloride takes part in the structure and so the exact structure would be $FeO_{0.833}(OH)_{1.167}Cl_{0.167}$ [158]. In fact, it is composed of an iron oxyhydroxide framework with tunnels, hollandite channels, that is filled with chloride ions [157,158].

The nanostructure of akaganeite is mostly formed in nanorods and can be synthesized by hydrolysis of an iron chloride solution. The reaction is performed at high temperatures, about 80 °C, and usually in the presence of surfactants, e.g., poly (ethylene oxide) (PEO) or polymers (e.g., polyethyleneimine) as a capping agent and structure director to control the particle size distribution and particles shape [159–161]. Control over the particles' size is possible by varying the concentration of the capping agent. In fact, smaller nanorods can be obtained by increasing the concentration of capping agent [159].

Akaganeite has aroused widespread interest as a new generation of adsorbent material owing to its outstanding uptake capacity for toxic anions and cations [157,158,162]. Akaganeite ion exchange capability strongly depends on the hydroxyl groups located outside of the hollandite channels [163]. Formation of oriented aggregations is a key feature of the akaganeite nanorods [161,164]. New synthesis techniques have been developed to fabricate mesoporous akaganeite nanorods with increased specific surface area and hierarchical scaffold structure which is formed through the nano-rods' aggregation [161]. These structures can provide even more uptake capability for akaganeite nanorods. Akaganeite nanorods can also be used as catalyst material. Among various iron oxide and iron oxide hydroxide nanostructures, akaganeite nanorods are the most effective catalysts for the reduction of 4-nitrotoluene using hydrazine hydrate as a reducing agent [83]. One of the promising applications of akaganeite nanorods is in the fabrication of magnetite nanorods. Converting akaganeite nanorods into magnetite is a common approach for the synthesis of ellipsoidal magnetite nanoparticles. This conversion can be performed through a reduction reaction by using a chemical reducing agent, such as hydrazine (Equation (6)) [160].

$$12\beta\text{-FeOOH} + N_2H_4 \rightarrow 4Fe_3O_4 + 8H_2O + N_2 \tag{6}$$

4.1.3. Lepidocrocite (γ-FeOOH)

Lepidocrocite is also called esmeraldite or hydrohematite. It is thermodynamically metastable and based on the aqueous condition, undergoes a phase transition to a more stable iron oxide structure [165,166]. This iron combination has anti-ferromagnetic properties. It can be formed in crystalline or amorphous structures from iron oxidation phenomena, such as atmospheric oxidation or sea corrosion [167]. Its crystal system and structure are orthorhombic, and its crystal habit is normally string-shaped [168]. Its appearance has a ruby red to orange-brown color, has a semi-metallic sheen, and is found in rocks, soil, and rust. It has transparent crystals that form thin sheet-like structures [169]. This fine appearance captures attention of jewelry designers and takes lepidocrocite to the jewelry industry.

Nanoscale lepidocrocite can be formed by the oxidation of zero-valent iron nanostructures (ZVINs) in oxygenated aqueous matrixes [170]. By the oxidation of ZVINs, ferrous ions diffuse out of the structure and an oxidized layer will form on the zero-valent core. The complete oxidation of zero-valent core resulted in the collapse of the core-shell structure and complete variations in the structure and composition. The resulting structure was a reddish-brown powder of nanosheets and nanoneedles which are characteristic morphology of iron oxide nanostructures [170,171]. In another approach, lepidocrocite nanoparticles can be fabricated by adding sodium hydroxide (NaOH) to a solution of ferric ions to reach a pH of ~6 [167,168,172].

Like other iron-based nanostructures, nanoscale lepidocrocite has gained applications as adsorbent and catalyst. It can be employed to remove heavy metals (e.g., chromium and uranium) and metal complex dyes (e.g., Lanacron brown S-GL which is known as Acid Brown 298, LBS-GL) [167,172,173]. These nanostructures are able to catalyze degradation reactions. For instance, lepidocrocite nanoparticles were employed for the degradation of carbon tetrachloride (CCl_4) and it showed that the compound was degraded in a second-order reaction [171]. From a microbiological point of view, it has been shown that iron reducing bacteria (IRB) can employ nanoscale lepidocrocite as an electron acceptor when special compounds (e.g., hydrogen, formate, pyruvate, serine, lactate, ore N-acetylglucosamine) are utilized as electron donors [174].

4.1.4. Feroxyhyte (δ-FeOOH)

Feroxyhyte has a hexagonal closed-packed crystal system and is essentially a planar antiferromagnet [175,176]. It can be found naturally as yellowish-brown deposits in iron-manganese nodules at high pressure environments, such as a very deep ocean and gleysol. However, by exposure to surface environment or aerated conditions, it converts to goethite [177]. It is the only phase of anhydrous ferric oxide-hydroxide that has significant magnetic properties at room temperature [178,179]. Feroxyhyte nanoparticles are usually synthesized by the rapid oxidation of ferric hydroxide ($Fe(OH)_2$), whereas hydrogen peroxide and the resulting nanostructures can be in plate, needle, or sphere morphologies [176,178,180–182]. In addition, magnetite (Fe_3O_4) nanoparticles can be used as a precursor to fabricate feroxyhyte nanoparticles. In this technique, magnetite nanoparticles are rapidly oxidized in the air [178]. Irradiation-based techniques were also developed to fabricate these nanoparticles [180]. In these approaches γ-irradiation is used to reduce ferric ions to ferrous and then a white suspension of ferric hydroxide is resulted. The resulting ferric hydroxide can be oxidized rapidly while coming into contact with atmospheric oxygen [180].

Nanostructures of feroxyhyte have a promising potential in environmental sciences and wastewater treatment. These particles can activate hydrogen peroxide to generate reactive oxygen species (ROS) which are effective in degradation of organic molecules [181]. Feroxyhyte nanoparticles can also eliminate inorganic pollutants. It has been shown that mesoporous feroxyhyte nanoparticles and ferrous ions have a synergistic action for phosphate immobilization. Hydroxyl groups on the nanoparticles and in-situ generated ferric ions were found to promote phosphate immobilization. This δ-FeOOH/Fe(II) synergistic system can reached about 94% removal efficiency at acidic pH [183]. Doping of nanostructures with foreign metal ions, such as cupper and manganese, is another approach to enhance the catalytic and adsorbent properties of feroxyhyte nanoparticles [178,179,184]. These nanostructures are applicable in batch and continuous flow models to remove heavy metal (e.g., arsenic, cadmium, mercury, and nickel) pollutants from drinking water [185,186]. Feroxyhyte can be used as a coating for other nano-adsorbents, such as maghemite (γ-Fe2O3) nanoparticles, to improve their stability and recovery capacity. It has been shown that maghemite nanoparticles with various ratios of feroxyhyte coating have different potentials for chromate (VI) removal, and the optimal mass ratio of coating to core is 1.0 [187].

Feroxyhyte nanoparticles can be used as a photocatalyst for water splitting and hydrogen generation. Besides common nano size features such as the small particle size and high surface area, these nanoparticles have a small band gap energy and mesoporous structure that make feroxyhyte nanoparticles a suitable photocatalyst [188]. Like what we have in the remediation activities, foreign metal doping can also increase the photocatalytic activity of feroxyhyte nanoparticles. Nickel ions (Ni^{2+}) can be used to increase the conductivity and charge transfer in the particles. By loading $Ni(OH)_2$

on the surface of the resulting structure, charge separation will also improve. These modifications significantly enhance the photocatalytic activity of feroxyhyte nanoparticles [189].

Feroxyhyte nanoparticles behave superparamagnetically or ferromagnetically at ambient temperature [176,182]. Therefore, these particles can be considered in technical, medical, and biomedical practices as an alternative for previously employed magnetic nanoparticles. Experimental evidences showed that feroxyhyte nanoparticles can release heat under an alternative magnetic field and the heat production can be controlled by changing the mass of nanoparticles [182].

4.1.5. Schwertmannite

In 1994 schwertmannite was introduced as a novel iron oxyhydroxysulphate from the Pyhäisalmi sulphide mine, Province of Oulu, Finland. It was found in ochreous precipitates from acid, sulphate-rich waters [190]. Schwertmannite has a structure similar to akaganeite, the only difference being that schwertmannite has sulfate ions instead of chloride ions so it is called iron oxyhydroxysulfate mineral with the chemical formula of $Fe8O8(OH)8-2x(SO4)x \cdot nH2O$ ($1 \leq x \leq 1.75$) [191,192]. It is poorly crystalline nanometric material with tetragonal crystal system and paramagnetic behavior [193]. It appears as an opaque yellowish-brown and has a fibrous morphology under the electron microscope [190,191].

In nature, schwertmannite nanocrystals were formed through the oxidation of ferrous ions and subsequently ferric ions precipitation. Synthetic schwertmannite nanoparticles can also be fabricated by the chemical oxidation of ferrous ions. Usually ferrous ions are provided from an $FeSO_4$ solution and hydrogen peroxide is used as an oxidizing agent (Equation (7)). Precipitation occurred by hydrolysis of the resulting ferric ions. Ferric ions first form a complex intermediate with water molecules and then, at high concentrations of sulfate ions (SO_4^{2+}), transform into schwertmannite. The total reaction is illustrated in Equation (8) [192].

$$2Fe^{2+} + H_2O_2 + 2H^+ \rightarrow 2Fe^{3+} + 2H_2O \tag{7}$$

$$8Fe(SO_4) + 2O_2 + (4 + x)H_2O \rightarrow Fe_8O_8(OH)_x(SO_4)_y + (8 - y)SO_4^{2+} + 2(8 - y)H^+ \tag{8}$$

Biosynthesis (biogenic fabrication) is another technique for the fabrication of schwertmannite nanostructures. It is dependent on the biological ferrooxidation of ferrous ions by iron oxidizing microorganisms that are able to oxidize ferrous ions or elemental sulfur as a source of energy. These organisms can precipitate nano-sized schwertmannite at ambient temperature and acidic pH (~2–3) [194]. *Acidithiobacillus ferrooxidans* is the most known and studied bacteria in this family [195].

Schwertmannite was found to have considerable potential as an absorbent material, which is applicable for water treatment and soil remediation purposes. It has a promising potential to remove both mineral (e.g., arsenic, chromium, antimony, and fluoride) and organic contaminants [192]. Song et al. (2018) compared the characteristic features of chemically synthesized and biogenic schwertmannite nanostructures that are important for remediation purposes [196]. They show that chemical synthesis can fabricate more schwertmannite than biogenic methods. But biogenic methods provide larger nanoparticles with larger specific surface area. From a surface morphology point of view, chemically synthesized particles have a smooth surface, while the surface of the biogenic particles is chestnut shell-like with higher saturated adsorption capacity [196]. In addition, biogenic schwertmannite has a higher chemical stability. Synthetic schwertmannite is more sensitive to the aging process and conversion to the goethite [197]. Key features of the biogenic nanoparticles can even be improved by a simple heating. Heating up to 250 °C resulted in the significant increase in the particles' porosity, specific surface area, and efficiency for contaminants removal [198]. Biofabrication of schwertmannite has a critical point; adherence of prepared nanostructures to the reactor wall during the biosynthesis process resulted in deterioration of particles' structural characteristics. To eliminate the adhered schwertmannite, Zhang et al. introduced additional schwertmannite into the reactor system prior to schwertmannite biosynthesis. They show that by doing so, all biogenic schwertmannite nanostructures are in a suspended form [199].

4.2. Hydrated Ferric Oxyhydroxides

Ferric oxyhydroxide is usually encountered as hydrated forms with a $FeO(OH) \cdot nH_2O$ general formula. The monohydrate form of this compound is the most common form and is known as ferric hydroxide and its simplified formula is $Fe(OH)_3$ (Figure S1f). For the first time it was reported from the Proprietary Mine at Broken Hill, New South Wales, Australia and called bernalite in homage to the British crystallographer J. D. Bernal (1901–1971) [200]. Bernalite is structurally related to perovskites (a calcium titanium oxide mineral, $CaTiO_3$) and its small crystals, up to 3 mm, occurs as flattened pyramidal to pseudo-octahedral with slightly concave faces. Due to goethite inclusions, the crystals can be transparent to opaque with pale green (bottle-green) color [200,201].

Bernalite nanoparticles can be transformed from goethite nanorods at alkaline pH ($\sim$10) and in the presence of arsenate. Arsenate has a key role in the formation of bernalite and the abundance of bernalite is in direct relation with the arsenate concentration [202]. In the laboratorial experiments, ferric hydroxide can be easily obtained by increasing the pH of a ferric ions solution up to about 8 as in Equation (9) shows [203].

$$Fe(III) + 3OH^- \rightarrow Fe(OH)_3 \tag{9}$$

It can be a basic reaction to prepare a precursor for other iron-based nanoparticles. For instance, ferric hydroxide was used as a precursor for the generation of monodisperse ferromagnetic and superparamagnetic iron oxide nanoparticles [204]. In a similar manner, controlled dehydration of ferric hydroxide resulted in Fe_2O_3 [205]. In a novel approach, Thuy et al. (2014) employed the reaction for the fabrication of colloidal ferric hydroxide to remove heavy metals from water. They combined the coprecipitation of ferric hydroxide and flotation technique using gamma poly-glutamic acid (γ-PGA) for collecting Au, Bi, Co, Cd, Cu, In, Ir, Mn, Ni, Pb, Pd, Pt, Te, Sb, Rh, Ru, and Zn. To perform the precipitation reaction, they added a solution of sodium hydroxide to heavy metals contaminated water containing ferrous ions [206]. Coprecipitation of ferric hydroxide can also be performed in the presence of other nanoparticles, such as magnetite nanoparticles, and form a core-shell structure. Arefi et al. showed that the resulting core-shell nanostructure is applicable as a heterogeneous catalyst for the direct oxidative amidation of alcohols with amine hydrochloride salts [207].

5. Iron Hydroxides

Iron hydroxides can be found in two different structures as ferrous hydroxide and ferric hydroxide. Molecular structure of ferrous hydroxide is provided in Figure 1, and it can be denoted as $Fe(OH)_2$ or FeH_2O_2. Basically, it is a nearly white substance, but by exposure to oxygen, ferrous ions are oxidized to ferric and provide a greenish to reddish-brown color of ferric hydroxide. Chemically, ferrous hydroxide is produced as a gelatinous precipitate when ferrous ions are exposed to an alkali, such as hydroxide salts or ammonium (Equation (10)). In the synthesis of ferrous hydroxide nanoparticles, the employed alkali has a significant effect on the shape and surface features of the resulted nanostructure. It was shown that ammonium derived the reaction to make nanospheres with a porous surface, while hydroxide salts (e.g., NaOH) make smooth nanorods. It is obvious that the product is sensitive to the oxygen, and hence the reaction should be performed under an inert atmosphere [208].

$$Fe^{2+} + 2OH^- \rightarrow Fe(OH)_2 \tag{10}$$

Nanostructures of ferrous hydroxide are great activators for hydrogen peroxide to produce hydroxyl radicals. This potential provides the particles a Fenton-like catalyst property that can be employed to degrade organic pollutants as a Fenton-like activator of hydrogen peroxide [208]. The molecular structure of ferric hydroxide is provided in Figure 1. Ferric hydroxide was discussed in detail in Section 4.2.

The content of this review was summarized in tabular format as Supplementary Material and provided in Table S1 for a quick overview.

6. Conclusions

Iron as a well-known element from the ancient time, retaining an outstanding position in the human life and culture. Due to its different valency and combining power with oxygen atoms and water molecules, it comes in a colorful variety of compounds with diverse chemical structures and physical appearances. In the recent years, by taking iron compounds to the nano realm, we can improve the iron to achieve unpresented properties and applications. In this review 19 different compounds of iron, from pure iron to combined forms of iron with oxygen and water, were investigated for their structures, properties, synthesis routes, and applications as nanomaterials. Nanostructures of iron can be prepared by simple chemical reactions and even in some cases these structures are achievable via biologic pathways or biochemical reactions. Provided data demonstrate that iron-based nanomaterials can be employed in medical, biomedical, industrial, engineering, environmental, and agricultural sceneries. The discovery of the potential applications of iron-based nanomaterials only in its emerging phase and there will be a huge amount of investigations in this regard in the future.

Supplementary Materials: The following are available online at http://www.mdpi.com/2227-9717/8/9/1128/s1, Figure S1: Molecular structures of iron compounds, FeO (a), Fe_2O_3 (b), Fe_3O_4 (c), FeOOH (d), $Fe(OH)_2$ (e), and $Fe(OH)_3$ (f), Table S1: A tabular summery of the data that provided in the main manuscript.

Author Contributions: Conceptualization, S.-M.T. and A.E.; Software, S.-M.T.; Data Curation, S.-M.T. and A.E.; Writing—Original Draft Preparation, S.-M.T., A.E., and M.Z.; Writing—Review & Editing, S.-M.T., A.E., and A.B.; Visualization, S.-M.T. and A.E.; Supervision, A.E. and A.B.; Project Administration, A.E. and A.B.; Funding Acquisition, A.B. and A.E. All authors have read and agreed to the published version of the manuscript.

Funding: The APC was funded by The University of Waikato.

Acknowledgments: This work was financially supported by Shiraz University of Medical Sciences under a grant NO. 99-01-121-23506. The APC was funded by The University of Waikato, Hamilton 3240, New Zealand.

Conflicts of Interest: The authors declare no conflict of interest.

References

1. Iron Ore Mine Production by Country 2014–2018. Available online: https://www.statista.com/statistics/267380/iron-ore-mine-production-by-country/ (accessed on 30 March 2019).
2. Fan, M.; Yuan, P.; Chen, T.; He, H.; Yuan, A.; Chen, K.; Zhu, J.; Liu, D. Synthesis, characterization and size control of zerovalent iron nanoparticles anchored on montmorillonite. *Chin. Sci. Bull.* **2010**, *55*, 1092–1099. [CrossRef]
3. Xiaomin, N.; Xiaobo, S.; Huagui, Z.; Dongen, Z.; Dandan, Y.; Qingbiao, Z. Studies on the one-step preparation of iron nanoparticles in solution. *J. Cryst. Growth* **2005**, *275*, 548–553. [CrossRef]
4. dos Santos, F.S.; Lago, F.R.; Yokoyama, L.; Fonseca, F.V. Synthesis and characterization of zero-valent iron nanoparticles supported on SBA-15. *J. Mater. Res. Technol.* **2017**, *6*, 178–183. [CrossRef]
5. Ebrahiminezhad, A.; Zare-Hoseinabadi, A.; Berenjian, A.; Ghasemi, Y. Green synthesis and characterization of zero-valent iron nanoparticles using stinging nettle (*Urtica dioica*) leaf extract. *Green Proc. Synth.* **2017**, *6*, 469–475. [CrossRef]
6. Taghizadeh, S.-M.; Berenjian, A.; Taghizadeh, S.; Ghasemi, Y.; Taherpour, A.; Sarmah, A.K.; Ebrahiminezhad, A. One-put green synthesis of multifunctional silver iron core-shell nanostructure with antimicrobial and catalytic properties. *Ind. Crop. Prod.* **2019**, *130*, 230–236. [CrossRef]
7. Akbari, A.; Mohamadzadeh, F. New method of synthesis of stable zero valent iron nanoparticles (nZVI) by chelating agent diethylene triamine penta acetic acid (DTPA) and removal of radioactive uranium from ground water by using Iron nanoparticle. *J. Nanostruct.* **2012**, *2*, 175–181. [CrossRef]
8. Sun, Y.-P.; Li, X.-q.; Cao, J.; Zhang, W.-x.; Wang, H.P. Characterization of zero-valent iron nanoparticles. *Adv. Colloid Interface Sci.* **2006**, *120*, 47–56. [CrossRef]

9. Üzüm, Ç.; Shahwan, T.; Eroğlu, A.E.; Hallam, K.R.; Scott, T.B.; Lieberwirth, I. Synthesis and characterization of kaolinite-supported zero-valent iron nanoparticles and their application for the removal of aqueous Cu^{2+} and Co^{2+} ions. *Appl. Clay Sci.* **2009**, *43*, 172–181. [CrossRef]

10. Wu, Y.-N.; Yang, L.-X.; Shi, X.-Y.; Li, I.-C.; Biazik, J.M.; Ratinac, K.R.; Chen, D.-H.; Thordarson, P.; Shieh, D.-B.; Braet, F. The selective growth inhibition of oral cancer by iron core-gold shell nanoparticles through mitochondria-mediated autophagy. *Biomaterials* **2011**, *32*, 4565–4573. [CrossRef]

11. Yang, L.-X.; Wu, Y.-N.; Wang, P.-W.; Huang, K.-J.; Su, W.-C.; Shieh, D.-B. Silver-coated zero-valent iron nanoparticles enhance cancer therapy in mice through lysosome-dependent dual programed cell death pathways: Triggering simultaneous apoptosis and autophagy only in cancerous cells. *J. Mater. Chem. B* **2020**, *8*, 4122–4131. [CrossRef]

12. Ebrahiminezhad, A.; Zare-Hoseinabadi, A.; Sarmah, A.K.; Taghizadeh, S.; Ghasemi, Y.; Berenjian, A. Plant-Mediated synthesis and applications of iron nanoparticles. *Mol. Biotechnol.* **2018**, *60*, 154–168. [CrossRef] [PubMed]

13. Huang, R.; Wang, X.; Xing, B. Removal of labile arsenic from flooded paddy soils with a novel extractive column loaded with quartz-supported nanoscale zero-valent iron. *Environ. Pollut.* **2019**, *255*, 113249. [CrossRef] [PubMed]

14. Liu, H.; Li, P.; Yu, H.; Zhang, T.; Qiu, F. Controlled fabrication of functionalized nanoscale zero-valent iron/celluloses composite with silicon as protective layer for arsenic removal. *Chem. Eng. Res. Des.* **2019**, *151*, 242–251. [CrossRef]

15. Tso, C.-p.; Shih, Y.-h. Effect of carboxylic acids on the properties of zerovalent iron toward adsorption and degradation of trichloroethylene. *J. Environ. Manag.* **2018**, *206*, 817–825. [CrossRef] [PubMed]

16. Ertosun, F.M.; Cellat, K.; Eren, O.; Gül, Ş.; Kuşvuran, E.; Şen, F. Comparison of nanoscale zero-valent iron, fenton, and photo-fenton processes for degradation of pesticide 2, 4-dichlorophenoxyacetic acid in aqueous solution. *SN Appl. Sci.* **2019**, *1*, 1491. [CrossRef]

17. Shojaei, S.; Shojaei, S. Optimization of process variables by the application of response surface methodology for dye removal using nanoscale zero-valent iron. *Int. J. Environ. Sci. Technol.* **2019**, *16*, 4601–4610. [CrossRef]

18. Ji, H.; Zhu, Y.; Duan, J.; Liu, W.; Zhao, D. Reductive immobilization and long-term remobilization of radioactive pertechnetate using bio-macromolecules stabilized zero valent iron nanoparticles. *Chin. Chem. Lett.* **2019**, *30*, 2163–2168. [CrossRef]

19. Lee, C.; Kim, J.Y.; Lee, W.I.; Nelson, K.L.; Yoon, J.; Sedlak, D.L. Bactericidal effect of zero-valent iron nanoparticles on *Escherichia coli*. *Environ. Sci. Technol.* **2008**, *42*, 4927–4933. [CrossRef]

20. Cornell, R.M.; Schwertmann, U. *The Iron OXIDES: Structure, Properties, Reactions, Occurrences and Uses*; John Wiley & Sons: Hoboken, NJ, USA, 2003.

21. Redl, F.X.; Black, C.T.; Papaefthymiou, G.C.; Sandstrom, R.L.; Yin, M.; Zeng, H.; Murray, C.B.; O'Brien, S.P. Magnetic, electronic, and structural characterization of nonstoichiometric iron oxides at the nanoscale. *J. Am. Chem. Soc.* **2004**, *126*, 14583–14599. [CrossRef]

22. Guntlin, C.P.; Ochsenbein, S.T.; Wörle, M.; Erni, R.; Kravchyk, K.V.; Kovalenko, M.V. Popcorn-shaped Fe_xO (Wüstite) nanoparticles from a single-source precursor: Colloidal synthesis and magnetic properties. *Chem. Mater.* **2018**, *30*, 1249–1256. [CrossRef]

23. Martinez, A.I.; Garcia-Lobato, M.; Perry, D.L. Study of the properties of iron oxide nanostructures. In *Research in Nanotechnology Developments*; Barrañón, A., Ed.; Nova Science Publishers, Inc.: Hauppauge, NY, USA, 2009; Volume 19, pp. 184–193.

24. Strobel, R.; Pratsinis, S.E. Direct synthesis of maghemite, magnetite and wustite nanoparticles by flame spray pyrolysis. *Adv. Powder Technol.* **2009**, *20*, 190–194. [CrossRef]

25. Chen, C.-J.; Chiang, R.-K.; Lai, H.-Y.; Lin, C.-R. Characterization of monodisperse wüstite nanoparticles following partial oxidation. *J. Phys. Chem. C* **2010**, *114*, 4258–4263. [CrossRef]

26. Gheisari, M.; Mozaffari, M.; Acet, M.; Amighian, J. Preparation and investigation of magnetic properties of wüstite nanoparticles. *J. Magn. Magn. Mater.* **2008**, *320*, 2618–2621. [CrossRef]

27. Papaefthymiou, G.; Redl, F.; Black, C.; Sandstrom, R.; Yin, M.; Murray, C.; O'Brien, S. Hybrid magnetic nanoparticles derived from wüstite disproportionation reactions at the nanoscale. In *ICAME 2005*; Springer: Berlin/Heidelberg, Germany, 2006; pp. 239–245. [CrossRef]

28. Ahmadipour, M.; Rao, K.V.; Rajendar, V. Formation of nanoscale $Mg_{(x)}Fe_{(1-x)}O$ (x = 0.1, 0.2, 0.4) structure by solution combustion: Effect of fuel to oxidizer ratio. *J. Nanomater.* **2012**, *2012*, 163909. [CrossRef]

29. Chen, C.-J.; Chiang, R.-K.; Kamali, S.; Wang, S.-L. Synthesis and controllable oxidation of monodisperse cobalt-doped wüstite nanoparticles and their core–shell stability and exchange-bias stabilization. *Nanoscale* **2015**, *7*, 14332–14343. [CrossRef]

30. Situ, S.F.; Samia, A.C.S. Highly efficient antibacterial iron oxide@carbon nanochains from wustite precursor nanoparticles. *ACS Appl. Mater. Interfaces* **2014**, *6*, 20154–20163. [CrossRef]

31. Chavez-Galan, J.; Almanza, R. Solar filters based on iron oxides used as efficient windows for energy savings. *Sol. Energy* **2007**, *81*, 13–19. [CrossRef]

32. Campos, E.A.; Pinto, D.V.B.S.; Oliveira, J.I.S.d.; Mattos, E.d.C.; Dutra, R.d.C.L. Synthesis, characterization and applications of iron oxide nanoparticles-a short review. *J. Aerosp. Technol. Manag.* **2015**, *7*, 267–276. [CrossRef]

33. Lassoued, A.; Dkhil, B.; Gadri, A.; Ammar, S. Control of the shape and size of iron oxide (α-Fe$_2$O$_3$) nanoparticles synthesized through the chemical precipitation method. *Results Phys.* **2017**, *7*, 3007–3015. [CrossRef]

34. Tadic, M.; Panjan, M.; Damnjanovic, V.; Milosevic, I. Magnetic properties of hematite (α-Fe$_2$O$_3$) nanoparticles prepared by hydrothermal synthesis method. *Appl. Surf. Sci.* **2014**, *320*, 183–187. [CrossRef]

35. Kefeni, K.K.; Msagati, T.A.; Nkambule, T.T.; Mamba, B.B. Synthesis and application of hematite nanoparticles for acid mine drainage treatment. *J. Environ. Chem. Eng.* **2018**, *6*, 1865–1874. [CrossRef]

36. Asoufi, H.M.; Al-Antary, T.M.; Awwad, A.M. Green route for synthesis hematite (α-Fe$_2$O$_3$) nanoparticles: Toxicity effect on the green peach aphid, *Myzus persicae* (Sulzer). *Environ. Nanotechnol. Monit. Manag.* **2018**, *9*, 107–111. [CrossRef]

37. Narayanan, K.B.; Han, S.S. One-pot green synthesis of hematite (α-Fe$_2$O$_3$) nanoparticles by ultrasonic irradiation and their in vitro cytotoxicity on human keratinocytes CRL-2310. *J. Cluster Sci.* **2016**, *27*, 1763–1775. [CrossRef]

38. Taghavi, F.S.; Ramazani, A.; Golfar, Z.; Woo, J.S. Green synthesis of α-Fe$_2$O$_3$ (hematite) nanoparticles using tragacanth Gel. *J. Appl. Chem. Res.* **2017**, *11*, 19–27.

39. Ali, H.R.; Nassar, H.N.; El-Gendy, N.S. Green synthesis of α-Fe$_2$O$_3$ using *Citrus reticulum* peels extract and water decontamination from different organic pollutants. *Energ. Source Part A* **2017**, *39*, 1425–1434. [CrossRef]

40. Ahmmad, B.; Leonard, K.; Islam, M.S.; Kurawaki, J.; Muruganandham, M.; Ohkubo, T.; Kuroda, Y. Green synthesis of mesoporous hematite (α-Fe$_2$O$_3$) nanoparticles and their photocatalytic activity. *Adv. Powder Technol.* **2013**, *24*, 160–167. [CrossRef]

41. Rufus, A.; Sreeju, N.; Philip, D. Synthesis of biogenic hematite (α-Fe$_2$O$_3$) nanoparticles for antibacterial and nanofluid applications. *RSC Adv.* **2016**, *6*, 94206–94217. [CrossRef]

42. Supattarasakda, K.; Petcharoen, K.; Permpool, T.; Sirivat, A.; Lerdwijitjarud, W.J.P.t. Control of hematite nanoparticle size and shape by the chemical precipitation method. *Powder Technol.* **2013**, *249*, 353–359. [CrossRef]

43. Anthony, J.W.; Bideaux, R.A.; Bladh, K.W.; Nichols, M.C. *Handbook of Mineralogy*; Mineralogical Society of America: Chantilly, VA, USA, 2004.

44. Basavegowda, N.; Mishra, K.; Lee, Y.R. Synthesis, characterization, and catalytic applications of hematite (α-Fe$_2$O$_3$) nanoparticles as reusable nanocatalyst. *Adv. Nat. Sci.* **2017**, *8*, 025017. [CrossRef]

45. Nazeer, A.A.; Udhayakumar, S.; Mani, S.; Dhanapal, M.; Vijaykumar, S.D. Surface modification of Fe$_2$O$_3$ and MgO nanoparticles with agrowastes for the treatment of chlorosis in *Glycine max*. *Nano Converg.* **2018**, *5*, 23. [CrossRef]

46. Danno, T.; Nakatsuka, D.; Kusano, Y.; Asaoka, H.; Nakanishi, M.; Fujii, T.; Ikeda, Y.; Takada, J. Crystal structure of β-Fe$_2$O$_3$ and topotactic phase transformation to α-Fe$_2$O$_3$. *Cryst. Growth Des.* **2013**, *13*, 770–774. [CrossRef]

47. Zhang, N.; Guo, Y.; Wang, X.; Zhang, S.; Li, Z.; Zou, Z. Beta-Fe$_2$O$_3$ nanoparticle-assembled film for photoelectrochemical water splitting. *Dalton Trans.* **2017**, *46*, 10673–10677. [CrossRef]

48. Kumar, A.; Singhal, A. Synthesis of colloidal β-Fe$_2$O$_3$ nanostructures—Influence of addition of Co^{2+} on their morphology and magnetic behavior. *Nanotechnology* **2007**, *18*, 475703. [CrossRef]

49. Nurdin, I.; Johan, M.; Yaacob, I.; Ang, B.; Andriyana, A. Synthesis, characterisation and stability of superparamagnetic maghemite nanoparticle suspension. *Mater. Res. Innov.* **2014**, *18*, 200–203. [CrossRef]

50. Aliahmad, M.; Moghaddam, N.N. Synthesis of maghemite (γ-Fe$_2$O$_3$) nanoparticles by thermal-decomposition of magnetite (Fe$_3$O$_4$) nanoparticles. *Mater. Sci.-Poland* **2013**, *31*, 264–268. [CrossRef]

51. Múzquiz-Ramos, E.; Guerrero-Chávez, V.; Macías-Martínez, B.; López-Badillo, C.; García-Cerda, L. Synthesis and characterization of maghemite nanoparticles for hyperthermia applications. *Ceram. Int.* **2015**, *41*, 397–402. [CrossRef]

52. Guivar, J.A.R.; Sadrollahi, E.; Menzel, D.; Fernandes, E.G.R.; López, E.O.; Torres, M.M.; Arsuaga, J.M.; Arencibia, A.; Litterst, F.J. Magnetic, structural and surface properties of functionalized maghemite nanoparticles for copper and lead adsorption. *RSC Adv.* **2017**, *7*, 28763–28779. [CrossRef]

53. Frison, R.; Cernuto, G.; Cervellino, A.; Zaharko, O.; Colonna, G.M.; Guagliardi, A.; Masciocchi, N. Magnetite–maghemite nanoparticles in the 5–15 nm range: Correlating the core–shell composition and the surface structure to the magnetic properties. a total scattering study. *Chem. Mater.* **2013**, *25*, 4820–4827. [CrossRef]

54. Kera, N.H.; Bhaumik, M.; Pillay, K.; Ray, S.S.; Maity, A. Selective removal of toxic Cr(VI) from aqueous solution by adsorption combined with reduction at a magnetic nanocomposite surface. *J. Colloid Interface Sci.* **2017**, *503*, 214–228. [CrossRef]

55. Prado, Y.; Daffé, N.; Michel, A.; Georgelin, T.; Yaacoub, N.; Greneche, J.-M.; Choueikani, F.; Otero, E.; Ohresser, P.; Arrio, M.-A. Enhancing the magnetic anisotropy of maghemite nanoparticles via the surface coordination of molecular complexes. *Nat. Commun.* **2015**, *6*, 10139. [CrossRef]

56. Mazarío, E.; Helal, A.S.; Stemper, J.; Mayoral, A.; Decorse, P.; Chevillot-Biraud, A.; Novak, S.; Perruchot, C.; Lion, C.; Losno, R. Maghemite nanoparticles bearing di(amidoxime) groups for the extraction of uranium from wastewaters. *AIP Adv.* **2017**, *7*, 056702. [CrossRef]

57. Palmqvist, N.M.; Seisenbaeva, G.A.; Svedlindh, P.; Kessler, V.G. Maghemite nanoparticles acts as nanozymes, improving growth and abiotic stress tolerance in *Brassica napus*. *Nanoscale Res. Lett.* **2017**, *12*, 631. [CrossRef] [PubMed]

58. Tavakkoli, M.; Kallio, T.; Reynaud, O.; Nasibulin, A.G.; Sainio, J.; Jiang, H.; Kauppinen, E.I.; Laasonen, K. Maghemite nanoparticles decorated on carbon nanotubes as efficient electrocatalysts for the oxygen evolution reaction. *J. Mater. Chem. A* **2016**, *4*, 5216–5222. [CrossRef]

59. Hu, J.; Chen, G.; Lo, I.M. Removal and recovery of Cr(VI) from wastewater by maghemite nanoparticles. *Water Res.* **2005**, *39*, 4528–4536. [CrossRef]

60. Jiang, W.; Pelaez, M.; Dionysiou, D.D.; Entezari, M.H.; Tsoutsou, D.; O'Shea, K. Chromium(VI) removal by maghemite nanoparticles. *Chem. Eng. J.* **2013**, *222*, 527–533. [CrossRef]

61. Avram, A.; Radoi, A.; Schiopu, V.; Avram, M.; Gavrila, H. Synthesis and characterization of γ-Fe$_2$O$_3$ nanoparticles for applications in magnetic hyperthermia. *Synthesis* **2011**, *10*, 1.

62. Mercante, L.; Melo, W.; Granada, M.; Troiani, H.; Macedo, W.; Ardison, J.; Vaz, M.; Novak, M. Magnetic properties of nanoscale crystalline maghemite obtained by a new synthetic route. *J. Magn. Magn. Mater.* **2012**, *324*, 3029–3033. [CrossRef]

63. Kuchma, E.A.; Zolotukhin, P.V.; Belanova, A.A.; Soldatov, M.A.; Lastovina, T.A.; Kubrin, S.P.; Nikolsky, A.V.; Mirmikova, L.I.; Soldatov, A.V. Low toxic maghemite nanoparticles for theranostic applications. *Int. J. Nanomed.* **2017**, *12*, 6365–6371. [CrossRef]

64. Kumar, N.; Kulkarni, K.; Behera, L.; Verma, V. Preparation and characterization of maghemite nanoparticles from mild steel for magnetically guided drug therapy. *J. Mater. Sci. Mater. Med.* **2017**, *28*, 116. [CrossRef]

65. Blanc-Beguin, F.; Nabily, S.; Gieraltowski, J.; Turzo, A.; Querellou, S.; Salaun, P. Cytotoxicity and GMI bio-sensor detection of maghemite nanoparticles internalized into cells. *J. Magn. Magn. Mater.* **2009**, *321*, 192–197. [CrossRef]

66. Perlstein, B.; Lublin-Tennenbaum, T.; Marom, I.; Margel, S. Synthesis and characterization of functionalized magnetic maghemite nanoparticles with fluorescent probe capabilities for biological applications. *J. Biomed. Mater. Res. B Appl. Biomater.* **2010**, *92*, 353–360. [CrossRef] [PubMed]

67. Hai, J.; Piraux, H.; Mazario, E.; Volatron, J.; Ha-Duong, N.; Decorse, P.; Lomas, J.; Verbeke, P.; Ammar, S.; Wilhelm, C. Maghemite nanoparticles coated with human serum albumin: Combining targeting by the iron-acquisition pathway and potential in photothermal therapies. *J. Mater. Chem. B* **2017**, *5*, 3154–3162. [CrossRef] [PubMed]

68. González, B.; Ruiz-Hernández, E.; Feito, M.J.; de Laorden, C.L.; Arcos, D.; Ramírez-Santillán, C.; Matesanz, C.; Portolés, M.T.; Vallet-Regí, M. Covalently bonded dendrimer-maghemite nanosystems: Nonviral vectors for in vitro gene magnetofection. *J. Mater. Chem.* **2011**, *21*, 4598–4604. [CrossRef]

69. Hai, J.; Piraux, H.; Mazario, E.; Volatron, J.; Ha-Duong, N.; Decorse, P.; Espinosa, A.; Whilem, C.; Verbeke, P.; Gazeau, F. Multifunctionality of maghemite nanoparticles functionalized by HSA for drug delivery. In Proceedings of the 2017 IEEE International Magnetics Conference (INTERMAG), Dublin, Ireland, 24–28 April 2017; pp. 1–2.

70. Štarha, P.; Stavárek, M.; Tuček, J.; Trávníček, Z. 4-Aminobenzoic acid-coated maghemite nanoparticles as potential anticancer drug magnetic carriers: A case study on highly cytotoxic cisplatin-like complexes involving 7-azaindoles. *Molecules* **2014**, *19*, 1622–1634. [CrossRef] [PubMed]

71. Thünemann, A.F.; Schütt, D.; Kaufner, L.; Pison, U.; Möhwald, H. Maghemite nanoparticles protectively coated with poly (ethylene imine) and poly (ethylene oxide)-b lock-poly (glutamic acid). *Langmuir* **2006**, *22*, 2351–2357. [CrossRef]

72. Liu, J.; Detrembleur, C.; Debuigne, A.; De Pauw-Gillet, M.-C.; Mornet, S.; Vander Elst, L.; Laurent, S.; Duguet, E.; Jérôme, C. Glucose-, pH-and thermo-responsive nanogels crosslinked by functional superparamagnetic maghemite nanoparticles as innovative drug delivery systems. *J. Mater. Chem. B* **2014**, *2*, 1009–1023. [CrossRef]

73. Blyakhman, F.A.; Safronov, A.P.; Zubarev, A.Y.; Shklyar, T.F.; Makeyev, O.G.; Makarova, E.B.; Melekhin, V.V.; Larrañaga, A.; Kurlyandskaya, G.V. Polyacrylamide ferrogels with embedded maghemite nanoparticles for biomedical engineering. *Results Phys.* **2017**, *7*, 3624–3633. [CrossRef]

74. Kim, J.-J.; Singh, R.K.; Seo, S.-J.; Kim, T.-H.; Kim, J.-H.; Lee, E.-J.; Kim, H.-W. Magnetic scaffolds of polycaprolactone with functionalized magnetite nanoparticles: Physicochemical, mechanical, and biological properties effective for bone regeneration. *RSC Adv.* **2014**, *4*, 17325–17336. [CrossRef]

75. Ngadiman, N.H.A.; Idris, A.; Irfan, M.; Kurniawan, D.; Yusof, N.M.; Nasiri, R. γ-Fe$_2$O$_3$ nanoparticles filled polyvinyl alcohol as potential biomaterial for tissue engineering scaffold. *J. Mech. Behav. Biomed. Mater.* **2015**, *49*, 90–104. [CrossRef]

76. Fallahiarezoudar, E.; Ahmadipourroudposht, M.; Idris, A.; Yusof, N.M. Optimization and development of Maghemite (γ-Fe$_2$O$_3$) filled poly-L-lactic acid (PLLA)/thermoplastic polyurethane (TPU) electrospun nanofibers using Taguchi orthogonal array for tissue engineering heart valve. *Mater. Sci. Eng. C* **2017**, *76*, 616–627. [CrossRef]

77. Chirita, M.; Grozescu, I.; Taubert, L.; Radulescu, H.; Princz, E. Fe$_2$O$_3$-nanoparticles, physical properties and their photochemical and photoelectrochemical applications. *Chem. Bull.* **2009**, *54*, 1–8.

78. Tucek, J.; Zboril, R.; Namai, A.; Ohkoshi, S.-i. ε-Fe$_2$O$_3$: An advanced nanomaterial exhibiting giant coercive field, millimeter-wave ferromagnetic resonance, and magnetoelectric coupling. *Chem. Mater.* **2010**, *22*, 6483–6505. [CrossRef]

79. López-Sánchez, J.; Muñoz-Noval, A.; Serrano, A.; Abuín, M.; de la Figuera, J.; Marco, J.; Pérez, L.; Carmona, N.; de la Fuente, O.R. Growth, structure and magnetism of ε-Fe$_2$O$_3$ in nanoparticle form. *RSC Adv.* **2016**, *6*, 46380–46387. [CrossRef]

80. Ohkoshi, S.-i.; Namai, A.; Yamaoka, T.; Yoshikiyo, M.; Imoto, K.; Nasu, T.; Anan, S.; Umeta, Y.; Nakagawa, K.; Tokoro, H. Mesoscopic bar magnet based on ε-Fe$_2$O$_3$ hard ferrite. *Sci. Rep.* **2016**, *6*, 27212. [CrossRef]

81. Tuček, J.; Machala, L.; Ono, S.; Namai, A.; Yoshikiyo, M.; Imoto, K.; Tokoro, H.; Ohkoshi, S.-i.; Zbořil, R. Zeta-Fe$_2$O$_3$–A new stable polymorph in iron(III) oxide family. *Sci. Rep.* **2015**, *5*, 15091. [CrossRef]

82. Lynch, S.; Batty, L.; Byrne, P. Environmental risk of metal mining contaminated river bank sediment at redox-transitional zones. *Minerals* **2014**, *4*, 52–73. [CrossRef]

83. Mohapatra, M.; Anand, S. Synthesis and applications of nano-structured iron oxides/hydroxides—A review. *Int. J. Eng. Sci. Technol.* **2010**, *2*, 127–146. [CrossRef]

84. Eusterhues, K.; Wagner, F.E.; Häusler, W.; Hanzlik, M.; Knicker, H.; Totsche, K.U.; Kögel-Knabner, I.; Schwertmann, U. Characterization of ferrihydrite-soil organic matter coprecipitates by X-ray diffraction and Mossbauer spectroscopy. *Environ. Sci. Technol.* **2008**, *42*, 7891–7897. [CrossRef]

85. Poulson, R.L.; Johnson, C.M.; Beard, B.L. Iron isotope exchange kinetics at the nanoparticulate ferrihydrite surface. *Am. Mineral.* **2005**, *90*, 758–763. [CrossRef]

86. Chasteen, N.D.; Harrison, P.M. Mineralization in ferritin: An efficient means of iron storage. *J. Struct. Biol.* **1999**, *126*, 182–194. [CrossRef]

87. Liu, G.; Debnath, S.; Paul, K.W.; Han, W.; Hausner, D.B.; Hosein, H.-A.; Michel, F.M.; Parise, J.B.; Sparks, D.L.; Strongin, D.R. Characterization and surface reactivity of ferrihydrite nanoparticles assembled in ferritin. *Langmuir* **2006**, *22*, 9313–9321. [CrossRef] [PubMed]

88. Baldi, F.; Marchetto, D.; Battistel, D.; Daniele, S.; Faleri, C.; De Castro, C.; Lanzetta, R. Iron-binding characterization and polysaccharide production by *Klebsiella oxytoca* strain isolated from mine acid drainage. *J. Appl. Microbiol.* **2009**, *107*, 1241–1250. [CrossRef] [PubMed]

89. Baldi, F.; Marchetto, D.; Paganelli, S.; Piccolo, O. Bio-generated metal binding polysaccharides as catalysts for synthetic applications and organic pollutant transformations. *N. Biotechnol.* **2011**, *29*, 74–78. [CrossRef] [PubMed]

90. Baldi, F.; Minacci, A.; Pepi, M.; Scozzafava, A. Gel sequestration of heavy metals by *Klebsiella oxytoca* isolated from iron mat. *FEMS Microbiol. Ecol.* **2001**, *36*, 169–174. [CrossRef] [PubMed]

91. Kianpour, S.; Ebrahiminezhad, A.; Mohkam, M.; Tamaddon, A.M.; Dehshahri, A.; Heidari, R.; Ghasemi, Y. Physicochemical and biological characteristics of the nanostructured polysaccharide-iron hydrogel produced by microorganism *Klebsiella oxytoca*. *J. Basic Microbiol.* **2016**, *2016*, 132–140. [CrossRef]

92. Kianpour, S.; Ebrahiminezhad, A.; Deyhimi, M.; Negahdaripour, M.; Raee, M.j.; Mohkam, M.; Rezaee, H.; Irajie, C.; Berenjian, A.; Ghasemi, Y. Structural characterization of polysaccharide-coated iron oxide nanoparticles produced by *Staphylococcus warneri*, isolated from a thermal spring. *J. Basic Microbiol.* **2019**, *2019*, 1–10. [CrossRef]

93. Kianpour, S.; Ebrahiminezhad, A.; Negahdaripour, M.; Mohkam, M.; Mohammadi, F.; Niknezhad, S.; Ghasemi, Y. Characterization of biogenic Fe (III)-binding exopolysaccharide nanoparticles produced by *Ralstonia* sp. SK03. *Biotechnol. Prog.* **2018**, *34*, 1167–1176. [CrossRef]

94. Ebrahiminezhad, A.; Zare, M.; Kiyanpour, S.; Berenjian, A.; Niknezhad, S.V.; Ghasemi, Y. Biosynthesis of xanthan gum coated iron nanoparticles by using *Xanthomonas campestris*. *IET Nanobiotechnol.* **2017**, *151*, 684–691. [CrossRef]

95. Stolyar, S.; Yaroslavtsev, R.; Bayukov, O.; Balaev, D.; Krasikov, A.; Iskhakov, R.; Vorotynov, A.; Ladygina, V.; Purtov, K.; Volochaev, M. Preparation, structure and magnetic properties of synthetic ferrihydrite nanoparticles. *J. Phys. Conf. Ser.* **2018**, *994*, 012003. [CrossRef]

96. Villacís-García, M.; Ugalde-Arzate, M.; Vaca-Escobar, K.; Villalobos, M.; Martínez-Villegas, N. Laboratory synthesis of goethite and ferrihydrite of controlled particle sizes. *Bol. Soc. Geol. Mex.* **2015**, *67*, 433–446. [CrossRef]

97. Sokolov, I.; Cherkasov, V.; Vasilyeva, A.; Bragina, V.; Nikitin, M. Paramagnetic colloidal ferrihydrite nanoparticles for MRI contrasting. *Colloids Surf. Physicochem. Eng. Aspects* **2018**, *539*, 46–52. [CrossRef]

98. Balaev, D.; Krasikov, A.; Dubrovskiy, A.; Popkov, S.; Stolyar, S.; Bayukov, O.; Iskhakov, R.; Ladygina, V.; Yaroslavtsev, R. Magnetic properties of heat treated bacterial ferrihydrite nanoparticles. *J. Magn. Magn. Mater.* **2016**, *410*, 171–180. [CrossRef]

99. Rout, K.; Mohapatra, M.; Anand, S. 2-Line ferrihydrite: Synthesis, characterization and its adsorption behaviour for removal of Pb(II), Cd(II), Cu(II) and Zn(II) from aqueous solutions. *Dalton Trans.* **2012**, *41*, 3302–3312. [CrossRef] [PubMed]

100. Antelo, J.; Arce, F.; Fiol, S. Arsenate and phosphate adsorption on ferrihydrite nanoparticles. synergetic interaction with calcium ions. *Chem. Geol.* **2015**, *410*, 53–62. [CrossRef]

101. Tosco, T.; Bosch, J.; Meckenstock, R.U.; Sethi, R. Transport of ferrihydrite nanoparticles in saturated porous media: Role of ionic strength and flow rate. *Environ. Sci. Technol.* **2012**, *46*, 4008–4015. [CrossRef]

102. Xiang, A.; Yan, W.; Koel, B.E.; Jaffé, P.R. Poly(acrylic acid) coating induced 2-line ferrihydrite nanoparticle transport in saturated porous media. *J. Nanopart. Res.* **2013**, *15*, 1705. [CrossRef]

103. Pariona, N.; Martinez, A.I.; Hdz-García, H.; Cruz, L.A.; Hernandez-Valdes, A. Effects of hematite and ferrihydrite nanoparticles on germination and growth of maize seedlings. *Saudi J. Biol. Sci.* **2017**, *24*, 1547–1554. [CrossRef]

104. Majidi, S.; Zeinali Sehrig, F.; Farkhani, S.M.; Soleymani Goloujeh, M.; Akbarzadeh, A. Current methods for synthesis of magnetic nanoparticles. *Artif. Cells Nanomed. Biotechnol.* **2016**, *44*, 722–734. [CrossRef]

105. Döpke, C.; Grothe, T.; Steblinski, P.; Klöcker, M.; Sabantina, L.; Kosmalska, D.; Blachowicz, T.; Ehrmann, A. Magnetic nanofiber mats for data storage and transfer. *Nanomaterials* **2019**, *9*, 92. [CrossRef] [PubMed]

106. Domingos, D.G.; Henriques, R.O.; Xavier, J.A.; Junior, N.L.; da Costa, R.H.R. Increasing activated sludge aggregation by magnetite nanoparticles addition. *Water Sci. Technol.* **2019**, *79*, 993–999. [CrossRef]

107. Alfaro, I.; Molina, L.; González, P.; Gaete, J.; Valenzuela, F.; Marco, J.F.; Sáez, C.; Basualto, C. Silica-coated magnetite nanoparticles functionalized with betaine and their use as an adsorbent for Mo(VI) and Re(VII) species from acidic aqueous solutions. *J. Ind. Eng. Chem.* **2019**, *78*, 271–283. [CrossRef]

108. Molina, L.; Gaete, J.; Alfaro, I.; Ide, V.; Valenzuela, F.; Parada, J.; Basualto, C. Synthesis and characterization of magnetite nanoparticles functionalized with organophosphorus compounds and its application as an adsorbent for La(III), Nd(III) and Pr(III) ions from aqueous solutions. *J. Mol. Liq.* **2019**, *275*, 178–191. [CrossRef]

109. Veisi, H.; Razeghi, S.; Mohammadi, P.; Hemmati, S. Silver nanoparticles decorated on thiol-modified magnetite nanoparticles (Fe_3O_4/SiO_2-Pr-S-Ag) as a recyclable nanocatalyst for degradation of organic dyes. *Mater. Sci. Eng. C* **2019**, *97*, 624–631. [CrossRef] [PubMed]

110. Guo, T.; Lin, M.; Huang, J.; Zhou, C.; Tian, W.; Yu, H.; Jiang, X.; Ye, J.; Shi, Y.; Xiao, Y. The recent advances of magnetic nanoparticles in Medicine. *J. Nanomater.* **2018**, *2018*, 7805147. [CrossRef]

111. Thanh, B.T.; Hai, T.H.; Van, P.H.; Minh Tung, L.; Lee, J. Detection of hepatitis B surface antigen by immunoassay using magnetite nanoparticles binding hepatitis B surface antibody. *Geosystem Eng.* **2019**, *22*, 206–213. [CrossRef]

112. Gupta, A.K.; Wells, S. Surface-modified superparamagnetic nanoparticles for drug delivery: Preparation, characterization, and cytotoxicity studies. *IEEE Trans. NanoBiosci.* **2004**, *3*, 66–73. [CrossRef]

113. Ebrahiminezhad, A.; Davaran, S.; Rasoul-Amini, S.; Barar, J.; Moghadam, M.; Ghasemi, Y. Synthesis, characterization and anti-*Listeria monocytogenes* effect of amino acid coated magnetite nanoparticles. *Curr. Nanosci.* **2012**, *8*, 868–874. [CrossRef]

114. Ebrahiminezhad, A.; Ghasemi, Y.; Rasoul-Amini, S.; Barar, J.; Davaran, S. Impact of amino-acid coating on the synthesis and characteristics of iron-oxide nanoparticles (IONs). *Bull. Korean Chem. Soc.* **2012**, *33*, 3957–3962. [CrossRef]

115. Ebrahiminezhad, A.; Ghasemi, Y.; Rasoul-Amini, S.; Barar, J.; Davaran, S. Preparation of novel magnetic fluorescent nanoparticles using amino acids. *Colloids Surf. B.* **2013**, *102*, 534–539. [CrossRef]

116. Ebrahiminezhad, A.; Rasoul-Amini, S.; Davaran, S.; Barar, J.; Ghasemi, Y. Impacts of iron oxide nanoparticles on the invasion power of *Listeria monocytogenes*. *Curr. Nanosci.* **2014**, *10*, 382–388. [CrossRef]

117. Ebrahiminezhad, A.; Rasoul-Amini, S.; Kouhpayeh, A.; Davaran, S.; Barar, J.; Ghasemi, Y. Impacts of amine functionalized iron oxide nanoparticles on HepG2 cell line. *Curr. Nanosci.* **2015**, *11*, 113–119. [CrossRef]

118. Ebrahiminezhad, A.; Varma, V.; Yang, S.; Berenjian, A. Magnetic immobilization of *Bacillus subtilis* natto cells for menaquinone-7 fermentation. *Appl. Microbiol. Biotechnol.* **2016**, *100*, 173–180. [CrossRef] [PubMed]

119. Ebrahiminezhad, A.; Varma, V.; Yang, S.; Ghasemi, Y.; Berenjian, A. Synthesis and application of amine functionalized iron oxide nanoparticles on menaquinone-7 fermentation: A step towards process intensification. *Nanomaterials* **2015**, *6*, 1. [CrossRef] [PubMed]

120. Gholami, A.; Rasoul-Amini, S.; Ebrahiminezhad, A.; Abootalebi, N.; Niroumand, U.; Ebrahimi, N.; Ghasemi, Y. Magnetic properties and antimicrobial effect of amino and lipoamino acid coated iron oxide nanoparticles. *Minerva Biotecnol.* **2016**, *28*, 177–186.

121. Gholami, A.; Rasoul-amini, S.; Ebrahiminezhad, A.; Seradj, S.H.; Ghasemi, Y. Lipoamino acid coated superparamagnetic iron oxide nanoparticles concentration and time dependently enhanced growth of human hepatocarcinoma cell line (Hep-G2). *J. Nanomater.* **2015**, *2015*, 451405. [CrossRef]

122. Berry, C.C.; Wells, S.; Charles, S.; Curtis, A.S.G. Dextran and albumin derivatised iron oxide nanoparticles: Influence on fibroblasts in vitro. *Biomaterials* **2003**, *24*, 4551–4557. [CrossRef]

123. Raee, M.J.; Ebrahiminezhad, A.; Gholami, A.; Ghoshoon, M.B.; Ghasemi, Y. Magnetic immobilization of recombinant *E. coli* producing extracellular asparaginase: An effective way to intensify downstream process. *Sep. Sci. Technol.* **2018**, *53*, 1397–1404. [CrossRef]

124. Ranmadugala, D.; Ebrahiminezhad, A.; Manley-Harris, M.; Ghasemi, Y.; Berenjian, A. Iron oxide nanoparticles in modern microbiology and biotechnology. *Crit. Rev. Microbiol.* **2017**, *43*, 493–507. [CrossRef]

125. Ranmadugala, D.; Ebrahiminezhad, A.; Manley-Harris, M.; Ghasemi, Y.; Berenjian, A. Magnetic immobilization of bacteria using iron oxide nanoparticles. *Biotechnol. Lett.* **2017**, *40*, 237–248. [CrossRef]

126. Seifan, M.; Sarmah, A.K.; Samani, A.K.; Ebrahiminezhad, A.; Ghasemi, Y.; Berenjian, A. Mechanical properties of bio self-healing concrete containing immobilized bacteria with iron oxide nanoparticles. *Appl. Microbiol. Biotechnol.* **2018**, *102*, 4489–4498. [CrossRef]

127. Seifan, M.; Ebrahiminezhad, A.; Ghasemi, Y.; Berenjian, A. Microbial calcium carbonate precipitation with high affinity to fill the concrete pore space: Nanobiotechnological approach. *Bioprocess Biosyst. Eng.* **2019**, *42*, 37–46. [CrossRef] [PubMed]

128. Seifan, M.; Ebrahiminezhad, A.; Ghasemi, Y.; Samani, A.K.; Berenjian, A. Amine-modified magnetic iron oxide nanoparticle as a promising carrier for application in bio self-healing concrete. *Appl. Microbiol. Biotechnol.* **2017**, *102*, 175–184. [CrossRef] [PubMed]

129. Seifan, M.; Ebrahiminezhad, A.; Ghasemi, Y.; Samani, A.K.; Berenjian, A. The role of magnetic iron oxide nanoparticles in the bacterially induced calcium carbonate precipitation. *Appl. Microbiol. Biotechnol.* **2018**, *102*, 3595–3606. [CrossRef] [PubMed]

130. Seifan, M.; Sarmah, A.K.; Ebrahiminezhad, A.; Ghasemi, Y.; Samani, A.K.; Berenjian, A. Bio-reinforced self-healing concrete using magnetic iron oxide nanoparticles. *Appl. Microbiol. Biotechnol.* **2018**, *102*, 2167–2178. [CrossRef] [PubMed]

131. Ernst, C.; Bartel, A.; Elferink, J.W.; Huhn, J.; Eschbach, E.; Schönfeld, K.; Feßler, A.T.; Oberheitmann, B.; Schwarz, S. Improved DNA extraction and purification with magnetic nanoparticles for the detection of methicillin-resistant *Staphylococcus aureus*. *Vet. Microbiol.* **2019**, *230*, 45–48. [CrossRef]

132. Haghighi, A.H.; Faghih, Z.; Khorasani, M.T.; Farjadian, F. Antibody conjugated onto surface modified magnetic nanoparticles for separation of HER2+ breast cancer cells. *J. Magn. Magn. Mater.* **2019**, *490*, 165479. [CrossRef]

133. Orhan, H.; Evli, S.; Dabanca, M.B.; Başbülbül, G.; Uygun, M.; Uygun, D.A. Bacteria killer enzyme attached magnetic nanoparticles. *Mater. Sci. Eng. C* **2019**, *94*, 558–564. [CrossRef]

134. Berovic, M.; Berlot, M.; Kralj, S.; Makovec, D. A new method for the rapid separation of magnetized yeast in sparkling wine. *Biochem. Eng. J.* **2014**, *88*, 77–84. [CrossRef]

135. Taghizadeh, S.-M.; Ebrahiminezhad, A.; Ghoshoon, M.B.; Dehshahri, A.; Berenjian, A.; Ghasemi, Y. Magnetic immobilization of *Pichia pastoris* cells for the production of recombinant human serum albumin. *Nanomaterials* **2020**, *10*, 111. [CrossRef]

136. Taghizadeh, S.M.; Berenjian, A.; Chew, K.W.; Show, P.L.; Mohd Zaid, H.F.; Ramezani, H.; Ghasemi, Y.; Raee, M.J.; Ebrahiminezhad, A. Impact of magnetic immobilization on the cell physiology of green unicellular algae *Chlorella vulgaris*. *Bioengineered* **2020**, *11*, 141–153. [CrossRef]

137. Ansari, F.; Grigoriev, P.; Libor, S.; Tothill, I.E.; Ramsden, J.J. DBT degradation enhancement by decorating *Rhodococcus erythropolis* IGST8 with magnetic Fe_3O_4 nanoparticles. *Biotechnol. Bioeng.* **2009**, *102*, 1505–1512. [CrossRef] [PubMed]

138. Ranmadugala, D.; Ebrahiminezhad, A.; Manley-Harris, M.; Ghasemi, Y.; Berenjian, A. Impact of 3–aminopropyltriethoxysilane-coated iron oxide nanoparticles on menaquinone-7 production using *B. subtilis*. *Nanomaterials* **2017**, *7*, 350. [CrossRef] [PubMed]

139. Ranmadugala, D.; Ebrahiminezhad, A.; Manley-Harris, M.; Ghasemi, Y.; Berenjian, A. The effect of iron oxide nanoparticles on *Bacillus subtilis* biofilm, growth and viability. *Process Biochem.* **2017**, *62*, 231–240. [CrossRef]

140. Ranmadugala, D.; Ebrahiminezhad, A.; Manley-Harris, M.; Ghasemi, Y.; Berenjian, A. Reduced biofilm formation in Menaquinone-7 production process by optimizing the composition of the cultivation medium. *Trends Pharma. Sci.* **2017**, *3*, 245–254.

141. Taghizadeh, S.-M.; Jafari, S.; Ahmad-Kiadaliri, T.; Mobasher, M.A.; Lal, N.; Raee, M.J.; Berenjian, A.; Ghasemi, Y.; Ebrahiminezhad, A. Magnetic immobilisation as a promising approach against bacteriophage infection. *Mater. Res. Express.* **2019**, *6*. [CrossRef]

142. Dieudonné, A.; Pignol, D.; Prévéral, S. Magnetosomes: Biogenic iron nanoparticles produced by environmental bacteria. *Appl. Microbiol. Biotechnol.* **2019**, *103*, 3637–3649. [CrossRef]

143. Yamazaki, T.; Suzuki, Y.; Kouduka, M.; Kawamura, N. Dependence of bacterial magnetosome morphology on chemical conditions in deep-sea sediments. *Earth. Planet. Sci. Lett.* **2019**, *513*, 135–143. [CrossRef]

144. Gorobets, S.; Gorobets, O.; Bulaievska, M.; Sharay, I. Detection of biogenic magnetic nanoparticles in ethmoid bones of migratory and non-migratory fishes. *SN Appl. Sci.* **2019**, *1*, 63. [CrossRef]

145. Gorobets, S.; Gorobets, O.; Magerman, A.; Gorobets, Y.; Sharay, I. Biogenic magnetic nanoparticles in plants. *arXiv* **2019**, arXiv:1901.07212.

146. Murros, K.E.; Salminen, J.; Wasiljeff, J.; Macías-Sánchez, E.; Soinne, L.S.; Faivre, D.; Valtonen, J.; Pohja, M.; Saari, P.; Pesonen, L. Magnetic nanoparticles in human cervical skin. *Front. Med.* **2019**, *6*, 123. [CrossRef]

147. Van de Walle, A.; Sangnier, A.P.; Abou-Hassan, A.; Curcio, A.; Hémadi, M.; Menguy, N.; Lalatonne, Y.; Luciani, N.; Wilhelm, C. Biosynthesis of magnetic nanoparticles from nano-degradation products revealed in human stem cells. *Proc. Natl. Acad. Sci. USA* **2019**, *116*, 4044–4053. [CrossRef] [PubMed]

148. Brok, E.; Frandsen, C.; Madsen, D.E.; Jacobsen, H.; Birk, J.; Lefmann, K.; Bendix, J.; Pedersen, K.; Boothroyd, C.; Berhe, A. Magnetic properties of ultra-small goethite nanoparticles. *J. Phys. D Appl. Phys.* **2014**, *47*, 365003. [CrossRef]

149. van der Zee, C.; Roberts, D.R.; Rancourt, D.G.; Slomp, C.P. Nanogoethite is the dominant reactive oxyhydroxide phase in lake and marine sediments. *Geology* **2003**, *31*, 993–996. [CrossRef]

150. Ghanbariasad, A.; Taghizadeh, S.-M.; Show, P.L.; Nomanbhay, S.; Berenjian, A.; Ghasemi, Y.; Ebrahiminezhad, A. Controlled synthesis of iron oxyhydroxide (FeOOH) nanoparticles using secretory compounds from *Chlorella vulgaris* microalgae. *Bioengineered* **2019**, *10*, 390–396. [CrossRef] [PubMed]

151. Nguyen, V.; Kynicky, J.; Ambrozova, P.; Adam, V. Microwave-Assisted Synthesis of Goethite Nanoparticles Used for Removal of Cr (VI) from Aqueous Solution. *Materials* **2017**, *10*, 783. [CrossRef] [PubMed]

152. Ghosh, M.K.; Poinern, G.E.J.; Issa, T.B.; Singh, P. Arsenic adsorption on goethite nanoparticles produced through hydrazine sulfate assisted synthesis method. *Korean J. Chem. Eng.* **2012**, *29*, 95–102. [CrossRef]

153. Kar, S.; Equeenuddin, S.M. Adsorption of hexavalent chromium using natural goethite: Isotherm, thermodynamic and kinetic study. *J. Geol. Soc. India* **2019**, *93*, 285–292. [CrossRef]

154. Adegoke, H.; Adekola, F.; Ashola, M. Adsorptive removal of hexavalent chromium using synthetic goethite nanoparticles. *Niger. J. Chem. Res.* **2018**, *23*, 20–38.

155. Tiraferri, A.; Hernandez, L.A.S.; Bianco, C.; Tosco, T.; Sethi, R. Colloidal behavior of goethite nanoparticles modified with humic acid and implications for aquifer reclamation. *J. Nanopart. Res.* **2017**, *19*, 107. [CrossRef]

156. Kalishyn, Y.; Bychko, I.; Polunkin, E.; Kameneva, T.; Strizhak, P. Enhancement in the oxidative stability of commercial gasoline fuel by the goethite nanoparticles. *ChemRxiv. Preprint* **2018**. [CrossRef]

157. Zhao, J.; Lin, W.; Chang, Q.; Li, W.; Lai, Y. Adsorptive characteristics of akaganeite and its environmental applications: A review. *Environ. Technol. Rev.* **2012**, *1*, 114–126. [CrossRef]

158. Kim, J.; Grey, C.P. ^{2}H and ^{7}Li solid-state MAS NMR study of local environments and lithium adsorption on the iron(III) oxyhydroxide, akaganeite (β-FeOOH). *Chem. Mater.* **2010**, *22*, 5453–5462. [CrossRef]

159. Mohapatra, J.; Mitra, A.; Tyagi, H.; Bahadur, D.; Aslam, M. Iron oxide nanorods as high-performance magnetic resonance imaging contrast agents. *Nanoscale* **2015**, *7*, 9174–9184. [CrossRef] [PubMed]

160. Kasparis, G.; Erdocio, A.S.; Tuffnell, J.M.; Thanh, N.T.K. Synthesis of size-tuneable β-FeOOH nanoellipsoids and a study of their morphological and compositional changes by reduction. *CrystEngComm* **2019**, *21*, 1293–1301. [CrossRef]

161. Yuan, Z.-Y.; Su, B.-L. Surfactant-assisted nanoparticle assembly of mesoporous β-FeOOH (akaganeite). *Chem. Phys. Lett.* **2003**, *381*, 710–714. [CrossRef]

162. Faria, M.C.; Rosemberg, R.S.; Bomfeti, C.A.; Monteiro, D.S.; Barbosa, F.; Oliveira, L.C.; Rodriguez, M.; Pereira, M.C.; Rodrigues, J.L. Arsenic removal from contaminated water by ultrafine δ-FeOOH adsorbents. *Chem. Eng. J.* **2014**, *237*, 47–54. [CrossRef]

163. Zhang, Y.-X.; Jia, Y. A facile solution approach for the synthesis of akaganéite (β-FeOOH) nanorods and their ion-exchange mechanism toward As(V) ions. *Appl. Surf. Sci.* **2014**, *290*, 102–106. [CrossRef]

164. Legg, B.A. The Formation and Aggregation of Iron Oxyhydroxide Nanoparticles in the Aqueous Environment. Ph.D. Thesis, UC Berkeley, Berkeley, CA, USA, 2013.

165. Majzlan, J.; Mazeina, L.; Navrotsky, A. Enthalpy of water adsorption and surface enthalpy of lepidocrocite (γ-FeOOH). *Geochim. Cosmochim. Acta* **2007**, *71*, 615–623. [CrossRef]

166. Vindedahl, A.M.; Strehlau, J.H.; Arnold, W.A.; Penn, R.L. Organic matter and iron oxide nanoparticles: Aggregation, interactions, and reactivity. *Environ. Sci. Nano* **2016**, *3*, 494–505. [CrossRef]

167. Navarro, G.; Acevedo, R.; Soto, A.; Herane, M. Synthesis and characterization of lepidocrocite and its potential applications in the adsorption of pollutant species. *J. Phys. Conf. Ser.* **2008**, *134*, 012023. [CrossRef]

168. Lee, G.; Kim, S.; Choi, B.; Huh, S.; Chang, Y.; Kim, B.; Park, J.; Oh, S. Magnetic properties of needle-like α-FeOOH and γ-FeOOH nanoparticles. *J. Korean Phys. Soc.* **2004**, *45*, 1019–1024.

169. Ewing, F. The crystal structure of lepidocrocite. *J. Chem. Phys.* **1935**, *3*, 420–424. [CrossRef]

170. Liu, A.; Liu, J.; Pan, B.; Zhang, W.-x. Formation of lepidocrocite (γ-FeOOH) from oxidation of nanoscale zero-valent iron (nZVI) in oxygenated water. *RSC Adv.* **2014**, *4*, 57377–57382. [CrossRef]

171. Agarwal, A.; Joshi, H.; Kumar, A. Synthesis, characterization and application of nano lepidocrocite and magnetite in the degradation of carbon tetrachloride. *S. Afr. J. Chem.* **2011**, *64*, 218–224.

172. Sheydaei, M.; Aber, S. Preparation of nano-lepidocrocite and an investigation of its ability to remove a metal complex dye. *Clean* **2013**, *41*, 890–898. [CrossRef]

173. Alexandratos, V.G.; Behrends, T.; Van Cappellen, P. Fate of adsorbed U(VI) during sulfidization of lepidocrocite and hematite. *Environ. Sci. Technol.* **2017**, *51*, 2140–2150. [CrossRef]

174. O'Loughlin, E.J.; Gorski, C.A.; Flynn, T.M.; Scherer, M.M. Electron donor utilization and secondary mineral formation during the bioreduction of lepidocrocite by *Shewanella putrefaciens* CN32. *Minerals* **2019**, *9*, 434. [CrossRef]

175. Koch, C.B.; Oxborrow, C.; Mørup, S.; Madsen, M.; Quinn, A.; Coey, J. Magnetic properties of feroxyhyte (δ-FeOOH). *Phys. Chem. Miner.* **1995**, *22*, 333–341. [CrossRef]

176. Chen, P.; Xu, K.; Li, X.; Guo, Y.; Zhou, D.; Zhao, J.; Wu, X.; Wu, C.; Xie, Y. Ultrathin nanosheets of feroxyhyte: A new two-dimensional material with robust ferromagnetic behavior. *Chem. Sci.* **2014**, *5*, 2251–2255. [CrossRef]

177. Chukhrov, F.; Zvyagin, B.; Gorshkov, A.; Yermilova, L.; Korovushkin, V.; Rudnitskaya, Y.S.; Yakubovskaya, N.Y. Feroxyhyte, a new modification of FeOOH. *Int. Geol. Rev.* **1977**, *19*, 873–890. [CrossRef]

178. Nishida, N.; Amagasa, S.; Kobayashi, Y.; Yamada, Y. Synthesis of Cu-doped δ-FeOOH nanoparticles by a wet chemical method. *J. Nanopart. Res.* **2018**, *20*, 181. [CrossRef]

179. Nishida, N.; Amagasa, S.; Ito, H.; Kobayashi, Y.; Yamada, Y. Manganese-doped feroxyhyte nano-urchins produced by chemical methods. *Hyperfine Interact.* **2018**, *239*, 33. [CrossRef]

180. Jurkin, T.; Štefanić, G.; Dražić, G.; Gotić, M. Synthesis route to δ-FeOOH nanodiscs. *Mater. Lett.* **2016**, *173*, 55–59. [CrossRef]

181. Pinto, I.S.; Pacheco, P.H.; Coelho, J.V.; Lorençon, E.; Ardisson, J.D.; Fabris, J.D.; de Souza, P.P.; Krambrock, K.W.; Oliveira, L.C.; Pereira, M.C. Nanostructured δ-FeOOH: An efficient Fenton-like catalyst for the oxidation of organics in water. *Appl. Catal. B* **2012**, *119*, 175–182. [CrossRef]

182. Chagas, P.; Da Silva, A.C.; Passamani, E.C.; Ardisson, J.D.; de Oliveira, L.C.A.; Fabris, J.D.; Paniago, R.M.; Monteiro, D.S.; Pereira, M.C. δ-FeOOH: A superparamagnetic material for controlled heat release under AC magnetic field. *J. Nanopart. Res.* **2013**, *15*, 1544. [CrossRef]

183. Li, Y.; Fu, F.; Cai, W.; Tang, B. Synergistic effect of mesoporous feroxyhyte nanoparticles and Fe(II) on phosphate immobilization: Adsorption and chemical precipitation. *Powder Technol.* **2019**, *345*, 786–795. [CrossRef]

184. Pinakidou, F.; Katsikini, M.; Paloura, E.; Simeonidis, K.; Mitraka, E.; Mitrakas, M. Monitoring the role of Mn and Fe in the As-removal efficiency of tetravalent manganese feroxyhyte nanoparticles from drinking water: An X-ray absorption spectroscopy study. *J. Colloid Interface Sci.* **2016**, *477*, 148–155. [CrossRef] [PubMed]

185. Kokkinos, E.; Soukakos, K.; Kostoglou, M.; Mitrakas, M. Cadmium, mercury, and nickel adsorption by tetravalent manganese feroxyhyte: Selectivity, kinetic modeling, and thermodynamic study. *Environ. Sci. Pollut. Res.* **2018**, *25*, 12263–12273. [CrossRef] [PubMed]

186. Tresintsi, S.; Simeonidis, K.; Estradé, S.; Martinez-Boubeta, C.; Vourlias, G.; Pinakidou, F.; Katsikini, M.; Paloura, E.C.; Stavropoulos, G.; Mitrakas, M. Tetravalent manganese feroxyhyte: A novel nanoadsorbent equally selective for As(III) and As(V) removal from drinking water. *Environ. Sci. Technol.* **2013**, *47*, 9699–9705. [CrossRef]

187. Hu, J.; Lo, I.M.; Chen, G. Performance and mechanism of chromate (VI) adsorption by δ-FeOOH-coated maghemite (γ-Fe$_2$O$_3$) nanoparticles. *Sep. Purif. Technol.* **2007**, *58*, 76–82. [CrossRef]

188. Pereira, M.C.; Garcia, E.M.; da Silva, A.C.; Lorençon, E.; Ardisson, J.D.; Murad, E.; Fabris, J.D.; Matencio, T.; de Castro Ramalho, T.; Rocha, M.V.J. Nanostructured δ-FeOOH: A novel photocatalyst for water splitting. *J. Mater. Chem.* **2011**, *21*, 10280–10282. [CrossRef]

189. da Silva Rocha, T.; Nascimento, E.S.; da Silva, A.C.; dos Santos Oliveira, H.; Garcia, E.M.; de Oliveira, L.C.A.; Monteiro, D.S.; Rodriguez, M.; Pereira, M.C. Enhanced photocatalytic hydrogen generation from water by Ni(OH)$_2$ loaded on Ni-doped δ-FeOOH nanoparticles obtained by one-step synthesis. *RSC Adv.* **2013**, *3*, 20308–20314. [CrossRef]

190. Bigham, J.; Carlson, L.; Murad, E. Schwertmannite, a new iron oxyhydroxysulphate from Pyhäsalmi, Finland, and other localities. *Mineral. Mag.* **1994**, *58*, 641–648. [CrossRef]

191. Sestu, M.; Navarra, G.; Carrero, S.; Valvidares, S.M.; Aquilanti, G.; Pérez-Lopez, R.; Fernandez-Martinez, A. Whole-nanoparticle atomistic modeling of the schwertmannite structure from total scattering data. *J. Appl. Crystallogr.* **2017**, *50*, 1617–1626. [CrossRef]

192. Zhang, Z.; Bi, X.; Li, X.; Zhao, Q.; Chen, H. Schwertmannite: Occurrence, properties, synthesis and application in environmental remediation. *RSC Adv.* **2018**, *8*, 33583–33599. [CrossRef]

193. Paikaray, S.; Schröder, C.; Peiffer, S. Schwertmannite stability in anoxic Fe(II)-rich aqueous solution. *Geochim. Cosmochim. Acta* **2017**, *217*, 292–305. [CrossRef]

194. Miyata, N.; Takahashi, A.; Fujii, T.; Hashimoto, H.; Takada, J. Biosynthesis of Schwertmannite and Goethite in a bioreactor with acidophilic Fe(II)-oxidizing betaproteobacterium strain GJ-E10. *Minerals* **2018**, *8*, 98. [CrossRef]

195. Xu, Y.; Yang, M.; Yao, T.; Xiong, H. Isolation, identification and arsenic-resistance of *Acidithiobacillus ferrooxidans* HX3 producing schwertmannite. *J. Environ. Sci.* **2014**, *26*, 1463–1470. [CrossRef]

196. Song, Y.; Liu, Y.; Wang, H. Comparison of the biological and chemical synthesis of schwertmannite at a consistent Fe^{2+} oxidation efficiency and the effect of extracellular polymeric substances of *Acidithiobacillus ferrooxidans* on biomineralization. *Materials* **2018**, *11*, 1739. [CrossRef]

197. Kumpulainen, S.; Räisänen, M.-L.; Von der Kammer, F.; Hofmann, T. Ageing of synthetic and natural schwertmannites at pH 2–8. *Clay Miner.* **2008**, *43*, 437–448. [CrossRef]

198. Qiao, X.; Liu, L.; Shi, J.; Zhou, L.; Guo, Y.; Ge, Y.; Fan, W.; Liu, F. Heating changes bio-schwertmannite microstructure and arsenic(III) removal efficiency. *Minerals* **2017**, *7*, 9. [CrossRef]

199. Zhang, J.; Shi, J.; Zhang, S.; Zhou, L.; Xu, J.; Ge, Y.; Fan, W.; Liu, F. Schwertmannite adherence to the reactor wall during the bio-synthesis process and deterioration of its structural characteristics and arsenic(III) removal efficiency. *Minerals* **2017**, *7*, 64. [CrossRef]

200. Birch, W.D.; Pring, A.; Reller, A.; Schmalle, H.W. Bernalite, $Fe(OH)_3$, a new mineral from Broken Hill, New South Wales: Description and structure. *Am. Mineral.* **1993**, *78*, 827–834.

201. Welch, M.; Crichton, W.; Ross, N. Compression of the perovskite-related mineral bernalite $Fe(OH)_3$ to 9 GPa and a reappraisal of its structure. *Mineral. Mag.* **2005**, *69*, 309–315. [CrossRef]

202. Han, J.; Ro, H.-M. Identification of bernalite transformation and tridentate arsenate complex at nano-goethite under effects of drying, pH and surface loading. *Sci. Rep.* **2018**, *8*, 8369. [CrossRef]

203. Grundl, T.; Delwiche, J. Kinetics of ferric oxyhydroxide precipitation. *J. Contam. Hydrol.* **1993**, *14*, 71–87. [CrossRef]

204. Chen, Z. Synthesis of monodisperse organic-and water-soluble iron oxide nanoparticles using $Fe(OH)_3$ as precursor. *J. Exp. Nanosci.* **2014**, *9*, 406–414. [CrossRef]

205. Pinto, P.S.; Lanza, G.D.; Ardisson, J.D.; Lago, R.M. Controlled dehydration of $Fe(OH)_3$ to Fe_2O_3: Developing mesopores with complexing iron species for the adsorption of β-Lactam antibiotics. *J. Braz. Chem. Soc.* **2019**, *30*, 310–317. [CrossRef]

206. Thuy, L.T.X.; Cuong, L.P.; Yabutani, T. Removal of trace metals from aqueous solution by $Fe(OH)_3$ coprecipitation and flotation using poly-glutamic acid. *Int. Res. J. Pure Appl. Chem.* **2014**, *4*, 797–804. [CrossRef]

207. Arefi, M.; Saberi, D.; Karimi, M.; Heydari, A. Superparamagnetic $Fe(OH)_3@Fe_3O_4$ nanoparticles: An efficient and recoverable catalyst for tandem oxidative amidation of alcohols with amine hydrochloride salts. *ACS Comb. Sci.* **2015**, *17*, 341–347. [CrossRef]

208. Yan, J.; Tang, H.; Lin, Z.; Anjum, M.N.; Zhu, L. Efficient degradation of organic pollutants with ferrous hydroxide colloids as heterogeneous Fenton-like activator of hydrogen peroxide. *Chemosphere* **2012**, *87*, 111–117. [CrossRef]

 *processes*

Article

Effects of Ethanol Concentration on Organosolv Lignin Precipitation and Aggregation from *Miscanthus x giganteus*

Muhammad Hazwan Hamzah [1,2,3,*] , **Steve Bowra** [4] **and Philip Cox** [5]

1 Department of Biological and Agricultural Engineering, Faculty of Engineering, Universiti Putra Malaysia, UPM Serdang 43400, Selangor, Malaysia
2 SMART Farming Technology Research Centre, Faculty of Engineering, Universiti Putra Malaysia, UPM Serdang 43400, Selangor, Malaysia
3 School of Chemical Engineering, University of Birmingham, Edgbaston B15 2TT, UK
4 Phytatec (UK) Ltd., Plas Gogerddan, Aberystwyth SY23 3EB, UK; steve.bowra@phytatec.com
5 Chemical Engineering, University of Wolverhampton, Wulfruna Street, Wolverhampton WV1 1LY, UK; P.W.Cox@wlv.ac.uk
* Correspondence: hazwanhamzah@upm.edu.my; Tel.: +603-97696422

Received: 23 April 2020; Accepted: 19 June 2020; Published: 16 July 2020

Abstract: This work assesses the behavior of organosolv lignin aggregates derived from *Miscanthus x giganteus* using different ethanol concentrations (10%, 25%, 50%, and 75% by volume). The percentage of lignin recovery was found to decrease from 75.8% to 71.4% and 25.1%, as the ethanol concentration was increased from 10% to 25% and 50%, respectively. Increasing the ethanol concentration further to 75% led to zero recovery. The purity of the precipitated lignin was consistently found to be ≥90%. Lignin derived from the dried supernatant obtained at 50% ethanol concentration resulted in high lignin purity (51.6%) in comparison with the other ethanol concentrations used. Fourier transform infrared spectroscopy analysis showed that the precipitated lignin and dried supernatant at 50% ethanol concentration possessed the highest peak intensity apportioned to wavenumber of lignin as compared to that of at 25% and 10% ethanol concentrations, and the results linked with the percentage of lignin purity. The results of particle size analysis for precipitated lignin demonstrated particle sizes of 306, 392, and 2050 nm for 10%, 25%, and 50% ethanol concentrations, respectively, and the remaining supernatant with average particle sizes of 1598, 1197, and 875 nm, respectively. These results were verified with the morphology of lignin macromolecules in scanning electron microscopy images. Results of the particle size distribution of lignin revealed that the overall size of lignin aggregates decreased with decreasing ethanol concentration. In summary, these findings suggest that ethanol concentration affected the behavior of lignin aggregates in water–ethanol solution.

Keywords: lignocellulosic; organosolv; lignin; aggregates; purity; concentration

1. Introduction

The demand for finite resources such as fossil fuels and natural gas are growing in most countries in the world in spite of the current energy crisis. The increasing awareness of the need for renewable and sustainable sources of energy has driven interest in lignocellulosic second-generation bioethanol. *Miscanthus* sp., a genus comprising of about 25 species has been proven as one of the biomass crops having high biomass energy potentials [1]. *Miscanthus x giganteus (MxG)*, a hybrid between *Miscanthus sinensis* and *Miscanthus sacchrisflorus*, is a promising high-yield lignocellulosic biomass crop that is currently used as a solid fuel used in co-firing power stations [2]. When compared with other genotypes, *MxG* has a wide range of potential benefits including the possibly unique and exclusive trait for adaptation to climate and environmental conditions, low levels of nutrients needed, soil carbon sequestration, and ease of harvesting and handling [3–6]. *MxG* also has an efficient rhizome system, which plays a key role as a nutrient reserve for the annual shoot growth in the growing season [7]. *Miscanthus* sp., which can be obtained mainly from the Europe and the United States, has enormous potential to become a feedstock to support second-generation bioethanol production. Similarly, the concept of biorefining has been identified as a method whereby cellulose, the substrate for ethanol production, can be recovered along with other renewable bio-based chemical building blocks, hemicellulose and lignin, thus potentially improving the overall economy of bioethanol production.

Within the integrated biorefinery concept, an area of research that has gained significant interest is the processing of lignin which, as mentioned earlier, is the second most abundant natural polymer after cellulose and hemicellulose [8]. Currently, lignin is traditionally viewed as the by-product of paper and pulp processes. However, with the emergence of second generation bioethanol production, large quantities of lignin will be created. Lignin isolated via different extraction methods can vary widely in terms of chemical composition and molecular structure. The differences also affect the physical properties such as solubility and particle size. Therefore, in the context growing interest in developing value added uses for lignin, this work focuses on characterizing lignin extracted using a modified organosolv method [9], particularly emphasizing the formation of aggregates. There is insufficient information available that describes the association behavior of lignin in solution, which depends on the solvent properties and lignin structure [10].

Understanding and interpreting the assembly of lignin macromolecules in solution are relevant and significant, as the heterogeneity and complexity of lignin structure, unique chemical reactivity, and the unknown molecular characteristics of lignin become the greatest bottlenecks in the utilization of lignin in producing various useful and renewable materials in the industry [11–13]. Previously, numerous studies on lignin aggregates have been conducted in conjunction with different types of methods and sources of lignin such as the aggregation and assembly of alkali lignin using iodine as probe [14], the impact of lignin source on its self-assembly in d-DMSO solution [10], and the aggregation of acetylated lignins in N,N-dimetylacetamide in the presence of salts [15]. A crucial physicochemical property of organosolv lignin which becomes the subject matter for this work is the higher solubility of organosolv lignin in organic solvents [16]. Organosolv lignin has a tendency to aggregate in most solvents, affecting the lignin recovery process, the biodegradation processes, and the preparation of lignin-based materials.

In this work, the resulting soluble lignin extract from delignification process was fractionated according to different ethanol solubilities under centrifugation, thereby generating two fractions: the precipitated fraction and supernatant fraction. Characterization of the resulting fractions was carried out by Klason lignin assay, Fourier transform infrared spectroscopy (FTIR), scanning electron miscroscopy (SEM), and particle size analysis. Specifically, the study compared the purity, lignin recovery, chemical structure, and especially particle size and morphology characterization for the resulting fractions.

2. Materials and Methods

2.1. Materials

The lignocellulosic biomass used was air-dried. The *MxG* was provided by the Institute of Biological, Environmental and Rural Sciences (IBERS, UK) in collaboration with Phytatec (UK) Ltd. The biomass was harvested in Aberystwyth, Wales, United Kingdom and kept in a cool, dry, and dark place throughout the study. Nitrogen (compressed oxygen free nitrogen, BOC, UK), and carbon dioxide (vapor withdrawal, BOC, UK) had ≥99.8% purity. Sulfuric acid (72%) (Fluka-Sigma Aldrich, UK) and absolute ethanol (Fisher Scientific, UK) were used as reagents.

2.2. Methodology

2.2.1. Biomass Preparation

Prior to hydrolysis, the *MxG* was mixed in distilled water and then warmed to 50 °C to soften the grass. The mixture was then soaked for 20 min to rehydrate the grass. The mixture was milled for 3 min in a domestic blender to reduce the particle size of the material. The grinding conditions of the temperature, soaking time, grinding time, and the solid-to-liquid ratio were previously optimized to yield an average particle size of 500 μm [17].

The *MxG* slurry was placed inside the reactor directly after the sample preparation at 120 °C. The sequentially processed *MxG* obtained at 120 °C was used for biomass hydrolysis at 180 °C and 200 °C. Subsequently, the *MxG* was mixed in water and a 1:1 ethanol-water solution for 180 °C and 200 °C, respectively, by warming to 50 °C and for a wetting time of 5 min prior to each hydrolysis step.

2.2.2. Biomass Hydrolysis

The steps taken in extracting lignin in this work are outlined in Figure 1.

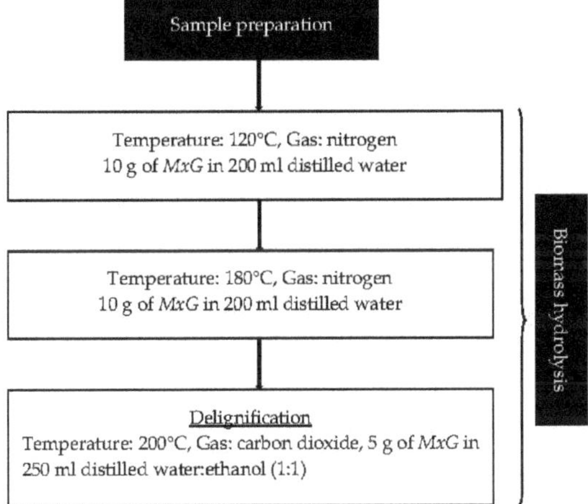

Figure 1. Steps taken in lignin extraction.

The *MxG* slurry was transferred to a 500 mL stirred pressure vessel (Alloy C276, Parr, IL, USA). The reactor was closed and pressurized with desired gas to 50 bar. The set point temperature was increased to the set temperature and was kept stable during the reaction time by a controller (4386, Parr). After the reaction, the reactor temperature was decreased through a cooling system with a cooling coil inside the pressure vessel in which a coolant flew at an initial −7 °C. When the temperature fell below 50 °C, the reactor was depressurized slowly to atmospheric pressure before the reactor was opened. Finally, the solution and solid fibers were separated using a laboratory test sieve (BS410-1 size 45 μm, Endecotts Ltd., England) for biomass hydrolysis at 120 °C and 180 °C. The solid fibers were placed in the drying cabinet (65 °C) until it reached constant weight. The liquid fraction or filtrate for biomass hydrolysis at 200 °C was recovered by vacuum filtration through a Pyrex sintered disc of porosity 2, rinsed with mixture of distilled water: ethanol (1:1).

The *MxG* was treated through a three-stage temperature profile sequential batch extraction method adapted from Hamzah et al. [9] to differentially separate extractives, hemicellulose, cellulose, and lignin. The first step applied subcritical water (SCW) at 120 °C with an equilibrium time of 30 min and 50 bar of nitrogen gas to remove water-soluble extractives that could have interfered with the isolation and later analytical steps. The second step used a SCW at a regime of 180 °C and a reaction time of

30 min under 50 bar of nitrogen gas to hydrolyze hemicelluloses prior to delignification. The final step involved lignin extraction via a SCW with associated modifiers using a 1:1 ethanol–water mixture at 200 °C, a reaction time of 60 min, and 50 bar of carbon dioxide gas. The nitrogen gas was used in SCW to maintain the constant pressure inside the reactor and to ensure that water remain as liquid [18]. In lignin extraction, carbon dioxide under pressure creates carbonic acid that serves as a catalyst for hydrolysis reaction [17,19].

2.2.3. Lignin Precipitation

The lignin precipitation method was adapted from Roque [17]. The 50% ethanol concentration of soluble lignin extract obtained after vacuum filtration was placed in a freezer at −20 °C for 2 h, after which the ethanol concentration was adjusted to either 10% and 25% by adding distilled water, and 75% by adding ethanol. The lignin was recovered using a J2-21 centrifuge (Beckman, Indianapolis, IN, USA) with a JA-10 rotor at 4 °C and at 10,000 revolutions per minute with 17,700 relative centrifugal force (RCF) for 10 min. The remaining supernatant was dried at 65 °C for further Klason lignin assay and FTIR analysis. The resulting precipitated lignin was air-dried and stored in 2 mL Eppendorf tubes at room temperature and later analyzed by a Klason lignin assay and FTIR analysis. The amount of precipitated lignin was compared to the total amount of lignin in the soluble fraction (precipitated lignin and dried supernatant), which gives the percentage of lignin recovery using Equation (1):

$$\text{Percentage of lignin recovery } (\%) = \frac{A_l}{A_l + A_s} \times 100\%, \tag{1}$$

where A_l is the amount of precipitated lignin (g) and A_s is the amount of lignin derived from dried supernatant (g).

2.2.4. Klason Lignin Determination

After the hydrolysis experiments, the precipitated lignin and dried supernatant was analyzed for lignin content using the Klason lignin assay following the Determination of Structural Carbohydrates and Lignin in Biomass Laboratory Analytical Procedure (NREL/TP-510-42618) [20].

2.2.5. FTIR Analysis

The FTIR analysis was carried out on the samples without any special pre-treatment. The IR spectra were determined using a spectrometer (FTIR-6300, JASCO, Easton, MD, USA) over a wavenumber ranging from 4000 cm^{-1} to 600 cm^{-1}. The resolution was 4 cm^{-1} and 32 scans were averaged. The precipitated lignin and dried supernatant of different ethanol concentration were analyzed for chemical structure characterization. Principle component analysis (PCA) was performed using the UnscramblerTM Version 10.3 software (CAMO). Two different pre-processing methods (smoothing followed by normalization) were performed on each of the three repeated spectrum measurements in the regions of 4000 cm^{-1} to 600 cm^{-1}. Analysis of FTIR spectra datasets by PCA determines the differences between spectra in terms of chemical structure and composition of the samples [21].

2.2.6. SEM Analysis

SEM images of lignin and dried supernatant were captured using a Philips XL30 FEG ESEM scanning electron microscope operating at 10 kV with various magnifications. Samples were coated with platinum for 120 s using Emscope SC500 sputter coater prior to analysis.

2.2.7. Particle Size Analysis

The precipitated lignin and dried supernatant were dispersed at 10 mg mL^{-1} based on initial concentration of ethanol-water solutions (10%, 25%, and 50% ethanol concentration). To achieve a good colloidal dispersion, the samples were ultrasound-treated for 10 s at room temperature using the 500W Fisher Scientific^{TM} Model 505 Sonic Dismembrator prior to measurement. The diameter measurement was carried out in triplicate and the average reading of the results was reported.

Malvern Zetasizer Nano ZS

Lignin particle size was performed according to the protocols described with modification [22,23]. Lignin particle size was analyzed at 23 °C. The obtained size distribution represented the dependencies of the relative intensity of the scattered light on the hydrodynamic diameter of lignin particles.

Malvern Mastersizer 2000

A refractive index of 1.6 [24] and absorption of 0.01 for lignin were used by the instrument to calculate the particle size distribution [25]. The mean particle size was reported in terms of D3,2 values. The D3,2 is the surface area mean diameter and refers to the diameter of a sphere equivalent volume to surface area of the particles in the sample.

2.2.8. Statistical Analysis

SPSS software (Version 22) was used to carry out statistical analysis. Post hoc test by Tukey's honest significant different (HSD) was conducted at $\alpha = 0.05$ to determine if the results obtained at each ethanol concentration were significantly different.

3. Results and Discussion

3.1. Percentage of Lignin Recovery

Figure 2 shows an increasing trend on percentage of lignin recovery as the ethanol co-solvent becomes more dilute. The post hoc test analysis illustrated that percentage of lignin recovery using 25% ethanol concentration (71.4%) did not differ significantly from the 10% ethanol concentration (75.8%), at a 95% confidence level. Lignin precipitation using 50% ethanol concentration only recovered 25.1% lignin. Further increasing the ethanol concentration to 75% led to zero recovery. The overall statistical analysis results were tabulated in Table 1.

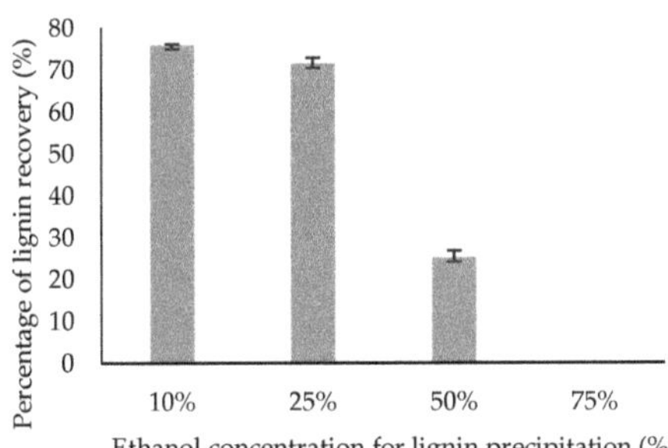

Figure 2. Percentage of lignin recovery at different ethanol concentration.

Table 1. Analysis of Tukeytest for the assessment of ethanol concentration.

Dependent Variable	Percentage of Ethanol Concentration	Comparison with Other Percentage of Ethanol Concentration	Sig.
Percentage of purity precipitated lignin	50%	25%	0.857
		10%	0.010 *
	25%	50%	0.857
		10%	0.010 *
	10%	50%	0.010 *
		25%	0.010 *
Percentage of purity dried supernatant	50%	25%	0.013 *
		10%	0.000 *
	25%	50%	0.000 *
		10%	0.525
	10%	50%	0.000 *
		25%	0.525
Percentage of lignin recovery	50%	25%	0.000 *
		10%	0.000 *
	25%	50%	0.000 *
		10%	0.100
	10%	50%	0.000 *
		25%	0.100

Note: Numbers with asterix (*) indicate statistically significant differences at the 0.05 level.

Critical precaution needs to be taken into consideration if water is added as the mixture of solution for delignification process. Based on the previous study within the research group, it is shown that the optimum delignification was achieved using ethanol:water ratio (1:1) [17] and similar concentration was used for delignification in this study. The optimal concentration of ethanol for delignification (1:1) is in agreement with the research conducted by Pasquini et al. [26]. The utilization of 100% ethanol is not preferred for delignification. The presence of both nucleophilic agents; water and ethanol produce a good solvent to promote the cleavage of lignin, and the capacity to dissolve the lignin fragments. With the addition of water, the nucleophilic agent stimulated the cleavage of the lignin but decreased the capability of the solvent to dissolve lignin in the delignification process [26]. High water content in the mixture of the solution for delignification may have demonstrated negative effects on delignification due to the nature of the hydrophobic biopolymer of lignin that could trigger adsorption of lignin fragments onto the surface of biomass fibers [27]. A suitable concentration of ethanol–water mixtures is desirable to avoid lignin re-precipitation onto the biomass fibers, thereby reducing the efficacy of delignification. Therefore, the best compromise between nucleophilicity and solubility of ethanol and water mixtures could promote good delignification.

Here, a new experiment looked at the addition of water to the soluble lignin extract was carried out after delignification for the purpose of lignin recovery study via centrifugation. In the lignin precipitation of ethanol pulping, the removal of lignin also depends on the capacity of aqueous ethanol solution to solubilise lignin fragments [26,28]. In a study with Alcell lignin and its solubility in ethanol–water mixtures, it was demonstrated that lignin solubility increased as the ethanol concentration increased until a maximum was reached at 70% ethanol concentration [29]. When lignin precipitation is conducted by diluting liquor with water that decreased the amount of organic solvent, the solubility of lignin decreased, thus more lignin was recovered [28,30,31]. The findings were in agreement with the finding of Sun et al. which states that addition of anti-solvent such as ethanol, 1-propanol, and 1-butanol decreases the lignin solubility in the resultant system, therefore leading to a higher lignin recovery [32]. In general, it seems that a 10% ethanol concentration could be proposed as the ethanol concentration for lignin recovery since there is no significant difference between lignin recovery using 25% and 10% ethanol addition.

3.2. Percentage of Lignin Purity

The resulting soluble lignin extract after delignification was fractionated according to different ethanol concentration under centrifugation thereby generating two fractions: precipitated and supernatant fraction. Both fractions were analyzed by Klason lignin assay to reveal the purity of lignin. A comparison between the three different ethanol concentrations is presented in Figure 3.

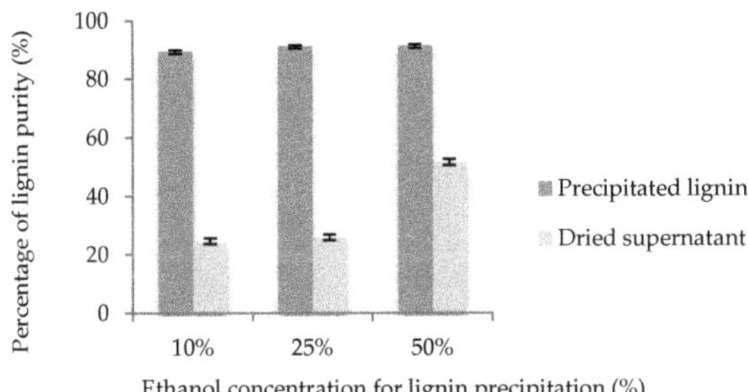

Figure 3. Percentage of lignin purity at different ethanol concentrations.

Overall, the purity of the obtained precipitated lignin consistently demonstrated a high purity (≥90%). Such high purity lignin has enormous possibilities to be used in several industrial applications such as polymer blends, adhesives, and corrosion inhibitors [12,16,33]. A few studies have also reported high purity of lignin from the organosolv method, for instance, 95.4% for sugarcane bagasse [34], 91.9% for *MxG* [35], as well as 96.5% for shrub willow [25].

As it can be seen from Figure 3, the purity of lignin derived from dried supernatant obtained at 50% ethanol concentration is considerably higher than that of at 25% and 10%. The high purity lignin derived from dried supernatant obtained at 50% ethanol concentration is due to less lignin that was precipitated as compared to other ethanol concentrations. Thus, more lignin remained with the impurities, as discussed in the FTIR analysis (Section 3.3). Considering that high dilution may lead to excessive energy costs if using 50% ethanol concentration, it is recommended to use 10% ethanol concentration for lignin precipitation process since there is no significant difference in the purity of the precipitated lignin using 50% and 25% ethanol concentrations.

3.3. FTIR Analysis

Figure 4a shows the PCA scores plot for the precipitated lignin and dried supernatant and it was elucidated that there were two definite clusters observed and were distinguishable within each other. Overall, the score plot for the dataset has functions, whose first two principles components explained 84% and 11% of the spectral variance, respectively. At the top are the spectra for the precipitated lignin at different ethanol concentration. A second cluster at the bottom consists of spectra for the dried supernatant. When comparison was made between similar types of spectra within samples, scores of precipitated lignin (10%_L_1 and 10%_L3, 25%_L_1 and 25%_L_3, 50%_L_1 and 50%_L_2) and scores of dried supernatant (10%_S_1 and 10%_S_2, 25%_S_1 and 25%_S_3, 50%_S_1 and 50%_S_2) were close to each other, indicating that the samples within similar type of spectra possess similar composition. Thus, a spectra was chosen only from the spectra with similar types to be analyzed for FTIR analysis.

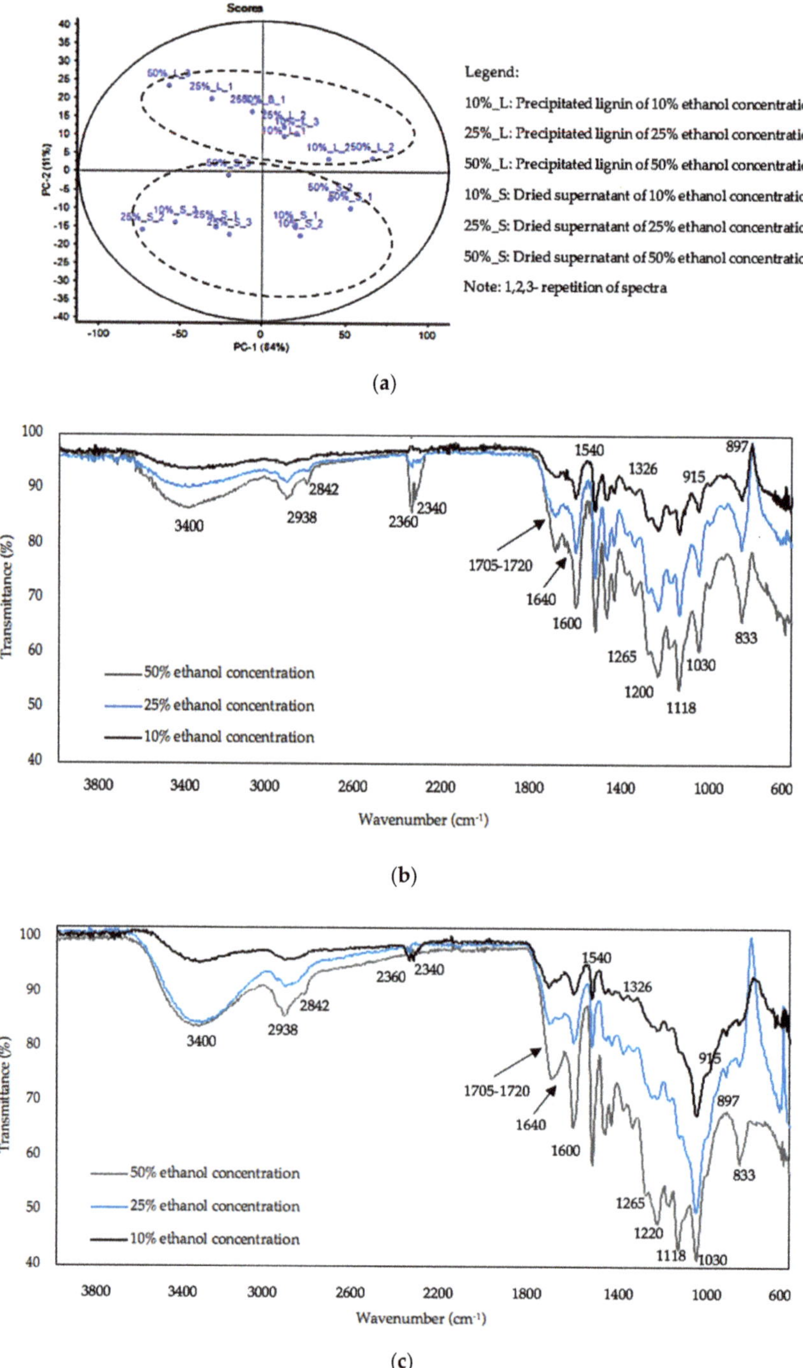

Figure 4. PCA and FTIR analysis at different ethanol concentrations. (**a**) PCA scores plot, (**b**) spectra of precipitated lignin, (**c**) spectra of dried supernatant.

The spectra of precipitated lignin at different ethanol concentration are shown in Figure 4b. When comparison was made between the three spectra at different ethanol concentrations, the spectra at 50% ethanol concentration had a high intensity peak as compared to the other two spectra at 25% and 10%, suggesting that a high intensity in peak indicates a high purity of lignin, especially at 3400 cm^{-1}. The peak at 3400 cm^{-1} indicated the presence of OH stretching vibrations in aromatic and aliphatic OH groups [36,37]. The abundance of hydroxyl groups, demonstrating that lignin recovered can be a good alternative to polyols in the production of lignin polymer composites through lignin depolymerisation and modification at higher bio-replacement ratios [38]. However, when the hydroxyl group peak is of interest, strong water absorption (around 3000–4000 cm^{-1}) could have influenced the results obtained and in turn, relatively led to the misinterpretation of the data. Therefore, the conclusion of the findings should be treated with caution, as the findings related to the availability of hydroxyl groups can be validated by various methods in future such as size exclusion chromatography and potentiometric titration, respectively.

The wavenumbers of 2938 cm^{-1} and 2842 cm^{-1} are attributed to CH stretching in aromatic methoxyl groups and in methyl and methylene groups of side chains [37]. An asymmetry and broadening of the peaks at 1705 cm^{-1} and 1600 cm^{-1} result from the weak absorption around 1640 cm^{-1} and may originate from both protein impurity and water associated with lignin, respectively [37]. The appearance of wavenumber at 1220 cm^{-1} is caused by the extraction process due to the hot water cleaved hemiacetal linkages, thus, liberating acids during biomass treatment which facilitate the breakage of ether linkages in biomass [39]. It is stated that the cleavage of O-acetyl groups and uronic acid substitutions on the hemicellulose released acids such as acetic and other organic acids which act as a catalyst, and catalyzed the formation and removal of oligosaccharides, and further hydrolyzed hemicellulose to monomeric sugars and aldehydes [40,41]. However, cellulose and hemicellulose appeared as contaminants as indicated by spectra wavenumbers at 897 and 1705 to 1720 cm^{-1}. A wavenumber of 897 cm^{-1} represents amorphous cellulose which aids hydrolysis of the cellulose to glucose if enzymatic hydrolysis occurs [42]. The wavenumbers of 1705 to 1720 cm^{-1} are attributed to ester carbonyl vibration in acetyl, feryloyl, and *p*-coumaryl groups in hemicelluloses [43]. Wavenumbers of 2340 to 2360 cm^{-1} were found in all spectra and are related to OH stretching from strong H-bonded-COOH [44].

In general, hardwoods or angiosperms are made up of guaiacyl (G) and syringl (S) units, while softwoods or gymnosperms contain only G units. Besides, grasses contain a variety of acidic guaiacyl units attached as esters and demonstrate more substitution of *p*-hydroxyphenyl units (H) such as ferulic, hydroxycinnamic, and *p*-coumaric acids [45]. Lignin extracted from *Miscanthus* sp. contains all three lignin monomers, G, H, and S units [7,46,47]. This data showed that the wavenumbers related to G, H, and S units could be at 1326 cm^{-1}(G-S units), 1265, 1030, 915 cm^{-1} (G units), 1118, 833 cm^{-1} (S units), 1705–1720 cm^{-1} (H units), and 2938, 2842 cm^{-1} for H–S units.

Supernatant obtained from the fractionation of the resulting soluble lignin extract at different ethanol concentration after centrifugation was dried for FTIR analysis (Figure 4c). In general, the spectra of 50% ethanol concentration had more broad intensity than 25% and 10% especially for wavenumbers apportioned to lignin. In fact, the supernatant has been proven to have the highest purity (51.6%). The peaks for the dried supernatant related to lignin including wavenumbers of 2842, 2340, 2360, 1640, and 1326 cm^{-1} were in weaker intensity as compared to peaks of precipitated lignin. The weak intensity and absence of peaks (915 cm^{-1}) related to lignin were due to more lignin precipitation and thus less lignin composition appeared in the dried supernatant. A peak of 1540 cm^{-1}, related to an aromatic ring stretching in lignin, was found in both spectra for the precipitated lignin and the dried supernatant [48]. In summary, from the spectra in Figure 4c, it is apparent that wavenumbers of 897 and 1705 to 1720 cm^{-1} related to the contamination of cellulose and hemicellulose were in high intensity and resulted in broader peaks at 50% ethanol concentration as compared to the 25% and 10% ethanol concentrations, thus the lignin derived from the supernatant had a lower purity level.

3.4. SEM Analysis

When comparing the three samples of precipitated lignin at different ethanol concentrations, SEM images showed similarity in the presence of the lignin macromolecule globule structure of spherical balls or droplets. The shape of the lignin macromolecule was not identical in terms of size, as shown in Figure 5a–c. The precipitated lignin at 50% ethanol concentration had a larger lignin macromolecule size than that of at 25% and 10%. On the other hand, the precipitated lignin at 25% and 10% ethanol concentrations exhibited a mixture of large and small lignin macromolecules. In addition, it formed more colloidal and amorphous lignin macromolecule structure. A representative of SEM images of dried supernatant is exhibited in Figure 5d–f at 50%, 25%, and 10% ethanol concentrations, respectively. SEM images at 50% ethanol concentration of dried supernatant revealed that a crystalline structure was observed in the dried supernatant as compared to that at 25% and 10%. The dried supernatant at 25% and 10% ethanol concentrations had less crystalline and smooth surfaces as compared to that of at 50%.

This is associated with the effect of surface area and particle size. Surface area is inversely proportional to particle size [49]. In a study of the effect of metal chlorides on the solubility of lignin in the black liquor of pre-hydrolysis kraft pulping, as the metal chlorides concentration increased, the particle size of lignin increased, hence the solubility of lignin decreased due to the increasing coagulation degree of lignin in the black liquor [50], thereby resulting in lower lignin precipitation. It is suggested that large particle size of lignin results in lower reaction surface area for the precipitation to take place. Therefore, less lignin is recovered.

The literature has emphasized the importance of ethanol concentration on lignin depolymerization. Findings revealed that there is a great increase in liquid residue from 41 to 65.5% by increasing ethanol concentration from 0 to 65 volume%. Nevertheless, the solid residue yields decreased steadily from 39% to 17% [51]. The high ethanol concentration exerted a negative impact on the recovery of solid residue in lignin depolymerization. Therefore, the study would be more convincing if the molecular weight of the lignin obtained at different ethanol concentrations using different established methods such as gel permeation chromatography and size exclusion chromatography in addition to structural morphology could be obtained. Currently, efforts to increase the use of lignin in biopolymer applications is related to degradation or deconstruction of lignin to small monomers by depolymerization and chemical modification of lignin to increase reactive sites into lignin molecules [52,53]. The available reactive hydroxyl groups in lignin demonstrate possibilities for chemical modifications prior to lignin valorization into valuable materials [54]. Thus, molecular weight determination will enable the SEM data to be verified by independent techniques.

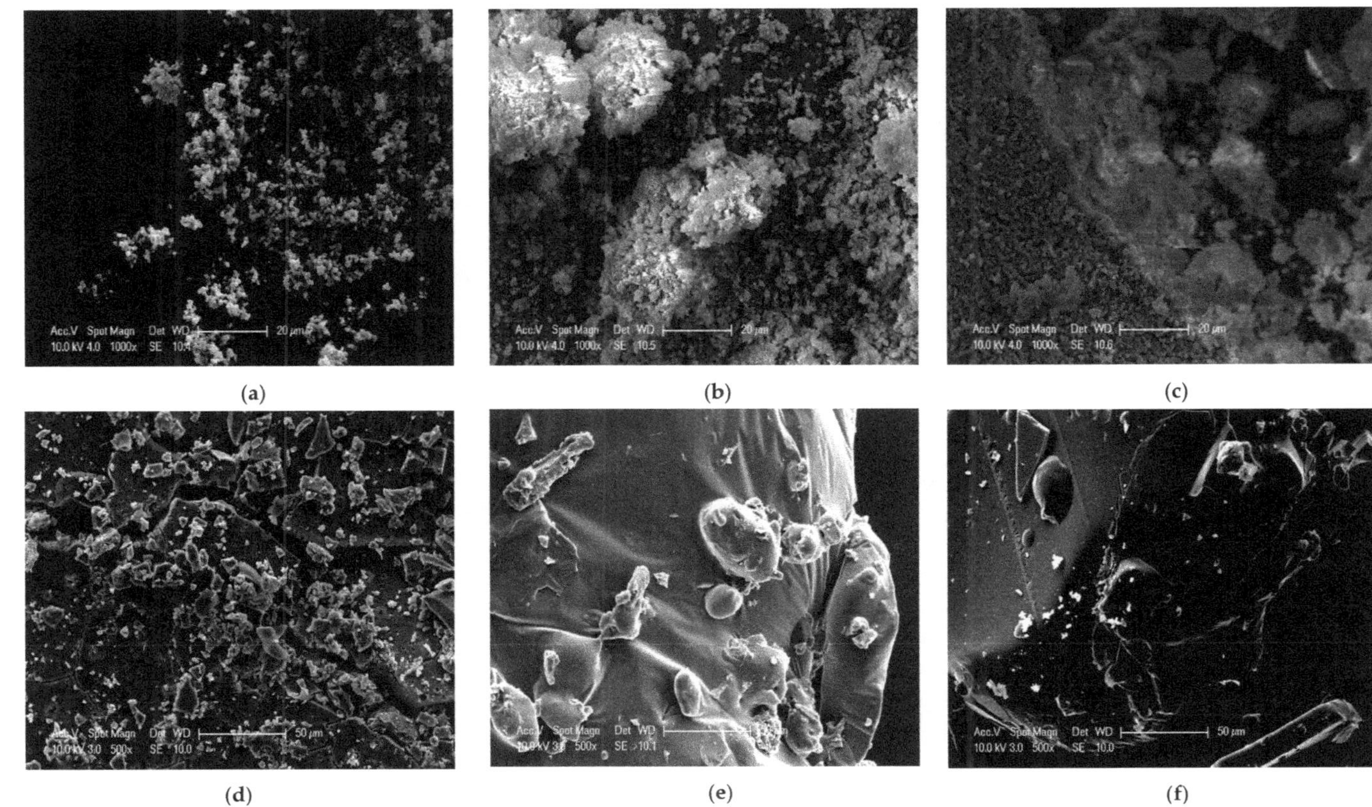

Figure 5. SEM images for precipitated lignin at (**a**) 50% ethanol concentration, (**b**) 25% ethanol concentration, (**c**) 10% ethanol concentration, and supernatant at (**d**) 50% ethanol concentration, (**e**) 25% ethanol concentration, (**f**) 10% ethanol concentration.

3.5. Particle Size Analysis

The data characterizing the particle size of lignin and the supernatant in different ethanol concentrations are presented in Table 2. A comparison between the three different ethanol concentrations using a Zetasizer illustrated that lignin precipitation at 50% ethanol concentration had the highest average particle size (2050.0 nm) as compared to that of at 25% (391.7 nm) and 10% (306.2 nm) and the remaining dried supernatant had the lowest average particle size (875.1 nm) as compared to that of the 25% (1197.3 nm) and 10% (1598.3 nm). The particle size of respective lignin macromolecules can be explained using the SEM image analysis (Figure 5), especially the SEM image of the precipitated lignin at 50% ethanol concentration that clearly shows that lignin macromolecules have regular, uniform, and large shapes as compared to that of the 25% and 10%. High average particle size of the precipitated lignin could also be due to the agglomeration of particles which reduces the surface area available for a precipitation reaction [55,56]. The surface area available during precipitation may influence the interaction of attractive and repulsive forces that further affects the lignin stability in solution and the lignin recovery process.

Table 2. Particle sizes of precipitated lignin and supernatant at different ethanol concentrations.

Sample	Ethanol Concentration (%)	Particle Size					
		Particle Diameter from Zetasizer Nano ZS (nm)		Particle Diameter from Mastersizer 2000 (μm)			
		Particle Size Distribution	Particle Size	$^1D_{v10}$	$^2D_{v50}$	$^3D_{v90}$	$^4D_{3,2}$
Lignin	50	342.0–1281.0 1281.0–6439.0	2050.0	5.5	98.2	257.7	18.3
	25	91.3–1718.0 3580.0–6439.0	391.7	7.0	36.2	131.7	8.6
	10	58.8–295.3 295.3–2669.0 3580.0–6439.0	306.2	3.8	67.8	300.2	11.1
Supernatant	50	68.1–295.3 295.3–1484.0 1484.0–6439	875.1	4.5	13.3	239.6	10.3
	25	825.0–2305.0	1197.3	3.0	9.3	86.3	6.9
	10	531.2–6439.0	1598.3	4.1	40.7	275.6	12.6

Note: D = diameter; v = volume. $^1D_{v10}$ = the maximum particle diameter below which 10% of the sample volume exists. $^2D_{v50}$ = the maximum particle diameter below which 50% of the sample volume exists. $^3D_{v90}$ = the maximum particle diameter below which 90% of the sample volume exists. $^4D_{3,2}$ = surface area mean diameter.

The 50% and 25% ethanol concentration had bimodal distributions whereas the 10% ethanol concentration of the precipitated lignin had multimodal distribution. The precipitated lignin at 25% and 10% ethanol concentration contains both nano- (<100 nm) and micro-size particles. The precipitated lignin at 50% ethanol concentration had only micro-size particles. The measured supernatant particles size at 10% and 25% ethanol concentration had a monomodal distribution whereas at 50% ethanol concentration, the supernatant showed a multimodal distribution. The supernatant of 25% and 10% ethanol concentration only had micro-particles whereby supernatant at 50% ethanol concentration contained both nano- and micro-particles. The resulting difference in particle size distributions for both precipitated lignin and supernatant may be due to the changes of physical and chemical properties; i.e., the samples in dried form, dispersed in the solution that cause the rearrangement of lignin macromolecules and the formation of new self-assembled structures [57].

Overall, the particle diameter of Mastersizer correlated with the patterns of particle size distribution of Zetasizer for both the precipitated lignin and dried supernatant. In general, reducing the ethanol concentration from 50% to 25% via Zetasizer results in a lower range of particle size distribution. A further reduction to 10% ethanol concentration showed that the distribution tends to move towards the right position of the distribution, demonstrating that the lignin particles tend to form a population of large particles within the precipitated lignin and dried supernatant.

Based on the data presented in Table 2, further distribution of Mastersizer results for precipitated lignin and dried supernatant are discussed. For precipitated lignin, reduction of 50% to 25% ethanol concentration showed that D_{v10} increased in size from 5.5 μm to 7.0 μm and decreased to 3.8 μm at 10% ethanol concentration. This indicates that the reduction of ethanol concentration from 50 to 10% ethanol concentration created large population of small particles. In contrast, D_{v50} and D_{v90} of the particle population exhibited different trends, whereby reducing ethanol concentration from 50% to 25% led to a descending trend in the size of lignin particles and ascending trend from 25% to 10% ethanol concentration. This showed that either re-aggregation of lignin particles occurred, or that lignin tends to form large particles at 10% ethanol concentration for D_{v50} and D_{v90} of the particle population (67.8 μm and 300.2 μm, respectively).

Similar findings were also observed for D_{v50} and D_{v90} of the particle population for supernatant, which showed an increment of lignin particle size at 10% ethanol concentration when compared to that of at 25% and 50% ethanol concentration. D_{v10} of the particle population showed different trends for supernatant. The particle size of 10% particle population (D_{v10}) decreased from 4.5 μm (50% ethanol concentration) to 3.0 μm (25% ethanol concentration) and increased to 4.1 μm (10% ethanol concentration). A possible explanation for various lignin aggregate behaviors may be the fact that lignin in an amphiphilic polymer contains both hydrophobic and hydrophilic segments besides possessing self-assembly behavior [58]. The hydrophilic segments of lignin macromolecules had an affinity to ethanol, whereas the hydrophobic segments dissociate lignin aggregate in ethanol-water mixture [58–61].

4. Conclusions

The focus of this study is primarily trying to understand the behavior of lignin aggregates and the study has raised important question on whether the effect of ethanol concentration influenced the behavior of lignin aggregates. In the present study, the effect of ethanol concentration was investigated on lignin recovery, particle size, chemical structure, and microscopy imaging properties. Findings showed that lignin recovery increased as the ethanol concentration decreased. Since there is no significant difference in the purity of precipitated lignin when using 50% and 25% ethanol concentration as compared to that of at 10%, it is suggested that 10% ethanol concentration is used for lignin recovery. Moreover, at 10% ethanol concentration, the purity of precipitated lignin was high ($\geq$90%). The results of lignin purity correlated well with the chemical structure analysis via FTIR. Both spectra of precipitated lignin and supernatant at 50% ethanol concentration had higher intensity peak than 25% and 10% especially for wavenumbers apportioned to lignin. The morphology images captured via SEM imaging revealed the findings of particle size analysis.

Overall, it was demonstrated that the lignin macromolecule agglomerates which form post extraction could be de-agglomerated by reducing the ethanol concentration from 50% to 10%. It is hypothesized that water had an influence on hygroscopic solvents i.e., different ethanol concentration that has different ability to attract and hold water molecules from the surrounding environment. Solvation of ethanol and water creates non-covalent interactions, such as hydrogen, Van der Waals and hydrophobic bonding that have a strong tendency to form aggregates with other molecules. The relationship between the solvent concentration and the resultant lignin macromolecule is complex. The present study is imperative and could facilitate an improved understanding on structural complexity of lignin for lignins obtained via SCW extraction.

Author Contributions: Conceptualization, M.H.H.; Formal analysis, M.H.H.; Investigation, M.H.H.; Methodology, M.H.H.; Supervision, S.B. and P.C.; Validation, S.B. and P.C.; Writing—original draft, M.H.H.; Writing—review and editing, P.C. All authors have read and agreed to the published version of the manuscript.

Funding: This research was funded by Ministry of Higher Education Malaysia and Universiti Putra Malaysia under Skim Latihan IPTA and the APC was funded by Universiti Putra Malaysia.

Acknowledgments: The authors would also like to thank Phytatec (U.K.) Ltd. for providing biomass feedstock.

Conflicts of Interest: The authors declare no conflict of interest.

References

1. Wahid, R.; Nielsen, S.F.; Hernandez, V.M.; Ward, A.J.; Gislum, R.; Jørgensen, U.; Møller, H.B. Methane production potential from *Miscanthus* sp.: Effect of harvesting time, genotypes and plant fractions. *Biosyst. Eng.* **2015**, *133*, 71–80. [CrossRef]

2. Lanzerstorfer, C. Combustion of miscanthus: Composition of the ash by particle size. *Energies* **2019**, *12*, 178. [CrossRef]

3. Kryževičienė, A. Cultivation of *Miscanthus x giganteus* for biofuel and its tolerance of Lithuania's climate. *Zemdirb. Agric.* **2011**, *98*, 267–274.

4. Wang, D.; Naidu, S.L.; Portis, A.R.; Moose, S.P.; Long, S.P. Can the cold tolerance of C4 photosynthesis in *Miscanthus x giganteus* relative to Zea mays be explained by differences in activities and thermal properties of Rubisco? *J. Exp. Bot.* **2008**, *59*, 1779–1787. [CrossRef]

5. Xi, Q.; Jezowski, S. Plant resources of Triarrhena and Miscanthus species in China and its meanning for Europe. *Plant Breed. Seed Sci.* **2004**, *49*, 63–75.

6. Jørgensen, U. Benefits versus risks of growing biofuel crops: The case of Miscanthus. *Curr. Opin. Environ. Sustain.* **2011**, *3*, 24–30. [CrossRef]

7. Lewandowski, I.; Clifton-Brown, J.C.; Scurlock, J.M.O.; Huisman, W. Miscanthus: European experience with a novel energy crop. *Biomass Bioenergy* **2000**, *19*, 209–227. [CrossRef]

8. Glasser, W.G. About Making Lignin Great Again—Some Lessons From the Past. *Front. Chem.* **2019**, *7*, 1–17. [CrossRef]

9. Hamzah, M.H.; Bowra, S.; Cox, P. Purity and structural composition of lignin isolated from *Miscanthus x giganteus* by sub-critical water extraction with associated modifiers. *J. Agric. Food Eng.* **2020**, *1*, 1–12.

10. Ratnaweera, D.R.; Saha, D.; Pingali, S.V.; Labbé, N.; Naskar, A.K.; Dadmun, M. The impact of lignin source on its self-assembly in solution. *RSC Adv.* **2015**, *5*, 67258–67266. [CrossRef]

11. Ganewatta, M.S.; Lokupitiya, H.N.; Tang, C. Lignin biopolymers in the age of controlled polymerization. *Polymers* **2019**, *11*, 1176. [CrossRef] [PubMed]

12. Lu, Y.; Lu, Y.C.; Hu, H.Q.; Xie, F.J.; Wei, X.Y.; Fan, X. Structural characterization of lignin and its degradation products with spectroscopic methods. *J. Spectrosc.* **2017**, 1–15. [CrossRef]

13. Vishtal, A.; Kraslawski, A. Challenges in industrial applications of technical lignins. *BioResources* **2011**, *6*, 3547–3568.

14. Deng, Y.; Feng, X.; Zhou, M.; Qian, Y.; Yu, H.; Qiu, X. Investigation of aggregation and assembly of alkali lignin using iodine as a probe. *Biomacromolecules* **2011**, *12*, 1116–1125. [CrossRef]

15. Clauss, M.M.; Weldin, D.L.; Frank, E.; Giebel, E.; Buchmeiser, M.R. Size-exclusion chromatography and aggregation studies of acetylated lignins in N,N-dimethylacetamide in the presence of salts. *Macromol. Chem. Phys.* **2015**, *216*, 2012–2019. [CrossRef]

16. Matsakas, L.; Karnaouri, A.; Cwirzen, A.; Rova, U.; Christakopoulos, P. Formation of lignin nanoparticles by combining organosolv pretreatment of birch biomass and homogenization processes. *Molecules* **2018**, *23*, 1822. [CrossRef]

17. Roque, R.M.N. Hydrolysis of Lignocellulosic Biomass by a Modified organosolv Method on a Biorefinery Perspective—Example of *Miscanthus x Giganteus*. Ph.D. Thesis, University of Birmingham, Birmingham, UK, 2013.

18. Mohan, M.; Banerjee, T.; Goud, V.V. Hydrolysis of bamboo biomass by subcritical water treatment. *Bioresour. Technol.* **2015**, *191*, 244–252. [CrossRef]

19. Rogalinski, T.; Liu, K.; Albrecht, T.; Brunner, G. Hydrolysis kinetics of biopolymers in subcritical water. *J. Supercrit. Fluids* **2008**, *46*, 335–341. [CrossRef]

20. Sluiter, A.; Hames, B.; Ruiz, R.; Scarlata, C.; Sluiter, J.; Templeton, D.; Crocker, D. *Determination of Structural Carbohydrates and Lignin in Biomass*; National Renewable Energy Laboratory: Golden, CO, USA, 2012.
21. Durak, T.; Depciuch, J. Effect of plant sample preparation and measuring methods on ATR-FTIR spectra results. *Environ. Exp. Bot.* **2020**, *169*, 103915. [CrossRef]
22. Šurina, I.; Jablonský, M.; Ház, A.; Sladková, A.; Briškárová, A.; Kačík, F.; Šima, J. Characterisation of non-wood lignin precipitated with sulphuric acid of various concentrations. *BioResources* **2015**, *10*, 1408–1423. [CrossRef]
23. Aleš, H.; Michal, J.; Lenka, D.; Alexandra, S.; Igor, Š. Thermal properties and size distribution of lignins precipitated with sulphuric acid. *Wood Res.* **2015**, *60*, 375–384.
24. Donaldson, L.A. Critical assessment of interference microscopy as a technique for measuring lignin distribution in cell walls. *N. Z. J. For. Sci.* **1985**, *15*, 349–360.
25. Stewart, H.E. Development of Food-Grade Microparticles From Lignin. Ph.D. Thesis, Massey University, Palmerston North, New Zealand, 2015.
26. Pasquini, D.; Pimenta, M.T.B.; Ferreira, L.H.; da Silva Curvelo, A.A. Extraction of lignin from sugar cane bagasse and *Pinus taeda* wood chips using ethanol–water mixtures and carbon dioxide at high pressures. *J. Supercrit. Fluids* **2005**, *36*, 31–39. [CrossRef]
27. Tu, Q.; Fu, S.; Zhan, H.; Chai, X.; Lucia, L.A. Kinetic modeling of formic acid pulping of bagasse. *J. Agric. Food Chem.* **2008**, *56*, 3097–3101. [CrossRef]
28. Xu, Y.; Li, K.; Zhang, M. Lignin precipitation on the pulp fibers in the ethanol-based organosolv pulping. *Colloids Surf. A Physicochem. Eng. Asp.* **2007**, *301*, 255–263. [CrossRef]
29. Ni, Y.; Hu, Q. Alcell lignin solubility in ethanol-water mixtures. *J. Appl. Polym. Sci.* **1995**, *57*, 1441–1446. [CrossRef]
30. Fernando, E.F. Lignin recovery from spent liquors from ethanol-water fractionation of sugar cane bagasse. *Cellul. Chem. Technol.* **2010**, *44*, 311–318.
31. Ortega, J.H. Process Design of Lignocellulosic Biomass Fractionation Into Cellulose, Hemicellulose and Lignin by Prehydrolysis and Organosolv Process. Master's Thesis, Wageningen University, Wageningen, The Netherlands, 2015.
32. Sun, J.; Dutta, T.; Parthasarathi, R.; Kim, K.H.; Tolic, N.; Chu, R.K.; Isern, N.G.; Cort, J.R.; Simmons, B.A.; Singh, S. Rapid room temperature solubilization and depolymerisation of polymeric lignin at high loadings. *Green Chem.* **2016**, *18*, 6012–6020. [CrossRef]
33. Hussin, M.H. Extraction, Modification and Characterization of Lignin From Oil Palm Fronds as Corrosion Inhibitors for Mild Steel in Acidic Solution. Ph.D. Thesis, Universite de Lorraine, Nancy and Metz, France, 2014.
34. Vallejos, M.E.; Felissia, F.E.; Curvelo, A.A.S.; Zambon, M.D.; Ramos, L.; Area, M.C. Chemical and physico-chemical characterisation of lignins obtained from ethanol-water fractionation of bagasse. *BioResources* **2011**, *6*, 1158–1171.
35. Bauer, S.; Sorek, H.; Mitchell, V.D.; Ibáñez, A.B.; Wemmer, D.E. Characterisation of *Miscanthus x giganteus* lignin isolated by ethanol organosolv process under reflux condition. *J. Agric. Food Chem.* **2012**, *60*, 8203–8212. [CrossRef]
36. Alriols, M.G.; García, A.; Llano-Ponte, R.; Labidi, J. Combined organosolv and ultrafiltration lignocellulosic biorefinery process. *Chem. Eng. J.* **2010**, *157*, 113–120. [CrossRef]
37. Boeriu, C.G.; Bravo, D.; Gosselink, R.J.A.; van Dam, J.E.G. Characterisation of structure-dependent functional properties of lignin with infrared spectroscopy. *Ind. Crops Prod.* **2004**, *20*, 205–218. [CrossRef]
38. Mahmood, N.; Yuan, Z.; Schmidt, J.; Xu, C. Depolymerisation of lignins and their applications for the preparation of polyols and rigid polyurethane foams: A review. *Renew. Sustain. Energy Rev.* **2016**, *60*, 317–329. [CrossRef]
39. Behera, S.; Arora, R.; Nandhagopal, N.; Kumar, S. Importance of chemical pretreatment for bioconversion of lignocellulosic biomass. *Renew. Sustain. Energy Rev.* **2014**, *36*, 91–106. [CrossRef]
40. Mosier, N.; Wyman, C.; Dale, B.; Elander, R.; Lee, Y.Y.; Holtzapple, M.; Ladisch, M. Features of promising technologies for pretreatment of lignocellulosic biomass. *Bioresour. Technol.* **2005**, *96*, 673–686. [CrossRef] [PubMed]
41. Xiao, L.-P.; Sun, Z.-J.; Shi, Z.-J.; Xu, F.; Sun, R. Impact of hot compressed water pretreatment on the structural changes of woody biomass for bioethanol production. *BioResources* **2011**, *6*, 1576–1598.

42. Mohamad Ibrahim, M.N.; Zakaria, N.; Sipaut, C.S.; Sulaiman, O.; Hashim, R. Chemical and thermal properties of lignins from oil palm biomass as a substitute for phenol in a phenol formaldehyde resin production. *Carbohydr. Polym.* **2011**, *86*, 112–119. [CrossRef]

43. Pandey, K.K. A study of chemical structure of soft and harwood and wood polymers by FTIR spectrscopy. *J. Appl. Polym. Sci.* **1999**, *71*, 1969–1975.

44. Davis, W.M.; Erickson, C.L.; Johnston, C.T.; Delfino, J.J.; Porter, J.E. Quantitative Fourier Transform Infrared spectroscopic investigation humic substance functional group composition. *Chemosphere* **1999**, *38*, 2913–2928. [CrossRef]

45. Holladay, J.E.; White, J.F.; Bozell, J.J.; Johnson, D. *Top Value-Added Chemicals from Biomass*; Pacific Northwest National Lab. (PNNL): Richland, WA, USA; National Renewable Energy Lab. (NREL): Golden, CO, USA, 2007.

46. Guo, F.; Shi, W.; Sun, W.; Li, X.; Wang, F.; Zhao, J.; Qu, Y. Differences in the adsorption of enzymes onto lignins from diverse types of lignocellulosic biomass and the underlying mechanism. *Biotechnol. Biofuels* **2014**, *7*, 38. [CrossRef]

47. Savy, D.; Cozzolino, V.; Vinci, G.; Nebbioso, A.; Piccolo, A. Water-Soluble Lignins from Different Bioenergy Crops Stimulate the Early Development of Maize (Zea mays, L.). *Molecules* **2015**, *20*, 19958–19970. [CrossRef] [PubMed]

48. Radotić, K.; Roduit, C.; Simonović, J.; Hornitschek, P.; Fankhauser, C.; Mutavdžić, D.; Steinbach, G.; Dietler, G.; Kasas, S. Atomic force microscopy stiffness tomography on living arabidopsis thaliana cells reveals the mechanical properties of surface and deep cell-wall layers during growth. *Biophys. J.* **2012**, *103*, 386–394. [CrossRef] [PubMed]

49. Kumar, A.; Kumar, J.; Bhaskar, T. Utilization of lignin: A sustainable and eco-friendly approach. *J. Energy Inst.* **2020**, *93*, 235–271. [CrossRef]

50. He, L.; Liu, Q.; Song, Y.; Deng, Y. Effects of metal chlorides on the solubility of lignin in the black liquor of prehydrolysis kraft pulping. *BioResources* **2014**, *9*, 4636–4642. [CrossRef]

51. Ye, Y.; Zhang, Y.; Fan, J.; Chang, J. Novel method for production of phenolics by combining lignin extraction with lignin depolymerisation in aqueous ethanol. *Ind. Eng. Chem. Res.* **2012**, *51*, 103–110. [CrossRef]

52. Xu, C.; Arneil, R.; Arancon, D.; Labidi, J.; Luque, R. Lignin depolymerisation strategies: Towards valuable chemicals and fuels. *Chem. Soc. Rev.* **2014**, *43*, 7485–7500. [CrossRef] [PubMed]

53. Matsushita, Y.; Yasuda, S. Reactivity of a condensed—Type lignin model compound in the Mannich reaction and preparation of cationic surfactant from sulfuric acid lignin. *J. Wood Sci.* **2003**, *49*, 166–171. [CrossRef]

54. Duval, A.; Lawoko, M. A review on lignin-based polymeric, micro- and nano-structured materials. *React. Funct. Polym.* **2014**, *85*, 78–96. [CrossRef]

55. Allen, E.; Smith, P.; Henshaw, J. *A Review of Particle Agglomeration*; AEA Technology: Dorset, UK, 2001.

56. Sayyar, S.; Abidin, Z.Z.; Yunus, R.; Muhammad, A. Extraction of oil from Jatropha seeds-optimisation and kinetics. *Am. J. Appl. Sci.* **2009**, *6*, 1390–1395. [CrossRef]

57. Shulga, G.; Vitolina, S. Lignin separated from the hydrolysate of the hydrothermal treatment of birch wood and its surface properties. *Cellul. Chem. Technol.* **2012**, *46*, 307–318.

58. Xiong, F.; Han, Y.; Wang, S.; Li, G.; Qin, T.; Chen, Y.; Chu, F. Preparation and formation mechanism of size-controlled lignin nanospheres by self-assembly. *Ind. Crops Prod.* **2017**, *100*, 146–152. [CrossRef]

59. Qian, Y.; Deng, Y.; Qiu, X.; Li, H.; Yang, D. Formation of uniform colloidal spheres from lignin, a renewable resource recovered from pulping spent liquor. *Green Chem.* **2014**, *16*, 2156. [CrossRef]

60. Li, H.; Deng, Y.; Liu, B.; Ren, Y.; Liang, J.; Qian, Y.; Qiu, X.; Li, C.; Zheng, D. Preparation of Nanocapsules via the Self-Assembly of Kraft Lignin: A Totally Green Process with Renewable Resources. *ACS Sustain. Chem. Eng.* **2016**, *4*, 1946–1953. [CrossRef]

61. Rao, X.; Liu, Y.; Zhang, Q.; Chen, W.; Liu, Y.; Yu, H. Assembly of Organosolv Lignin Residues into Submicron Spheres: The Effects of Granulating in Ethanol/Water Mixtures and Homogenisation. *ACS Omega* **2017**, *2*, 2858–2865. [CrossRef] [PubMed]

 *processes*

Article

Industrial Processes Management for a Sustainable Society: Global Research Analysis

Emilio Abad-Segura [1], Manuel E. Morales [2,3], Francisco Joaquín Cortés-García [4] and
Luis Jesús Belmonte-Ureña [1,2,*]

1   Department of Economics and Business, University of Almeria, 04120 Almeria, Spain; eas297@ual.es
2   European Center of Biotechnology and Bioeconomy, 3 Rue des Rouges-Terres, 51110 Pomacle, France
3   NEOMA BS, Campus Reims, 59 Rue Pierre Taittinger, 51100 Reims, France; manuel.morales@neoma-bs.fr
4   Faculty of Business and Management, Universidad Autónoma de Chile, 7500912 Santiago, Chile;
     franciscojoaquincortesgarcia@gmail.com
*   Correspondence: lbelmont@ual.es

Received: 7 May 2020; Accepted: 22 May 2020; Published: 24 May 2020

Abstract: Few decades ago, the development of the industrial sector was disconnected from society's protection. Negative effects awareness emerges from the current industrial processes through the Sustainable Development Goals (SDGs), considering the causal implications to build up a more sustainable society. The aim of this study is to analyze the state of the art in industrial processes management to obtain positive and sustainable effects on society. Thus, a bibliometric analysis of 1911 articles was set up during the 1988–2019 period, bringing up the authors' productivity indicators in the scientific field, that is, journals, authors, research institutions, and countries. We have identified environmental management; the impact assessments of industrial processes on the environment and its relation with a more sustainable society; as well as the study of the sustainable management of water resources as the related axes in the study of environmental protection with political, economic, and educational approaches. The growing trend of world scientific publications let us observe the relevance of industrial processes management in the implementation of efficient models to achieve sustainable societies. This research contributes to the academic, scientific, and social debate on decision-making both in public and private institutions, and in multidisciplinary groups.

Keywords: sustainable process; industry; green technology; effective management; sustainable society; scientific research

---

1. Introduction

Currently, the industrial processes used by responsible and competitive firms must incorporate design efficiency steps. Hence, the sustainability concept is engaged in a process where we can get better indicators with fewer resources [1,2].

In this context, these processes have to support sustainable development, guaranteeing the basic needs of the population, through the rational management of natural resources, and without compromising the sustainability of future societies. Indeed, industrial processes are considered sustainable when they are innovative and compose safety and waste management [3]. Therefore, a society will be sustainable if it is organized in such a way that can guarantee the citizens' and ecosystems' life, through generations [4].

In the overall process, it is necessary to understand that a change from a consumer society to a sustainable conservative society must focus on the wellbeing of the planet and on guarantying future generations. This can only be achieved through the respect of biophysical boundaries and the reduction in resources exploitation, which in turn is supposed to encourage social justice from the per capita consumption point of view [5].

Initiatives are supposed to raise awareness about the creation of a more sustainable society that encompasses the Sustainable Development Goals (SDGs) [6,7]. A sustainable society will provide a higher level of well-being in the environment, social and economic dimensions.

With the industrial revolution, global production has been one of the main economic activities of society, triggering a negative impact on the environment. On the other hand, human activity has brought environmental awareness in the median and the political and social agenda around the world [8,9]. Sustainability in this study refers to the reduction of negative impacts on environmental, social, and economic relations, in order to approach climate change, pollution, and resources' management [10–12].

The motivation of this study, well justified in the research, is the evolution of the knowledge base of the management of industrial processes with the firm intention of contributing to a society under a model that includes the dimensions of sustainability: environmental, economic, and social.

In the literature review, we have found enough material to answer the research questions defined in this study: (i) what is the knowledge structure of the industrial processes for a sustainable society; (ii) what are the most productive authors, institutions, and countries and; (iii) what are the current research axes in the field and the further perspectives.

Thereby, the aim of this study is the analysis of the current state of the art in management of industrial processes to increase sustainability during the 1988–2019 period.

We identified a sample of 1911 articles from scientific journals in the Scopus database from Elsevier to answer the research questions defined in our study. This study uses the bibliometric method to synthesize the current available knowledge on the management of industrial processes for a sustainable society, from the first article appearance in Scopus (1988) to the last year (2019).

The scientific production in the research field let us identify the main drivers, their current and potential trends, and the gaps in critical knowledge.

The main limit found in this study is the impossibility to know if the number of publications obey the community regulation issues, the demands of interest groups, or the needs required by a global society.

Finally, a set of research axes dedicated to the environmental protection with political, economic, and educational approaches are currently being explored to address environmental management concerns; the impact of industrial processes on the environment and their relationship with a more sustainable society; and the sustainable management of water resources and their effects on society. Therefore, we can conclude that industrial processes are implementing efficient and accountable models, giving rise to sustainability enhancement on society.

2. Research Scope

Once the research topic is identified, this section frames the conceptual background within some theoretical principles, and provides the definition used in the study, in order to avoid misinterpretations. Hence, it works as a guideline for the research and provides a framework for outcomes interpretation.

2.1. Backgrounds

The Industrial Revolution was the economic, social, and technological starting point; it began in the second half of the 18th century in the United Kingdom, and a few decades after spread to parts of western Europe and North America. In this period, the socioeconomic paradigm evolved from a rural economy, based mainly on agriculture and trade, to an urban, industrialized, and mechanized economy [13–15].

Therefore, the Industrial Revolution was a turning point in history, where both agricultural and industrial production grew exponentially, while reducing production time.

In the pre-industrial economy, most of the population subsisted thanks to agricultural work, mainly based on self-consumption, trade representing a small percentage of the economy, since productivity was low [16]. Furthermore, cities were small and barely developed, and absolutist monarchies were the government regimes that supported these pre-industrial societies.

The industrial economy triggered technological, socioeconomic, and cultural changes, affecting all structures of society. Overall, scholars addressing technological changes considered the use of new materials, energy sources, and machines, the ones addressing cultural ones inferred an increase in knowledge in the scientific, technical, and health areas, while the social scholars refer to population growth, especially in urban settings [17–19].

Technology currently holds a crosscutting impact on all sectors of the economy and society. The speed of changes is breeding a new industrial revolution, which pushes companies to rethink business models and the sustainability of their processes [20,21].

2.2. Framework

A series of theoretical principles, encompassing the industrial processes framework for a sustainable society, support this study. Once the literature review is completed, Table 1 presents the articles that support the main research objectives.

Table 1. Main articles reviewed in relation to the objective of the research topic.

Year	Article Title [Reference]	Journal	Author(s)
2019	Does corporate social responsibility affect the cost of equity in controversial industry sectors? [22]	*Review of Accounting and Finance*	Hmaittane, A.; Bouslah, K.; M'Zali, B.
2018	UN sustainable development goals: How can sustainable/green chemistry contribute? The view from the agrochemical industry [23]	*Current Opinion in Green and Sustainable Chemistry*	O'Riordan, T.J.C.
2015	Sustainable Development Titanium Industry [24]	*Economy in the Industry*	Kostygova, L.A.
2014	Green Process Engineering as the Key to Future Processes [25]	*Processes*	Patel, D.; Kellici, S.; Saha, B.
2012	Controversial issues in factors determining intra-industry trade [26]	*International Journal of Economics and Business Research*	Ferto, I.; Soós, K.A.
2012	An Overview of Cement production: How "green" and sustainable is the industry? [27]	*Environmental Management and Sustainable Development,*	Potgieter, J.H.
2006	Where is developing country industry in sustainable development planning? [28]	*Sustainable Development*	Luken, R.A.
2004	Striving for process excellence [process industry] [29]	*Manufacturing Engineer*	Venables, M.
1999	Sustainable development for industry and society [30]	*Building Research & Information*	Stigson, B.
1995	Green materials and green processes [31]	*Journal of Materials Research*	Szekely, J.; Laudise, R.

The industrial processes addressing the achievement of more efficient societies are sustained by the organizational theory, which provides the basis for insertion in the organizations. From the literature review, the theoretical framework of the impact of industrial processes on the environmental, economic, and social dimensions of societies is established.

The stakeholders' theory appears for the first time in a study published by Freeman in 1984, suggesting the change in the company's business models integrating the corporate social responsibility (CSR) concept. Although, the initial aim of these changes was in the environmental dimension, organizations have also showed interest in how to interact with stakeholders [32–35]. Therefore, the increased need to discover how to encourage corporate social responsibility matches the increasing social demands for the business sector to take responsibility for its social impacts and look after the firms' interests [36,37].

In this study, industry is considered as a set of operations carried out in order to obtain, transform, or transport products encompassed in the secondary sector. The secondary sector is based on the transformation of raw materials into products for intermediate or final consumption, so the industry understanding implies a production process, which will use a certain amount of labor and capital [38,39].

Moreover, industrial firms require resources that, even if not participating directly in the industrial process, make possible to obtain merchandises, which may be for final consumption, if they go straight to consumers, or intermediate goods, if they must cross through another industrial transformation before to be send to consumers.

We consider a process to be a sequence of steps logically arranged with specific aims, i.e., in a firm, a process refers to actions looking for higher efficiency in production, triggering an increase in its profitability by producing more with lower costs [40–42].

In this study, the industrial process is responsible for obtaining, transforming, or transporting one or more raw materials, in order to shift them into materials or products that satisfy the societal needs maximizing utility [43,44]. According to the use of natural resources, work characteristics and flexibility on changes, four kinds of processes can be identified: batch processes, continuous flow, piecework, and mass production [45–47].

The concept of sustainability was established in the Brundtland Report in 1987, to relate environment to development, defined as the possibility to meet the current needs without compromising the necessary resources of the future. Eleven years later, in 1998, the concept of sustainability was encompassed in the triple bottom line, where social, environmental, and economic dimensions are interconnected in the firms' understanding [48–51].

In 2015, the United Nations published the 2030 Agenda for Sustainable Development, with 17 Sustainable Development Goals (SDGs) and 169 targets, to verify the progress on sustainability, regarding the economic, social, and environmental dimensions. Sustainability influences organizations, mainly through the SDGs dissemination [52,53]. Thereby, the good practices on firms' sustainability need to be aligned with the SDGs indicators to improve accountability and compare information on their actions. Regarding the efficient management of industrial processes, companies should address to following SDGs: 7th goal: Affordable and clean energy; 9th goal: Industry, innovation and infrastructure; 11th goal: Sustainable cities and communities; 12th goal: Responsible production and consumption; and 17th goal: Partnership for the goals [54–57].

The industrial sector must know the 169 goals and identify which are the activities influenced by each of them, so, we can set specific objectives and indicators with which to measure the progress on each activity, resulting in better communication with society and stakeholders. Indeed, companies act as corporate citizens and they have the commitment to implement sustainability in their value chains, minimizing their environmental impact, achieving economic benefits, and improving society [58,59].

Furthermore, the industrial sector must target beyond the written goals defined in the 2030, because each SDG needs innovative solutions. Therefore, stable, sustainable, and equitable wellbeing and social justice is an essential requirement for facing poverty, although this will not be enough [60–62].

According to the United Nations Educational, Scientific and Cultural Organization (UNESCO), the principles that represent sustainability are interdependence, diversity, human rights, global equity and justice, rights of future generations, conservation, values and decisions about lifestyle, democracy and citizen participation, precautionary principle, and economic vitality. All those principles should be included in sustainability trainings [63–65].

The sustainable process takes into account the success of environmental policies based on functionality, efficiency, and sustainability. Companies with highly effective resources and materials management improve their performance and logistics when prioritizing regional production. Sustainability in management also encourage firms to minimize waste and scrap production and through control and review processes, to choose partner companies that have sustainable certifications and standards joining the socially responsible approach [66].

Contributing to the sustainability definition, the UN World Commission on Environment and Development identify that a sustainable society is the one that meets the current needs without compromising the ability of future generations to take care of their own needs [67–69].

In general, society joins sustainability seeking to improve the life quality of citizens, improve autonomy, and move through a common good. Hence, the sustainable principles in society indicate the planetary boundaries and capacity for resources, among which are: (i) conservation: buying durable products, consciously consuming, and recycling; (ii) recycling: use over and over, close loops to save energy, avoid pollution, protect habitats, and conserve resources; (iii) rational use of renewable resources; and (iv) population control, to optimize resources in relation to consumption [70–73].

2.3. Underlying Terminology

Other concepts have been identified in order to build an underlying conceptual structure in this thematic area, those concepts entail the basis of knowledge on resource management of industrial processes to contribute positively to a sustainable society.

Thus, the effective management concept integrates efficiency as a fundamental term in management systems, measuring the level in which we carry out activities and we achieve successful results according to the plans; how do we manage the interconnected functions to create a corporate policy, in addition to organizing, planning, controlling, and steering the organization's resources in order to achieve their objectives. In this sense, an organization must act effectively, understood as the achievement of results in accordance to the guiding plans [74,75]. We have tools to support organizations in achieving efficiency management of their activities, i.e., systems' management laying on the bases for effective management [76].

On the other hand, the term green technology, also known as green computing, or green IT, refers to the efficient use of computing resources, minimizing the environmental impact, maximizing their economic viability, and ensuring social responsibilities. This set of methods reduces the IT impact on the environment [77,78].

Green-tech protocols must affect practically everything from food to clothing, furniture, cleaning products, packaging, etc. Those protocols represent the only way in which we can expect a reduction or even the elimination of hazardous materials and dangerous habits for society and for the planet [79–81].

The eco-technologies group is composed by the techniques, raw materials, and production systems supported by a non-stop research work. Innovation allows the accomplishment of objectives that transform people's lives on a global scale, based on: (i) circular economy: use of resources, manufacturing-production, consumption, waste, and recycling; (ii) decrease: modify the way of consuming, reducing waste, the amount of energy used and pollution; and (iii) sustainability: finding a way to meet current needs without putting the resources of future generations at risk [82–86].

Furthermore, contributing to the framework of this research, the ISO 2600 assists companies and organizations engaged in a socially responsible operation, clarifying what social responsibility means and helps in the translation of principles into effective actions. The theoretical framework states that companies integrating respect of society and the environment recognize it as a critical success factor and move towards the sustainability assessment in particular and the overall performance in general [87–89].

3. Materials and Methods

Bibliometric analysis applies mathematical and statistical methods to scientific literature, with the aim of studying and analyzing scientific activity. The instruments used to measure the scientific activity are the bibliometric indicators that provide information on the scientific activity outcomes. Garfield introduced the bibliometric analysis in the mid-20th century; this method has been widespread in scientific research, for decades contributing to the revision, summary, and analysis of the state of the art across multiple disciplines [90–92]. Thereby, bibliometric analysis has evolved based on the critical thinking of sciences and the availability of scientific databases for researchers.

The aim of this study is to provide an outlook of the research dynamics of industrial process management to achieve a more sustainable society. We performed a quantitative bibliometric analysis to identify, organize, and analyze trends in the research topic. In recent decades, the bibliometric analysis has contributed to the revision of scientific knowledge and has been used successfully in different scientific fields [93–96].

In the Scopus database, we used a search query using the terms "sustainability", "management", "process", and "industry" to examine the subfields of the title, abstract, and keywords, in a 32 years period, from 1988 to 2019, as shown in other bibliometric works [97,98]. The analyzed period gathered the articles from the first article collected in the Scopus (1988) to the last full year (2019).

We got the sample of the analyzed articles through a search in April 2020. This search only included scientific articles, both in open and non-open access. Hence, the final sample included 1911 documents. The analyzed variables were the year of publication, journal, author, country of affiliation of the author, research institution where the author is affiliated, subject area, and keywords. In this study, the indicators of the analyzed scientific production were the distribution by years of the published articles, and the productivity of the authors, countries, and research institutions. In this sense, Figure 1 shows a diagram with the steps applied in research on industrial process management for a more sustainable society.

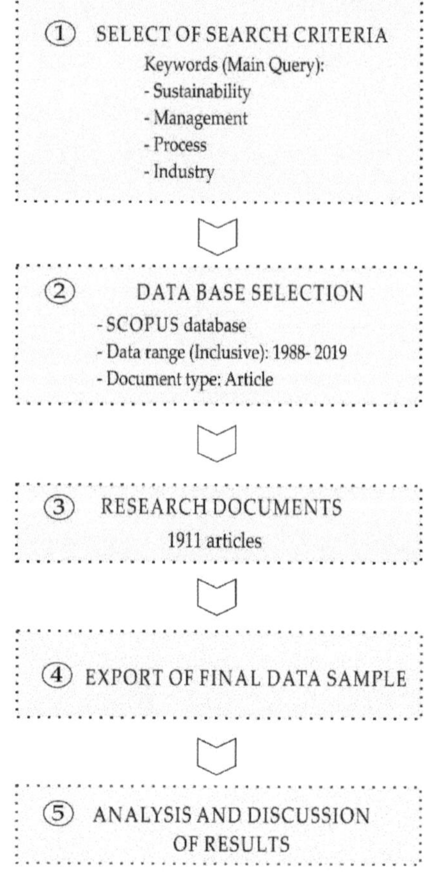

Figure 1. Flowchart of the methodology.

The quality indicators referring to the impact of the different agents used in this study were: the h-index, which allows detecting the most outstanding authors in the discipline, based on the number of citations received from their articles; citations' number; and the indicator that measures the quality of the scientific journals included in the Scopus database, the 2018 SCImago Journal Rank (SJR) [99–101].

The network indicators, which measure the collaborative links between authors, institutions, and countries, are included in the analysis to provide reliability and suitability in bibliometric analysis, using the analysis of co-authorship. Co-authorship of a technical document is an official declaration of the participation of two or more authors or organizations. Co-authorship analysis is widely used to understand and evaluate patterns of scientific collaboration. In co-authoring networks, nodes represent authors, organizations, or countries, who are connected when they share the authorship of an article.

The keywords analysis in our research topic has allowed the uncovering of the main issues or further perspectives of research topics or problems, based on the analysis of co-occurrences, since scientific articles can be reduced to a set of word appearances. With the analysis of co-occurrences, the proximity relationship of two or more terms in a text unit is established. The co-occurrence of two concepts is very high if they frequently appear together in one set of documents and rarely do so separately in the rest. Furthermore, the graphic representation of the co-occurrence networks allows them to be viewed [102–104].

We have used the software tool VOSviewer (version 1.6.10., Leiden University, Leiden, The Netherlands) for the analysis of these network indicators, which provides data about the interactions and contents assessments, in order to measure the activities of research networks [105,106].

The results obtained are worthy for scholars, practitioners, decision-makers, public authorities, and other stakeholders, because we are assessing the scientific production of a research field with strong influence over society.

4. Results and Discussion

4.1. Analysis of Scientific Production

The following subsection displays the main characteristics of scientific production, the evolution in the number of articles, and percentages of variation between periods, and, finally, the total number of journals where articles on this topic were published.

Hence, Table 2 shows the evolution of the main characteristics of the articles published on industrial processes for a sustainable society, from 1988 to 2019. The time horizon of the study is 32 years and it has been divided into four-year periods, in order to facilitate the analysis. In this time horizon, interest on the relationship between industrial processes and a sustainable model of society has increased, especially in the last eight years, as observed in the collected data.

Therefore, if in the first period (1988–1991) only 2 articles were published on this topic, in the last four-year period analyzed (2016–2019), the number attempt 845, that is, practically 425 times more. The number of publications increases in the last four-year period, where 44.20% of the total articles analyzed have been published. This exponential growth could be due to the initiative promoted by the United Nations in 2015, in addition to the definition of the 2030 Agenda and the Millennium Development Goals (MDGs), which propelled collaboration and the progressive increase in scientific publications [107,108]. During the last 8 years (2012–2019), 70% of the articles have been published (1339), and 2019 was the year with more publications, 271 articles.

On the other hand, the authors Buckley, Pass, and Prescott published in 1988 the first article on the research topic of management of industrial processes for a sustainable society, with the title "Measures of international competitiveness: A critical survey", in the *Journal of Marketing Management*. This article is classified in the "Business, Management, and Accounting" thematic area [109]. Likewise, it should be noted that the same article can be classified in more than one category, which will depend on the publisher and author.

Table 2. Main characteristics of scientific production (1998–2019).

Period	A	AU	C	TC	TC/A	J
1988–1991	2	4	2	205	102.5	2
1992–1995	2	4	1	63	31.5	2
1996–1999	38	73	13	764	20.1	35
2000–2003	84	224	34	2716	32.3	73
2004–2007	135	345	42	3415	25.3	105
2008–2011	311	879	53	11,919	38.3	226
2012–2015	494	1553	70	11,339	23.0	315
2016–2019	845	2813	82	5299	6.3	386

A: number of articles; AU: number of authors; C: number of countries; TC: number of citations in total articles; TC/A: number of citations per article; J: number of journals.

Figure 2 shows the evolution in the number of articles and the variation percentage between each four-year period studied. In addition, we highlight the increase percentage between the second and third periods analyzed (1800%), in 1992–1995 and 1996–1999, even at that time the scientific activity in this field of study was still very low. The percentage increase in the number of publications in the 2008–2011 period (130.40%), obey to the fact that it is the first four-year period with more than 300 articles (311).

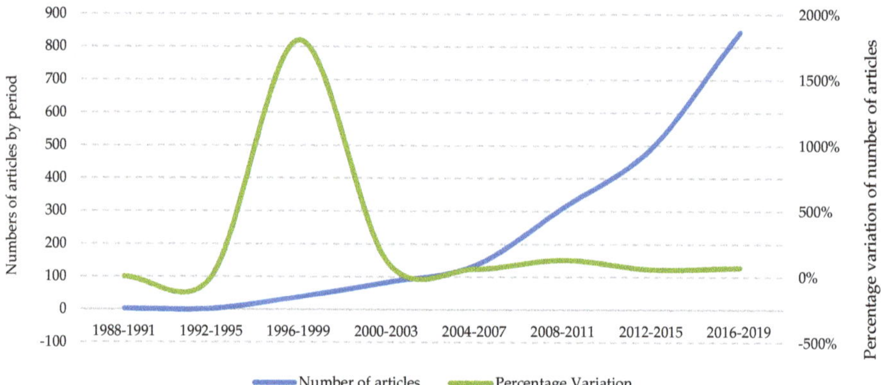

Figure 2. Evolution of the number of articles and percentage of variation between four-year periods.

The total amount of authors who have contributed to this research topic during the analyzed period was 5895. The total number of authors has also increased during this period, as does the volume of articles. In the last three years (2016–2019), 47.70% of the total amount of authors from the 32-year period is concentrated. In addition, we observe that the number of authors who published on the relationship between industrial processes and a sustainable society in the first period (1988–1991) was 4, increasing to 2813 between 2016–2019. The number of authors experience a higher increase than the number of published articles, because in the recent years, the average number of authors per article has also increased. In the four-year period 1988–1991, the average number of authors per article was 2 authors per article, while in the last period (2016–2019) it increased to 3.3, the highest average of authors per article in the studied time horizon.

In the analyzed period (1988–2018), the total number of countries that has at least one pubblication is 99. Thereby, the number of countries has increased from two in the first period (1988–1991) to 82 in the last one (2016–2019).

We account 205 citations for the 2 articles published between 1988–1991, then the number of citations grew exponentially, from the second period (1992–1995), with 63, to the sixth four-year period analyzed (2008–2011), with 11,919. Since 2011, the total number of citations decreased, as showed by the 2012–2015 and 2016–2019 periods, with 11,339 and 5229, respectively. The circumstance of the decrease in the number of citations is due to the fact that the most recent articles published will receive more citations in the coming years [110]. The decreasing in the overall number of citations is also related to the decreasing in the average number of citations per article, because it has decreased from the first four-year period (2001–2003), with 102.5, to the last period (2016–2018), with 6.3.

Finally, the total number of journals where "Industrial process management for a sustainable society" topics were published has been 864 throughout the 32 years period. Thus, the journals incidence increased from 2 in the first observed period to 386 in the last period (2016–2019), which represents 44.7% of the total.

4.2. Distribution of Publications by Subject Area and Journal

The following Section 4.2 displays the main thematic areas related to scientific production. Likewise, a comparison of growth trends according to the main thematic areas is presented and a ranking of the most prolific journals on this subject is shown.

During the analyzed timeframe, 1988–2019, we found 26 categories according to the Scopus classification, with studies related to industrial processes and sustainable society. We classified the sample of 1911 articles in the 26 subject areas, an article can be classified in more than one subject area, depending on the author and editor's interest.

Figure 3 presents the evolution over time of the 5 main themes, through the Scopus analytics that links articles of the sample (1988–2019) with the underpinning subject areas.

The most studied category throughout the studied period is "Environmental Science", with 22% of published articles (914) in this category. Followed, by "Business, Management, and Accounting" (14%, 583), "Engineering" (14%, 583), "Social Sciences" (11%, 467), and "Energy" (9%, 373) categories in order of importance. Therefore, the previous five categories represent 71% of the documents published from 1988 to 2019. The rest of the subject area categories do not exceed 3% of total articles published except for "Agricultural and Biological Sciences" (5.2%, 214), and "Economics, Econometrics, and Finance" (4%, 172).

Environmental Science, with an exponential growth, is the only subject area that published articles in all the analyzed periods, followed by Business, Management, and Accounting that almost published in all periods, except in 1992–1995.

The shared features identified in the "Environmental Science", "Business, Management, and Accounting" and "Engineering and Social Sciences" publications categories, is that the research topics are linked both conceptually and practically to management, technological processes, and the dynamics of industrialization, environmental impact, environmental sustainability, and the socio-economic systems that encompass the biophysical boundaries. Hence, it is recognized by authors who relate the impact of environmental and organizational practices of industries to business [111,112].

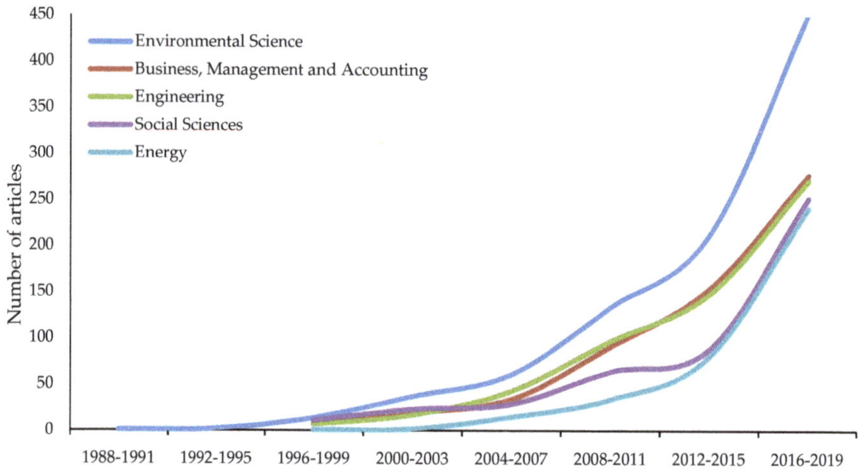

Figure 3. Comparison of growth trends of main subject areas (1988–2019).

Table 3 shows the characteristics of articles in the overall investigation of industrial processes for a sustainable society. We rank the 20 journals with most articles published, the journals belonging to the first quartile (Q1) of the 2018 SJR index stands out with 75% of total articles. In addition, the *International Journal of Production Economics*, has the highest impact factor, SJR, with 2.475 (Q1), followed by *Business Strategy and the Environment* journal, with 2.166 (Q1).

The *Journal of Cleaner Production* (137, 7.17%) and *Sustainability* (111, 5.81%) are the two journals that have published the most articles. The top 20 journals in this research has published 29.20% (558) of all articles. On the other hand, the *Journal of Cleaner Production* is the journal that has the most in the first position of the ranking, with 3 out of 8 periods analyzed, even when the *Sustainability* (99) journal attains the first position, with the highest number published articles, in the last period of the analysis (2016–2019). Furthermore, the *Journal of Cleaner Production* is well-recognized by the scientific community, standing out as the most cited per period (4142), although the *International Journal of Production Economics* holds the highest average of citations per article (185.22) per period and the highest h-index for published articles (38), followed by *Resources Conservation and Recycling* (18) in the second place, while the highest h-index for all subjects is presented by *Science of the Total Environment* (205).

We notice that the herein analyzed research has increased in the number of journals and authors over the years, as evidenced by the increase in the indicators of number of articles and journals [113,114]. The European origin journals stand out as the most relevant: the United Kingdom (8), Netherlands (4), Switzerland (1), Germany (1), and Italy (1); and the American ones (5).

Table 3. Top 20 prolific journals (1998–2019).

Journal	A	TC	TC/A	Ha	Hj	SJR(Q)	Country	R (A)							
								88–91	92–95	96–99	00–03	04–07	08–11	12–15	16–19
Journal of Cleaner Production	137	4142	30.23	38	150	1.620(Q1)	Netherlands	0	0	0	0	1(6)	1(8)	1(33)	2(90)
Sustainability (Switzerland)	111	609	5.49	15	53	0.549(Q2)	Switzerland	0	0	0	0	0	0	2(12)	1(99)
Resources Conservation and Recycling	40	810	20.25	18	103	1.541(Q1)	Netherlands	0	0	0	0	12(2)	5(5)	4(11)	3(22)
Journal of Environmental Management	31	564	18.19	13	146	1.206(Q1)	USA	0	0	0	7(2)	70(1)	3(6)	7(7)	5(15)
Science of the Total Environment	29	475	16.38	14	205	1.536(Q1)	Netherlands	0	0	0	0	0	148(1)	5(9)	4(19)
WIT Transactions on Ecology and the Environment	25	19	0.76	2	19	0.125(Q4)	UK	0	0	0	0	2(6)	4(6)	3(12)	119(1)
Business Strategy and the Environment	21	686	32.67	11	84	2.166(Q1)	USA	0	0	8(1)	0	5(3)	8(4)	20(3)	7(10)
Marine Policy	21	343	16.33	9	79	1.242(Q1)	UK	0	0	0	55(1)	11(2)	38(2)	11(6)	8(10)
Waste Management	18	603	33.50	15	127	1.523(Q1)	UK	0	0	0	0	0	46(2)	14(5)	6(11)
International Journal of Life Cycle Assessment	16	264	16.50	9	89	1.538(Q1)	Germany	0	0	0	0	57(1)	32(2)	9(6)	9(7)
Journal of Sustainable Tourism	14	264	18.86	7	83	1.365(Q1)	UK	0	0	2(2)	8(2)	0	36(2)	52(2)	12(6)
Journal of Construction Engineering and Management	13	560	43.08	12	95	1.044(Q1)	USA	0	0	0	0	10(2)	10(4)	26(3)	31(4)
Water Science and Technology	13	258	19.85	8	124	0.455(Q2)	UK	0	0	35(1)	0	14(2)	161(1)	6(8)	135(1)
International Journal of Production Research	12	619	51.58	10	115	1.585(Q1)	UK	0	0	0	0	0	34(2)	10(6)	27(4)
Journal of Industrial Ecology	11	479	43.55	9	85	1.486(Q1)	USA	0	0	0	0	0	2(8)	49(2)	180(1)
Chemical Engineering Transactions	10	30	3.00	3	29	0.273(Q3)	Italy	0	0	0	0	0	0	15(4)	10(6)
Management of Environmental Quality	10	88	8.80	5	29	0.358(Q3)	UK	0	0	0	0	3(4)	135(1)	27(3)	105(2)
Energy	9	309	34.33	7	158	2.048(Q1)	UK	0	0	13(1)	0	36(1)	13(3)	115(1)	36(3)
International Journal of Production Economics	9	1667	185.22	8	155	2.475(Q1)	Netherlands	0	0	0	0	0	130(1)	16(4)	26(4)
Journal of Management in Engineering	8	292	36.50	6	55	1.271(Q1)	USA	0	0	0	0	73(1)	0	8(7)	0

A: number of articles; R: rank position by number of articles in the four-year period; TC: number of citations; TC/A: number of citations by article; Ha: h-index in articles; Hj: h-index in journal; SJR(Q): Scimago Journal Rank 2018 (Quartile).

In 2008, Vachon and Klassen published the most cited article (848) "Environmental management and manufacturing performance: The role of collaboration in the supply chain" in the *International Journal of Production Economics*, included into the "Business, Management and Accounting" and "Economics, Econometrics and Finance" subject area categories [115]. This article was the only published article about this topic of analysis in the previous journal during the period (2008–2011).

4.3. Productivity of Authors, Institutions, and Countries

Section 4.3 displays the most productive authors and the cooperation between them based on co-authorship. As for the institutions, the most prolific are shown, the countries with the highest production and the network of cooperation among them based on co-authorship, and the most productive international collaborations.

Hence, Table 4 shows what we consider as the most important variables from articles written by the 12 most prolific authors on the topic of management of industrial processes for a sustainable society during the period 1988–2019. It is noteworthy to mention that seven authors in this ranking are European: Germany (3), Italy (2), and the United Kingdom (1), and 2 North Americans: the United States (1), and Canada (1).

The most productive author on the analyzed topic is the German, Seuring, from the Universität Kassel, with seven published articles, followed by the Italians, Colla, from the Scuola superiore di studi universitari e di perfezionamento Sant'Anna, and, Ulgiati, from Parthenope University of Naples, with six published articles each. Regarding citations, the German, Seuring from the Universität Kassel holds the first place with 583 citations, although the authors with the highest average number of citations per article are also Germans, Foerstl and Hartmann, with 107.40 and 103.60, respectively and not the Italians Colla and Ulgiati, accounting six articles each. In addition, Seuring also stands out with the highest h-index (7), followed by Ulgiati (6). All authors in this ranking published their first article from 2000, and seven show interest in this topic in 2019, publishing at least one article.

Likewise, the main thematic area (Environmental Sciences) associated with the contributions of each of the most productive authors has been identified, reflecting the motivations in this scientific field and the inclinations of the journals for the publication of these lines of research, such as the reviewed literature shows [116,117].

Table 4. Top 12 prolific authors (1988–2019).

Author	A	TC	TC/A	Institution	C	First A	Last A	h-index
Seuring, S.	7	583	83.29	Universität Kassel	Germany	2003	2015	7
Colla, V.	6	75	12.50	Scuola Superiore Sant'Anna	Italy	2014	2018	5
Ulgiati, S.	6	129	21.50	Parthenope University of Naples	Italy	2011	2019	6
Brent, A.C.	5	102	20.40	Victoria University of Wellington	New Zealand	2005	2010	5
Foerstl, K.	5	537	107.40	German Graduate School of Management and Law	Germany	2010	2018	5
Genovese, A.	5	334	66.80	Sheffield University Management School	UK	2015	2018	5
Govindan, K.	5	168	33.60	Syddansk Universitet	Denmark	2014	2019	4
Hartmann, E.	5	518	103.60	Friedrich-Alexander-Universität	Germany	2010	2019	4
Hewage, K.	5	205	41.00	The University of British Columbia	Canada	2011	2019	4
Huisingh, D.	5	360	72.00	The University of Tennessee	USA	2009	2019	5
Hussain, M.	5	2	0.40	Abu Dhabi University	United Arab Emirates	2017	2019	1
Raut, R.D.	5	21	4.20	National Institute of Industrial Engineering	India	2015	2019	3

A: number of articles; TC: number of citations; TC/A: number of citations by article; C: country; First A: First article; Last A: Last article; h-index: Hirsch index in this research topic.

Figure 4 shows the networking map between the main authors that published on the topic industrial processes for a sustainable society, based on the co-authorship analysis. In Figure 4, the different colors represent the networking clusters of articles production, while the size of the circle changes according to the number of articles published by author. The network shows high dispersion in the authors' collaboration during the analyzed period (1988–2019). This limited collaboration in the network influences the overall evolution of the social structure of this research topic. We note that the limited scientific collaboration between the authors, in some way involves a wide scope in the subject areas, in addition to promoting innovation, and facilitating access to various topics of study in the field of research [118,119].

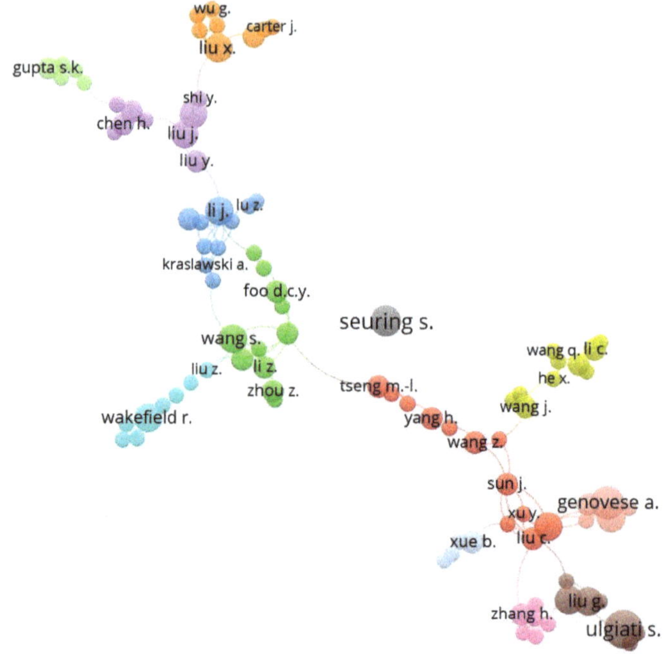

Figure 4. Network map of cooperation between authors based on co-authorship (1988–2019).

Table 5 shows the 10 institutions with the highest number of published articles during the period of study (1988–2019). We observe that the institutions in the ranking have a diverse origin, for example, 30% of the institutions are European (Imperial College London, Politecnico di Milano, and Wageningen University and Research Center), 30% from Asia (Chinese Academy of Sciences, Hong Kong Polytechnic University, and Beijing Normal University), and 20% from Australia (University of Queensland and University of Melbourne). The Chinese Academy of Sciences is the most productive institution with 21 articles, while the most cited institution (851) with the highest average of citations per article (50.06) is the University of British Columbia (851) from Canada. The University of British Columbia and the Imperial College London share the highest h-index (12) in this research topic.

On the other hand, Hong Kong Polytechnic University is the institution with the highest percentage of international collaboration (64.3%), although the international collaboration does not influence the number of citations, in comparison with articles written without international co-authorship. The Politecnico di Milano and the Wageningen University and Research Center are the only two institutions where international co-authorship outstands as a profitable practice, because the articles published by these two institutions have a much higher number of international citations.

Table 5. Top 10 prolific institutions (1988–2019).

Institution	C	A	TC	TC/A	h-index	IC (%)	TCIC	TCNIC
Chinese Academy of Sciences	China	21	302	14.38	11	33.3%	15.86	13.64
The University of British Columbia	Canada	17	851	50.06	12	41.2%	48.86	50.90
Imperial College London	UK	17	772	45.41	12	47.1%	42.13	48.33
University of Queensland	Australia	17	440	25.88	10	52.9%	29.33	22.00
Universidade de Sao Paulo - USP	Brazil	14	241	17.21	8	14.3%	14.00	17.75
Hong Kong Polytechnic University	Hong Kong	14	565	40.36	10	64.3%	25.11	67.80
Politecnico di Milano	Italy	14	322	23.00	11	50.0%	36.00	10.00
University of Melbourne	Australia	14	355	25.36	9	21.4%	11.33	29.18
Wageningen University and Research Centre	Netherlands	13	594	45.69	9	61.5%	64.25	16.00
Beijing Normal University	China	13	124	9.54	8	61.5%	10.88	7.40

C: country; A: number of articles; TC: number of citations; TC/A: number of citations by article; h-index: Hirsch index in research topic; IC: percentage of articles made with international collaboration; TCIC: number of citations by article made with international collaboration; TCNIC: number of citations by article made without international collaboration.

Table 6 shows the countries with the highest scientific production on industrial processes for a sustainable society, including the ones we consider the most characteristics.

In our study, the United States stands out as the country with most publications (289) and with the highest h-index, along with the United Kingdom (47). In addition, the United States has the highest number of citations (7481), and 32.25 average citations per article, positioning in the second place in the average citation rate, just behind Canada (43.96) with only 95 articles.

The amount of American and English publications stands out in the interest of those countries on the management of industrial processes for a sustainable society. The United States started publishing on the management of industrial processes for a sustainable society in 1996, and it has led the ranking of publication productivity on this topic in all periods, excluding the 2004–2017 period. China has the second position in articles productivity at the last period (2016–2019), with 90 items, representing the 68.2% of total production. We observe, what we think is a notworthy characteristic in the top 10 ranking of countries articles production, that Italy (2003), Brazil (2003), and Spain (2003), out of 10 countries, published its first article until the fourth period (2000–2003).

In short, the United States, the United Kingdom, Australia, China, and Italy, are the 5 most engaged countries on this research subject, with 50% (952) of all analyzed articles. This ranking coincides with the countries that produce the most scientific documents, and the excellence rate that includes the most cited articles in their area. The latter is an indicator of the high quality of research in these countries [120,121].

Table 6. Top 10 prolific countries in number of articles (1998–2019).

Country	A	TC	TC/A	h-index	R (A)							
					1988–1991	1992–1995	1996–1999	2000–2003	2004–2007	2008–2011	2012–2015	2016–2019
USA	289	6828	23.63	47	0	0	1(9)	1(17)	3(16)	1(62)	1(93)	1(92)
UK	232	7481	32.25	47	2(1)	0	2(8)	3(12)	1(20)	2(50)	3(52)	3(89)
Australia	189	3760	19.89	35	1(1)	0	5(2)	2(12)	2(19)	3(43)	2(53)	5(59)
China	132	1146	8.68	20	0	0	9(1)	19(1)	10(5)	11(8)	7(27)	2(90)
Italy	110	2236	20.33	25	0	0	0	22(1)	21(2)	9(9)	4(42)	7(56)
Brazil	107	1064	9.94	19	0	0	0	17(1)	9(6)	10(8)	5(36)	6(56)
Spain	101	2416	23.92	25	0	0	0	31(1)	13(4)	4(18)	8(27)	8(51)
India	100	1290	12.90	16	0	0	6(2)	6(4)	11(5)	12(8)	10(20)	4(61)
Germany	98	2822	28.80	28	0	0	11(1)	5(6)	4(10)	8(13)	6(29)	11(39)
Canada	95	4176	43.96	28	0	0	7(1)	4(7)	6(9)	5(17)	9(20)	10(41)

A: number of articles; R: rank position by number of articles in the four-year period; TC: number of citations; TC/A: number of citations by article; h-index: Hirsch index in research topic.

Table 6 shows the indicators regarding international collaboration ordered by the amount of scientific production in the observed period (1988–2019). We identify the international networks of scientific collaboration lead by China with 55.3% (45 countries), followed by Canada (50.5%, 28), Spain (45.5%, 38), and the United Kingdom (41.4%, 45). In our ranking, India is the country with the lowest percentage of international collaboration (17%, 8).

On the other hand, we observe in Table 7 that the number of articles' citations is higher where international co-authorship is entailed than those made without international collaborations, exception made by the United Kingdom, Germany, and Canada.

Table 7. Top 10 prolific countries and international collaboration (1998–2019).

Country	NC	Main Collaborators	IC (%)	TC/A	
				IC	NIC
USA	44	UK, Canada, China, Australia, Germany	33.6%	32.48	19.15
UK	45	USA, Spain, Germany, Italy, Canada	41.4%	29.72	34.03
Australia	36	China, UK, USA, Canada, Netherlands	39.2%	22.26	18.37
China	38	Australia, USA, UK, Hong Kong, Italy	55.3%	9.03	8.25
Italy	30	UK, Spain, USA, China, Germany	33.6%	33.35	13.73
Brazil	22	Canada, UK, USA, China, Portugal	21.5%	12.43	9.26
Spain	38	UK, Italy, Chile, Germany, USA	45.5%	30.57	18.36
India	8	USA, Denmark, UK, Australia, Canada	17.0%	38.76	7.60
Germany	26	UK, Switzerland, USA, Austria, Netherlands	39.8%	26.59	30.25
Canada	28	USA, UK, Australia, Brazil, China	50.5%	27.31	60.96

NC: number of collaborator countries; IC: percentage of articles made with international collaboration; TC/A: number of citations by article; IC: international collaboration; NIC: no international collaboration.

Figure 5 shows a networking map between countries based on the co-authorship analysis, where the different colors represent the different clusters build up by the groups of countries, while the size of the circle changes according to the number of articles published by country [122,123]. The VOSviewer software identifies eight sets of components.

Cluster 1 is the largest and includes 14 countries, led by Australia with 189 published articles. Cluster 1 is associated with Malaysia (65 articles), Iran (22), Mexico (21), Chile (18), Indonesia (18), Japan (16), Lithuania (11), Saudi Arabia (11), Pakistan (10), Egypt (9), Bangladesh (7), Kenya (6), and Nigeria (6). Cluster 2 has Italy (100) at the head and is associated with Brazil (107), Canada (95), Sweden (52), France (50), Portugal (26), Thailand (21), Norway (16), United Arab Emirates (11), Serbia (10), and Slovenia (7). Cluster 3 is led by Spain (101) and includes Finland (42), Russian Federation (22), Greece (18), New Zealand (18), Romania (14), Poland (13), and Hungary (12). Cluster 4 is leaded by Germany (98), and includes the Netherlands (60), Switzerland (35), Austria (22), Belgium (20), Ireland (17), and Colombia (9). Cluster 5 is leaded by the United States (289) and includes Turkey (38), South Korea (26), Singapore (9), Qatar (6), and Croatia (5). Cluster 6 has China (132) at the head and cooperates with South Africa (47), Taiwan (32), Hong Kong (23), and the Philippines (8). Cluster 7 is led by the United Kingdom (232), and collaborates with Czech Republic (8), and Slovakia (7). Finally, cluster 8 is composed by India (100) and Denmark (29).

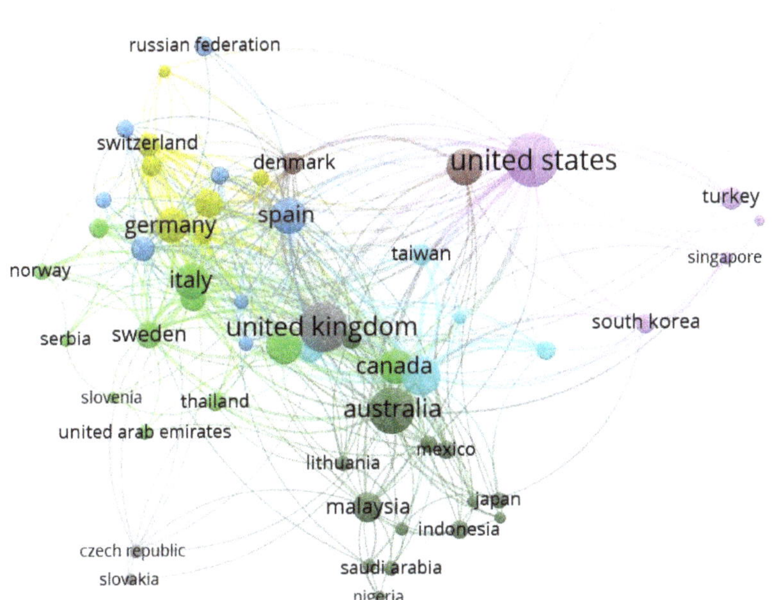

Figure 5. Network of cooperation between countries based on co-authorship (1988–2019).

4.4. Keywords Analysis

Section 4.4 displays an analysis of the most relevant keywords, the keyword network map based on co-occurrence and their evolution, during the period analyzed (1988–2019).

Therefore, Table 8 lists the 20 most frequently used keywords in the sample of 1911 articles during the observed period (1988–2019) divided in eight four-year periods within the 32-year timeframe. The most outstanding terms are Sustainability (864 articles, 45.2%) and Sustainable Development (757, 39.6%). In the literature review, both terms are considered as synonyms [124], then if we observe them together, they rank first in 6 out of 8 four-year periods (Sustainability: 2000–2003, 2008–2011, 2016–2019; and Sustainable Development: 1996–1999, 2004–2007, 2012–2015). We identify the keywords Environmental Management (262, 13.7%), Decision Making (235, 12.3%), and Environmental Impact (212, 11.1%) listed in order of importance.

Indeed, we identify four thematic axes according to the keywords semantics. Hence, the first group is associated with sustainability in processes (Sustainability, Sustainble Development, Life Cycle, Recycling, and Life Cycle Assessment); the second is related to the environmental dimension of industry (Environmental Management, Environmental Impact, Environmental Sustainability, and Environmental Protection); the third with the management of industrial processes (Decision Making, Waste Management, Supply Chain Management, Water Management, Project Management, Economics, and Stakeholder); and the fourth with the industry itself (Construction Industry, Industry, and Innovation).

Table 8. Top 20 keywords (1988–2019).

Keyword	1988–2019		1988–1991		1992–1995		1996–1999		2000–2003		2004–2007		2008–2011		2012–2015		2016–2019	
	(A)	%	R (A)	%	R (A)	%	R (A)	%	R (A)	%	R (A)	%	R (A)	%	R (A)	%	R (A)	%
Sustainability	864	45.2%	0	–	28(1)	50.0%	6(5)	13.2%	1(39)	46.4%	2(47)	34.8%	1(154)	49.5%	2(206)	31.2%	1(412)	18.2%
Sustainable Development	757	39.6%	14(1)	50.0%	0	–	1(10)	26.3%	2(28)	33.3%	1(51)	37.8%	2(123)	39.5%	1(213)	24.9%	2(331)	14.6%
Environmental Management	262	13.7%	8(1)	50.0%	3(2)	100.0%	12(3)	7.9%	10(7)	8.3%	10(11)	8.1%	4(42)	13.5%	4(72)	8.5%	4(124)	5.0%
Decision Making	235	12.3%	0	–	0	–	0	–	7(8)	9.5%	11(10)	7.4%	6(35)	11.3%	8(52)	7.1%	3(130)	4.1%
Environmental Impact	212	11.1%	0	–	0	–	2(9)	23.7%	3(15)	17.9%	3(20)	14.8%	9(26)	8.4%	5(63)	5.3%	8(79)	3.1%
Environmental Sustainability	182	9.5%	0	–	0	–	0	–	33(4)	4.8%	18(7)	5.2%	11(22)	7.1%	7(53)	4.5%	7(96)	2.6%
Waste Management	169	8.8%	0	–	29(1)	50.0%	44(2)	5.3%	18(6)	7.1%	4(18)	13.3%	8(32)	10.3%	10(38)	6.5%	10(72)	3.8%
Supply Chain Management	149	7.8%	0	–	0	–	0	–	0	–	104(3)	2.2%	22(16)	5.1%	15(31)	3.2%	6(98)	1.9%
Life Cycle	147	7.7%	0	–	0	–	0	–	38(4)	4.8%	26(6)	4.4%	13(20)	6.4%	11(37)	4.0%	9(79)	2.4%
Construction Industry	143	7.5%	0	–	0	–	89(1)	2.6%	27(4)	4.8%	8(12)	8.9%	5(35)	11.3%	13(31)	7.1%	11(60)	4.1%
Industry	132	6.9%	0	–	0	–	0	–	6(9)	10.7%	36(5)	3.7%	7(33)	10.6%	6(60)	6.7%	48(24)	3.9%
Recycling	132	6.9%	0	–	19(1)	50.0%	5(5)	13.2%	41(4)	4.8%	7(13)	9.6%	10(25)	8.0%	19(27)	5.1%	13(57)	3.0%
Environmental Protection	113	5.9%	0	–	12(1)	50.0%	4(5)	13.2%	5(10)	11.9%	16(8)	5.9%	27(13)	4.2%	12(36)	2.6%	20(40)	1.5%
Water Management	108	5.7%	0	–	0	–	47(2)	5.3%	13(7)	8.3%	67(4)	3.0%	15(19)	6.1%	9(39)	3.8%	25(37)	2.2%
Project Management	98	5.1%	0	–	0	–	0	–	0	–	9(12)	8.9%	12(21)	6.8%	17(28)	4.3%	24(37)	2.5%
Priority Journal	89	4.7%	0	–	0	–	0	–	128(2)	2.4%	98(3)	2.2%	39(11)	3.5%	16(29)	2.2%	16(43)	1.3%
Innovation	87	4.6%	0	–	0	–	0	–	0	–	20(7)	5.2%	21(16)	5.1%	31(20)	3.2%	15(43)	1.9%
Economics	83	4.3%	0	–	0	–	3(5)	13.2%	20(5)	6.0%	81(3)	2.2%	36(11)	3.5%	18(27)	2.2%	29(32)	1.3%
Life Cycle Assessment	83	4.3%	0	–	0	–	0	–	111(2)	2.4%	54(4)	3.0%	50(10)	3.2%	27(21)	2.0%	14(45)	1.2%
Stakeholder	80	4.2%	0	–	0	–	0	–	0	–	21(7)	5.2%	18(17)	5.5%	20(26)	3.4%	39(29)	2.0%

A: number of articles; R: rank position by number of articles in the four-year period; %: percentage of articles in which it appears; (–): not data.

Figure 6 represents the keywords network map of the 1911 articles on the research topic. The keyword network maps is build up through the co-occurrences method and the different colors of nodes identify the clusters, while their size changes according to the keywords' frequency. We identify four main lines of research: "Environmental Protection", "Environmental Management", "Environmental Impact", and "Water Management" [35,39,64,74,125,126]. We set up this clusters according to the keywords associated to most articles in the sample, using the VOSviewer software.

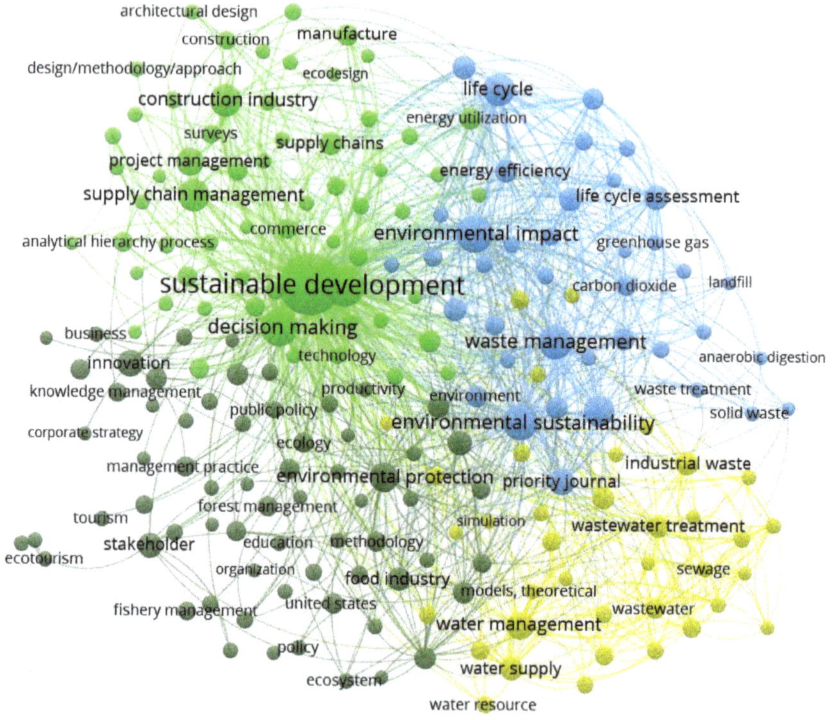

Figure 6. Keywords network map based on co-occurrence (1998–2019).

Cluster 1 (grey) is about the thematic axe "Sustainability in processes". Cluster 1 is the bigger group and it has the 32% of the total keywords, Environmental Protection leads Cluster 1 with (113 articles) followed by Innovation (87 articles), Economics (83), Stakeholder (80), Food Industry (68), Economic and Social Effects (58), Agriculture (55), Environmental Economics (54), Conservation of Natural Resources (52), Forestry (45), Corporate Social Responsibility (44), Ecology (44), Forest Management (43), Implementation Process (39), Land Use (38), Governance Approach (33), Case Study (32), Public Policy (32), Strategic Approach (31), Strategic Planning (31), Biodiversity (30), Developing Countries (30), Industrial Ecology (30), Ecotourism (27), Education (27), Policy (26), Economic Aspect (25), or Ecosystem (25) among others.

We associate cluster 1 with the "Environmental protection" research line. This research line enables a political, economic, educational, and touristic approach. The political approach considers the relation to normative regulation; the economic one considers the costs and the consumption; the educational approach studies training at different academic levels; and the touristic approach considers tourism as an outstanding sector for a more sustainable society [127,128].

Cluster 2 (green) is about the thematic axe "Environmental dimension of industry". Cluster 2 has the highest centrality and includes the 30% of total keywords. Cluster 2 encompasses the keywords Environmental Management (262), Decision Making (235), Supply Chain Management (149), Construction Industry (143), Project Management (98), Supply Chains (62), Manufacture (60), Pollution Control (57), Energy Utilization (52), Analytical Hierarchy Process (51), Manufacturing (51), Risk Management (49), Commerce (46), Risk Assessment (44), Surveys (42), Decision Support Systems (39), Construction (37), Industrial Economics (36), Performance Assessment (36), Information Management (35), Sustainability Assessment (35), Sensitivity Analysis (34), Sustainable Supply Chains (34), Cleaner Production (33), Technology (33), Buildings (31), Quality Control (31), Environmental Technology (29), Hierarchical Systems (29), Product Design (29), Manufacturing Industries (28), Product Development (28), Costs (27), Industrial Management (27), Textile Industry (27), Environmental Regulations (26), Industrial Research (26), and Benchmarking (25).

We associate cluster 2 with the "Environmental management" research line. This research line studies the process that allows the reduction of environmental impacts and the efficiency progress to achieve economic, environmental, and operational improvements of organizations [129,130].

Cluster 3 (blue) is about the thematic axe "Industrial processes". Cluster 3 gathers 19% of all keywords. It is associated with the following keywords: Environmental Sustainability (182), Waste Management (169), Life Cycle (147), Recycling (132), Life Cycle Assessment (83), Life Cycle Analysis (72), Environmental Performance (70), Climate Change (63), Wastewater Treatment (58), Life Cycle Assessment (LCA) (56), Environmental Impact Assessment (55), Waste Disposal (54), Environment (52), Carbon Dioxide (48), Optimization (44), Greenhouse Gases (37), Investments (37), Carbon Footprint (36), Emission Control (32), Biomass (31), Economic Analysis (31), Circular Economy (30), Global Warming (30), Carbon (27), Cost Benefit Analysis (27), Greenhouse Gas (27), Gas Emissions (26), Solid Waste (26), Biofuel (25), and Energy Management (25).

We associate cluster 3 with the "Environmental impact" research line. This research line studies the impact of industrial processes on the environment and its relationship with a more sustainable society. The scientific production of this thematic axis deals with the direct impact of the industry on nature, by the territorial development, the use of natural resources and waste production [131–134].

Finally, cluster 4 (yellow) is holds the "Industry" thematic axe. Cluster 4 is the smallest one, composed by 18% of keywords and head the "Water Management" research line. Cluster 4 is associated with the following keywords: Water Supply (71), Industrial Waste (62), Wastewater (45), Water Conservation (41), Waste Water Management (39), Water Resources (38), Energy Conservation (36), Water Quality (36), Chemical Industry (35), Water (32), Pollution (31), Water Resource (31), Water Pollution (30), Water Treatment (30), Waste Water (29), Sewage (28), Waste Treatment (27), Effluent (26), Effluents (25), and Energy (25).

Water management is a research line that studies the sustainable water resources management and its effects on society. We define this research line as the effort to improve the understanding on how to promote a collaborative management of water, land, and resources, with the aim to maximize wellbeing and social justice, without compromising the sustainable functionality of ecosystems [135,136].

The four previous mentioned thematic axes encompass concepts related to the industrial processes management to achieve a more sustainable society, for example, sustainable development in globalized times and the causal effects of industrial management in society. There is a consolidation in the development of research that has led to an increase in the number of contributions, in relation to the implementation of management and implementation of sustainable policies in industrial processes. Similarly, there is an evolution in terminology that is accompanied by the emergence of new lines of research, such as those expressed below and in line with the results obtained [137,138].

4.5. Future Research Perspectives

Figure 7 shows the changes over time and stability of each keyword network, because we can identify the time period in which they have been studied, according to the research sample observed in the 1988–2019 period.

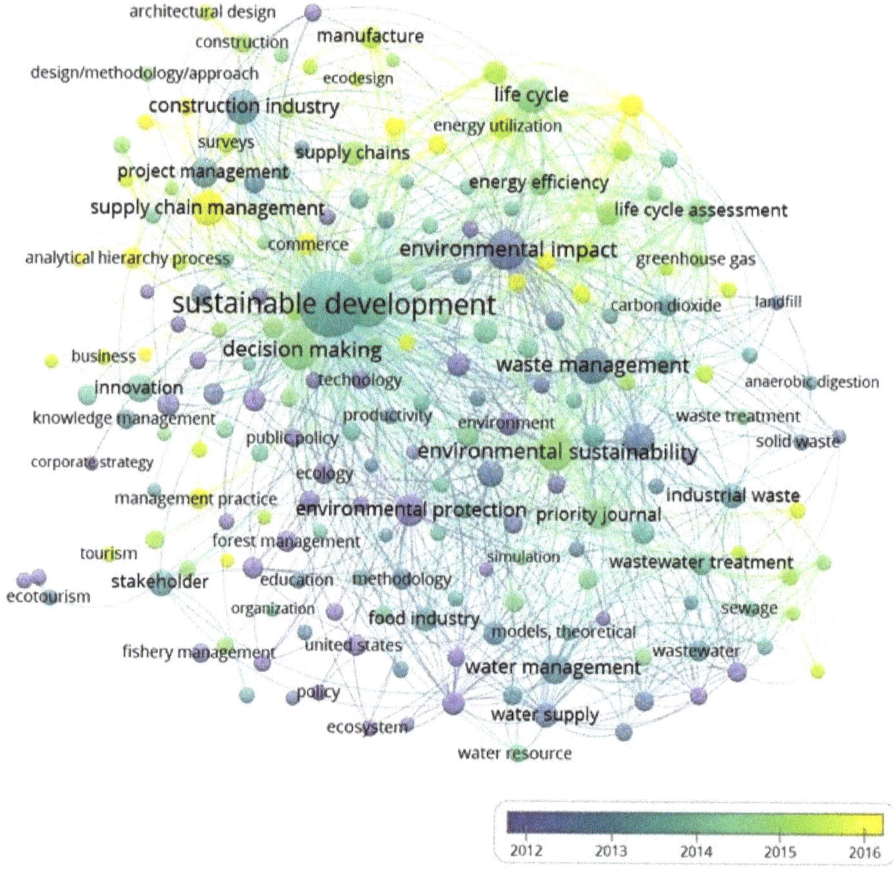

Figure 7. Evolution of keywords network map based on co-occurrence (1998–2019).

We assume a significative increase in the keywords related to the research topic in the 2012–2015 period, influenced by the exponential increase in the articles publication on management of industrial processes for a more sustainable society. The keywords diversity in the bibliometric analysis brings out an idea of the range of researchs topics in the field of study. On the other hand, we also observe the emergence of another keywords group from 2016. The new keyword group encompass concepts like green cluster and environmental management to improve firms' environmental image and social responsibility coming from public authorities and civil society pressures.

Hence, Figure 7 let us to improve our understanding on the keywords relevance over time. So, the early bird keywords in the timeline have strong influence over the most recent keywords.

Research moves forward worldwide, incorporating new concepts and strategies that open new research lines. Indeed, we are witnessing the emergence of keywords such as social welfare, attracting consumer, or competitiveness that bridge the gap between Industry 4.0 and sustainability. The outcoming group of keywords pave the way for a new research line where the impact of Industry 4.0 on the triple bottom line will be explored [139,140]. The previous mentioned research line seeks to better understand how this dynamic relationship between economic development (job creation, cost reduction, and risk management); environmental responsibility (energy and resources conservation, renewable energy consumption, recycling, packaging reduction and decarbonization); and social welfare (working condition regulations, community enhancement, and the development of social responsibility in products and services).

On the other hand, we also identify controversial keywords related with the research topic in the analyzed literature, keywords like agrochemical, pesticide, tobacco, defense industry, bioengineering, robotics, or conflict mineral exploitation. These terms are connected to a new research line that adresses controversial issues at industries, provocative industries by its definition of international trade agreements, or by the unfair labor and wage conditions resulting from the high mobility of global capitals [141,142].

Moreover, we identify a concurrent research line that explores the regional and social perspectives of the "sin stock sectors". These sectors include the alcohol, tobacco, gambling, sex and firearms industry of which characteristics changes around the world. So, concerning this research line, the studies' interests turn around the behavior of ethical and socially responsible investments regarding ethically questionable activities, assuming the social benefit as the final goal [143,144].

We recognize "the concept design of sustainable industrial process" as a new thematic axe encompassing keywords such as ecodesign, architectural design, or construction industry. Currently, a large number of companies include sustainable development within their commercial strategies, therefore industrial engineers seek to implement this strategy in the design of manufacturing process.

Likewise, artificial intelligence (AI) and sustainable development concepts set up the development of a new research line, aiming to understand how current technology will affect the planet through climate change and biodiversity loss. Therefore, previous research line studies how AI applications bring environmental benefits or not, considering tradeoffs, spillover, and rebound effects on water quality, air pollution, deforestation, soil quality degradation, and biodiversity [145,146].

5. Conclusions

The aim of this study is to analyze the research trends of scientific production on industrial processes management for a more sustainable society, during the last 32 years. We drew up a bibliometric analysis of 1911 articles from the Elsevier Scopus database to depict the evolution of scientific production, thematic areas, journals, authors, institutions, the most productive countries, and the main research lines based on the keyword network analysis.

Scientific production increased every year during the 1988–2019 period, especially in the last period (2016–2019), where 845 documents have been published, accounting for the 44.2% of total articles.

We observe that most articles of our research topic are in the following four subject areas: Environmental Science; Business, Management and Accounting; Engineering; and Social Sciences.

The outstanding diversity of subject areas in the articles sample demonstrates the cross-sectional approach of sustainability, which is linked with environmental, economic, social, scientific, and technological disciplines. Furthermore, the most productive journals are the *Journal of Cleaner Production* and *Sustainability*, together gathering 12.98% of total articles published. We observe that 75% of journals are in the first quartile of journals' ranking Scopus. The *Journal of Cleaner Production* is the most cited, and it has the highest h-index, but the *International Journal of Production Economics* has the highest average citations per article.

We notice that the 60% of authors are Europeans, therefore the most prolific authors, the most cited, and the author's with highest average citations per article belong to German research institutes.

The countries that publish the most in this research area are the United States and the United Kingdom, Australia, China, and Italy, with 50% of the articles sample. Hong Kong has the highest percentage of international collaboration regarding articles' publication; even when this international exposition is not translated into higher citations rate, in comparison with the articles without international collaboration. The Netherlands and Italy are the only two countries that take advantage of the international co-authorship getting higher international citations.

In addition, we set up four thematic axes resulting from the industrial process management of industrial for a more sustainable society. The first thematic axis studies environmental protection, taking different approaches (political, economic, educational, or touristic); the second thematic axis addresses the environmental management; the third analyses the impact of environmental industrial processes on a more sustainable society; and, finally, the fourth thematic axis studies the sustainable management of water resources and its causal effects on society.

Likewise, we observed that the research field of industrial processes management for a sustainable society is gaining more attention from scholars and practitioners over the years. New research topics emerge, within the boundaries of interest of this study, to analyze the impact of Industry 4.0 on the sustainability triple bottom line. New research subjects address controversial issues at industries, provocative industries by its definition of international trade agreements, or by the unfair labor and wage conditions resulting from the high mobility of global capitals. Furthermore, we identify a concurrent research line that explores the behavior of ethical and socially responsible investments regarding non-ethical activities. On the other hand, new studies should address the concept design of sustainable industrial process, implemented in the manufacturing processes design. We identify the need for studies seeking to understand the relationship between artificial intelligence and sustainable development, to assess how current technological advances will affect the planet, due to challenges such as climate change and the loss of biodiversity.

Innovation in this research field has been identified based on the morphology of the clusters of authors, institutions, countries, and keywords, and the intensity of the relationships that develop in them. The results obtained are a complement to the area of knowledge of industrial processes and allow establishing the relationship between science and technology and supporting the decision-making process.

This study identifies some limitations that could be addresses in further research studies. In our study, the outcome is conditioned by the methodology we define, the database chosen, the keywords selected to extract the articles sample; the timeframe; and the analyzed variables. Therefore, the use of a different database and variables will bring out different outcomes, complementing the analysis and opening the floor for comparative studies.

Finally, we observed in the recent years that international research on industrial processes management for a more sustainable society shows an upward trend.

Author Contributions: Conceptualization, E.A.-S., M.E.M. and L.J.B.-U.; Investigation, E.A.-S., M.E.M. and L.J.B.U.; Methodology, M.E.M. and L.J.B.-U.; Resources, E.A.-S., M.E.M., F.J.C.-G. and L.J.B.-U.; Software, M.E.M. and L.J.B.-U.; Supervision, E.A.-S., M.E.M. and L.J.B.-U.; Validation, M.E.M., F.J.C.-G. and L.J.B.-U.; Writing—original draft, E.A.-S., M.E.M., F.J.C.-G. and L.J.B.-U.; Writing—review & editing, E.A.-S., M.E.M. and L.J.B.-U; formal analysis, data curation, E.A.-S. and L.J.B.-U.; project administration, E.A.-S., M.E.M., F.J.C.-G. and L.J.B.-U.; funding acquisition, L.J.B.-U. All authors have read and agreed to the published version of the manuscript.

Funding: This research received no external funding.

Conflicts of Interest: The authors declare no conflict of interest.

References

1. Swain, R.B.; Yang-Wallentin, F. Achieving sustainable development goals: Predicaments and strategies. *Int. J. Sustain. Dev. World Ecology* **2019**, *27*, 96–106. [CrossRef]
2. Xu, G. A Method of Establishing Process Specifications in Process Industry Based on Statistical Process Control. *J. Mech. Eng.* **2019**, *55*, 208. [CrossRef]
3. Xia, X.; Zhang, C. The Impact of Authorized Remanufacturing on Sustainable Remanufacturing. *Processes* **2019**, *7*, 663. [CrossRef]
4. Yao, T.; Huang, Z.; Zhao, W. Are smart cities more ecologically efficient? Evidence from China. *Sustain. Cities Soc.* **2020**, 102008. [CrossRef]
5. Jacobides, M.G.; Kudina, A. How Industry Architectures Shape Firm Success when Expanding in Emerging Economies. *Glob. Strategy J.* **2013**, *3*, 150–170. [CrossRef]
6. Lee, S. Role of social and solidarity economy in localizing the sustainable development goals. *International J. Sustain. Dev. World Ecology* **2019**, *27*, 65–71. [CrossRef]
7. Vladimirova, K.; Le Blanc, D. Exploring Links between Education and Sustainable Development Goals Through the Lens of UN Flagship Reports. *Sustain. Dev.* **2016**, *24*, 254–271. [CrossRef]
8. Hickel, J. The contradiction of the sustainable development goals: Growth versus ecology on a finite planet. *Sustain. Dev.* **2019**, *27*, 873–884. [CrossRef]
9. Gude, V.G. Sustainable chemistry and chemical processes for a sustainable future. *Resour. Effic. Technol.* **2017**, *3*, 249–251. [CrossRef]
10. Lanciano, E.; Saleilles, S. Small firms in the sustainable transformation of food industry: Entangling entrepreneurship and activism in grassroots innovation processes. *Sociol. DEL Lav.* **2017**, *147*, 111–127. [CrossRef]
11. Kyzas, G.; Matis, K. The Flotation Process Can Go Green. *Processes* **2019**, *7*, 138. [CrossRef]
12. Torres, J.; Valera, D.L.; Belmonte, L.J.; Herrero-Sánchez, C. Economic and social sustainability through organic agriculture: Study of the restructuring of the citrus sector in the "bajo andarax" district (Spain). *Sustainability* **2016**, *8*, 918. [CrossRef]
13. Humphries, J.; Schneider, B. Spinning the industrial revolution. *Econ. Hist. Rev.* **2018**, *72*, 126–155. [CrossRef]
14. Greasley, D.; Oxley, L. Causality and the First Industrial Revolution. *Ind. Corp. Chang.* **1988**, *7*, 33–47. [CrossRef]
15. Boyer, G.R.; Mokyr, J. The Economics of the Industrial Revolution. *Ind. Labor Relat. Rev.* **1986**, *39*, 604. [CrossRef]
16. Landsberger, H.A.; Hartwell, R.M. The Industrial Revolution. *Soc. Forces* **1974**, *53*, 144. [CrossRef]
17. Jackson, R.V. Rates of Industrial Growth during the Industrial Revolution. *Econ. Hist. Rev.* **1992**, *45*, 1. [CrossRef]
18. Loureiro, A. There is a fourth industrial revolution: The digital revolution. *Worldw. Hosp. Tour. Themes* **2018**, *10*, 740–744. [CrossRef]
19. Rhyne, E.; Otero, M. Microfinance Matures: Opportunities, Risks, and Obstacles for an Emerging Global Industry. *Innov. Technol. Gov. Glob.* **2007**, *2*, 91–114. [CrossRef]
20. Darkow, I.-L.; Von Der Gracht, H.A. Scenarios for the future of the European process industry - the case of the chemical industry. *Eur. J. Futures Res.* **2013**, 1. [CrossRef]
21. Jänicke, M. "Green growth": From a growing eco-industry to economic sustainability. *Energy Policy* **2012**, *48*, 13–21. [CrossRef]
22. Hmaittane, A.; Bouslah, K.; M'Zali, B. Does corporate social responsibility affect the cost of equity in controversial industry sectors? *Rev. Account. Financ.* **2019**, *18*, 635–662. [CrossRef]
23. O'Riordan, T.J.C. UN sustainable development goals: How can sustainable/green chemistry contribute? The view from the agrochemical industry. *Curr. Opin. Green Sustain. Chem.* **2018**, *13*, 172–173. [CrossRef]
24. Kostygova, L.A. Sustainable Development Titanium Industry. *Econ. Ind.* **2015**, *4*, 20. [CrossRef]
25. Patel, D.; Kellici, S.; Saha, B. Green Process Engineering as the Key to Future Processes. *Processes* **2014**, *2*, 311–332. [CrossRef]

26. Ferto, I.; Soós, K.A. Controversial issues in factors determining intra-industry trade. *Int. J. Econ. Bus. Res.* **2012**, *4*, 132. [CrossRef]

27. Potgieter, J.H. An Overview of Cement production: How "green" and sustainable is the industry? *Environ. Manag. Sustain. Dev.* **2012**, 1. [CrossRef]

28. Luken, R.A. Where is developing country industry in sustainable development planning? *Sustain. Dev.* **2006**, *14*, 46–61. [CrossRef]

29. Venables, M. Striving for process excellence [process industry]. *Manuf. Eng.* **2004**, *83*, 40–43. [CrossRef]

30. Stigson, B. Sustainable development for industry and society. *Build. Res. Inf.* **1999**, *27*, 424–430. [CrossRef]

31. Szekely, J.; Laudise, R. Green materials and green processes. *J. Mater. Res.* **1995**, *10*, 485–486. [CrossRef]

32. Freeman, R.E.; Phillips, R.A. Stakeholder Theory: A Libertarian Defense. *Bus. Ethics Q.* **2002**, *12*, 331–349. [CrossRef]

33. Smith, P.A. Stakeholder Engagement Framework. *Inf. Secur. Int. J.* **2017**, *38*, 35–45. [CrossRef]

34. Schaltegger, S.; Hörisch, J.; Freeman, R.E. Business Cases for Sustainability: A Stakeholder Theory Perspective. *Organ. Environ.* **2017**, *32*, 191–212. [CrossRef]

35. Plaza Úbeda, J.A.; Jiménez, J.D.B.; Belmonte-Ureña, L.J. Stakeholders, environmental management and performance: An integrated approach. *Cuad. De Econ. Y Dir. De La Empresa* **2011**, *14*, 151–161. [CrossRef]

36. Cots, E.G. Stakeholder social capital: A new approach to stakeholder theory. *Bus. Ethics A Eur. Rev.* **2011**, *20*, 328–341. [CrossRef]

37. Freeman, E.; Moutchnik, A. Stakeholder management and CSR: Questions and answers. *Uwf Umw.* **2013**, *21*, 5–9. [CrossRef]

38. Al-Kassem, A.H. Recruitment and Selection Practices in Business Process Outsourcing Industry. *Arch. Bus. Res.* **2017**, 5. [CrossRef]

39. Styhre, A.; Ingelgård, A.; Roth, J. Gendering knowledge: The practices of knowledge management in the pharmaceutical industry. *Knowl. Process. Manag.* **2001**, *8*, 65–74. [CrossRef]

40. Plaut, J. Industry environmental processes: Beyond compliance. *Technol. Soc.* **1998**, *20*, 469–479. [CrossRef]

41. Canabarro, N.; Soares, J.F.; Anchieta, C.G.; Kelling, C.S.; Mazutti, M.A. Thermochemical processes for biofuels production from biomass. *Sustain. Chem. Process.* **2013**, *1*, 22. [CrossRef]

42. Roddeck, W. Automation in forming processes. *Comput. Ind.* **1979**, *1*, 111–116. [CrossRef]

43. Rosenthal, K.; Lütz, S. Recent developments and challenges of biocatalytic processes in the pharmaceutical industry. *Curr. Opin. Green Sustain. Chem.* **2018**, *11*, 58–64. [CrossRef]

44. Sahu, O.; Rao, D.G.; Thangavel, A.; Ponnappan, S. Treatment of sugar industry wastewater using a combination of thermal and electrocoagulation processes. *Int. J. Sustain. Eng.* **2017**, *11*, 16–25. [CrossRef]

45. Huang, Y.; Fan, L. Artificial intelligence for waste minimization in the process industry. *Comput. Ind.* **1993**, *22*, 117–128. [CrossRef]

46. Leppänen, P.; Kraslawski, A.; Avramenko, Y. Innovation in Organization of Design Process in Pulp and Paper Industry Projects. *Knowl. Process. Manag.* **2011**, *18*, 133–149. [CrossRef]

47. Zhang, L.; Wang, J.; Wen, H.; Fu, Z.; Li, X. Operating performance, industry agglomeration and its spatial characteristics of Chinese photovoltaic industry. *Renew. Sustain. Energy Rev.* **2016**, *65*, 373–386. [CrossRef]

48. McManus, P. Defining sustainable development for our common future: A history of the World Commission on Environment and Development (Brundtland Commission). *Aust. Geogr.* **2014**, *45*, 559–561. [CrossRef]

49. Holden, E.; Linnerud, K.; Banister, D. The Imperatives of Sustainable Development. *Sustain. Dev.* **2016**, *25*, 213–226. [CrossRef]

50. Brundtland, G.H. World summit on sustainable development. *BMJ* **2002**, *325*, 399–400. [CrossRef]

51. Holden, E.; Linnerud, K.; Banister, D. Sustainable development: Our Common Future revisited. *Global Environ. Chang.* **2014**, *26*, 130–139. [CrossRef]

52. Dalampira, E.; Nastis, S.A. Mapping Sustainable Development Goals: A network analysis framework. *Sustain. Dev.* **2019**, *28*, 46–55. [CrossRef]

53. Wysokińska, Z. Millenium Development Goals/UN and Sustainable Development Goals/UN as Instruments for Realising Sustainable Development Concept in the Global Economy. *Comp. Econ. Res.* **2017**, *20*, 101–118. [CrossRef]

54. Fedulova, L. Development trends and implementation of digital technologies for sustainable development goals. *Environ. Econ. Sustain. Dev.* **2020**, *7*, 6–14. [CrossRef]

55. Rosen, M.A. How Can We Achieve the UN Sustainable Development Goals? *Eur. J. Sustain. Dev. Res.* **2017**, *1*, 1–4. [CrossRef]

56. Haliscelik, E.; Soytaş, M.A. Sustainable development from millennium 2015 to Sustainable Development Goals 2030. *Sustain. Dev.* **2019**, *27*, 545–572. [CrossRef]

57. Luken, R.; Mörec, U.; Meinert, T. Data quality and feasibility issues with industry-related Sustainable Development Goal targets for Sub-Saharan African countries. *Sustain. Dev.* **2019**, *28*, 91–100. [CrossRef]

58. Kopnina, H. The victims of unsustainability: A challenge to sustainable development goals. *Int. J. Sustain. Dev. World Ecol.* **2015**, *23*, 113–121. [CrossRef]

59. Zhu, P. Studies on Sustainable Development of Ecological Sports Tourism Resources and Its Industry. *J. Sustain. Dev.* **2009**, *2*. [CrossRef]

60. Jones, P.; Comfort, D. Sustainability and the UK Waste Management Industry. *Eur. J. Sustain. Dev. Res.* **2018**, *2*. [CrossRef]

61. Vivoda, V.; Kemp, D. How do national mining industry associations compare on sustainable development? *Extr. Ind. Soc.* **2019**, *6*, 22–28. [CrossRef]

62. Williams, S.; Robinson, J. Measuring sustainability: An evaluation framework for sustainability transition experiments. *Environ. Sci. Policy* **2020**, *103*, 58–66. [CrossRef]

63. Henriques, M.H.; Brilha, J.B. UNESCO Global Geoparks: A strategy towards global understanding and sustainability. *Episodes* **2017**, *40*, 349–355. [CrossRef]

64. Paskova, M.; Zelenka, J. Sustainability Management of Unesco Global Geoparks. *Sustain. Geosci. Geotourism* **2018**, *2*, 44–64. [CrossRef]

65. Agbedahin, A. Sustainable development, Education for Sustainable Development, and the 2030 Agenda for Sustainable Development: Emergence, efficacy, eminence, and future. *Sustain. Dev.* **2019**, *27*, 669–680. [CrossRef]

66. Bapat, S.S.; Aichele, C.P.; A High, K. Development of a sustainable process for the production of polymer grade lactic acid. *Sustain. Chem. Process.* **2014**, *2*, 3. [CrossRef]

67. E Watanabe, M. The United Nations Sustainable Development Goals. *BioScience* **2020**, *70*, 205–212. [CrossRef]

68. Alawneh, R.; Ghazali, F.E.M.; Ali, H.; Sadullah, A.F. A Novel framework for integrating United Nations Sustainable Development Goals into sustainable non-residential building assessment and management in Jordan. *Sustain. Cities Soc.* **2019**, *49*, 101612. [CrossRef]

69. Lee, K.H.; Noh, J.; Khim, J.S. The Blue Economy and the United Nations' sustainable development goals: Challenges and opportunities. *Environ. Int.* **2020**, *137*, 105528. [CrossRef]

70. Balta, M.T.; Dincer, I.; Hepbasli, A. Development of sustainable energy options for buildings in a sustainable society. *Sustain. Cities Soc.* **2011**, *1*, 72–80. [CrossRef]

71. Deakin, M.; Reid, A. Sustainable urban development: Use of the environmental assessment methods. *Sustain. Cities Soc.* **2014**, *10*, 39–48. [CrossRef]

72. Naganathan, H.; Chong, W.K. Evaluation of state sustainable transportation performances (SSTP) using sustainable indicators. *Sustain. Cities Soc.* **2017**, *35*, 799–815. [CrossRef]

73. Atanda, J. Developing a social sustainability assessment framework. *Sustain. Cities Soc.* **2019**, *44*, 237–252. [CrossRef]

74. Singla, L.; Vashist, S. Efficient Workload Management in Cloud Computing. *Int. J. Sci. Res.* **2012**, *3*, 154–157. [CrossRef]

75. Shavkun, I.; Dybchinska, Y. Efficient manager: Creative dimension. *Manag. Entrep. Trends Dev.* **2019**, *2*, 47–59. [CrossRef]

76. Baldenius, T.; Reichelstein, S. Incentives for Efficient Inventory Management: The Role of Historical Cost. *Manag. Sci.* **2005**, *51*, 1032–1045. [CrossRef]

77. Garcia-Verdugo, E.; Altava, B.; Burguete, M.I.; Lozano, P.; Luis, S.V. Ionic liquids and continuous flow processes: A good marriage to design sustainable processes. *Green Chem.* **2015**, *17*, 2693–2713. [CrossRef]

78. Keane, M.A. Advances in greener separation processes? Case study: Recovery of chlorinated aromatic compounds. *Green Chem.* **2003**, *5*, 309. [CrossRef]

79. Salimova, T.; Guskova, N.; Krakovskaya, I.; Sirota, E. From industry 4.0 to Society 5.0: Challenges for sustainable competitiveness of Russian industry. *IOP Conf. Series: Mater. Sci. Eng.* **2019**, *497*, 012090. [CrossRef]

80. Valera, D.L.; Belmonte, L.J.; Molina-Aiz, F.D.; López, A.; Camacho, F. The greenhouses of Almería, Spain: Technological analysis and profitability. *Acta Hortic.* **2017**, *1170*, 219–226. [CrossRef]
81. Makris, D.P. Green extraction processes for the efficient recovery of bioactive polyphenols from wine industry solid wastes – Recent progress. *Curr. Opin. Green Sustain. Chem.* **2018**, *13*, 50–55. [CrossRef]
82. Khalil, M.E. Exploring Inclusiveness in Green Hotels for Sustainable Development in Egypt. *Int. J. Ind. Sustain. Dev.* **2020**, *1*, 15–23. [CrossRef]
83. Guo, J.; Mao, H.; Wang, T. Ecological Industry: A Sustainable Economy Developing Pattern. *J. Sustain. Dev.* **2010**, *3*. [CrossRef]
84. Chen, C.-W. Improving Circular Economy Business Models: Opportunities for Business and Innovation: A new framework for businesses to create a truly circular economy. *Johns. Matthey Technol. Rev.* **2020**, *64*, 48–58. [CrossRef]
85. Olabi, A.G. Circular economy and renewable energy. *Energy* **2019**, *181*, 450–454. [CrossRef]
86. Bolger, K.; Doyon, A. Circular cities: Exploring local government strategies to facilitate a circular economy. *Eur. Plan. Stud.* **2019**, *27*, 2184–2205. [CrossRef]
87. ISO 26000 and SDGS. Available online: https://www.iso.org/publication/PUB100401.html (accessed on 2 May 2020).
88. Hahn, R. ISO 26000 and the Standardization of Strategic Management Processes for Sustainability and Corporate Social Responsibility. *Bus. Strat. Environ.* **2012**, *22*, 442–455. [CrossRef]
89. ISO 26000:2010. Available online: https://www.iso.org/iso-26000-social-responsibility.html (accessed on 3 May 2020).
90. Price, D.D.S. The analysis of scientometric matrices for policy implications. *Scientometrics* **1981**, *3*, 47–53. [CrossRef]
91. Garfield, E. Derek Price and the Practical World of Scientometrics. *Sci. Technol. Hum. Values* **1988**, *13*, 349–350. [CrossRef]
92. Prathap, G. Eugene Garfield: From the metrics of science to the science of metrics. *Scientometrics* **2017**, *114*, 637–650. [CrossRef]
93. Abad-Segura, E.; González-Zamar, M.-D. Global Research Trends in Financial Transactions. *Mathematics* **2020**, *8*, 614. [CrossRef]
94. González-Zamar, M.-D.; Jiménez, L.O.; Ayala, A.S.; Abad-Segura, E. The Impact of the University Classroom on Managing the Socio-Educational Well-being: A Global Study. *Int. J. Environ. Res. Public Heal.* **2020**, *17*, 931. [CrossRef] [PubMed]
95. Belmonte-Ureña, L.J.; Garrido-Cardenas, J.A.; Camacho-Ferre, F. Analysis of World Research on Grafting in Horticultural Plants. *HortScience* **2020**, *55*, 112–120. [CrossRef]
96. Abad-Segura, E.; González-Zamar, M.-D.; Infante-Moro, J.C.; Ruipérez García, G. Sustainable Management of Digital Transformation in Higher Education: Global Research Trends. *Sustainability* **2020**, *12*, 2107. [CrossRef]
97. Duque-Acevedo, M.; Belmonte-Ureña, L.J.; Cortés-García, F.J.; Camacho-Ferre, F. Agricultural waste: Review of the evolution, approaches and perspectives on alternative uses. *Glob. Ecol. Conserv.* **2020**, *22*. [CrossRef]
98. Abad-Segura, E.; Cortés-García, F.J.; Belmonte-Ureña, L.J. The sustainable approach to corporate social responsibility: A global analysis and future trends. *Sustainability* **2019**, *11*, 5382. [CrossRef]
99. Leydesdorff, L.; Van Den Besselaar, P. Scientometrics and communication theory: Towards theoretically informed indicators. *Scientometrics* **1997**, *38*, 155–174. [CrossRef]
100. Abad-Segura, E.; González-Zamar, M.D. Effects of Financial Education and Financial Literacy on Creative Entrepreneurship: A Worldwide Research. *Educ. Sci.* **2019**, *9*, 238. [CrossRef]
101. Prathap, G. Quantity, quality, and consistency as bibliometric indicators. *J. Assoc. Inf. Sci. Technol.* **2013**, *65*, 214. [CrossRef]
102. Su, H.-N.; Lee, P.-C. Mapping knowledge structure by keyword co-occurrence: A first look at journal papers in Technology Foresight. *Scientometrics* **2010**, *85*, 65–79. [CrossRef]
103. Lee, C.I.S.; Felps, W.; Baruch, Y. Mapping Career Studies: A Bibliometric Analysis. *Acad. Manag. Proc.* **2014**, *2014*, 14214. [CrossRef]
104. Ponomariov, B.; Boardman, C. What is co-authorship? *Scientometrics* **2016**, *109*, 1939–1963. [CrossRef]
105. Van Eck, N.J.; Waltman, L. Software survey: VOSviewer, a computer program for bibliometric mapping. *Scientometrics* **2009**, *84*, 523–538. [CrossRef] [PubMed]

106. Van Eck, N.J.; Waltman, L.; Noyons, E.C.M.; Buter, R.K. Automatic term identification for bibliometric mapping. *Scientometrics* **2010**, *82*, 581–596. [CrossRef] [PubMed]

107. Fontana, L.B.; Oldekop, J.A. The sustainable development goals: The bumpy road ahead. *World Dev.* **2020**, *127*, 104770. [CrossRef]

108. Dang, H.-A.H.; Serajuddin, U. Tracking the sustainable development goals: Emerging measurement challenges and further reflections. *World Dev.* **2020**, *127*, 104570. [CrossRef]

109. Buckley, P.J.; Pass, C.L.; Prescott, K. Measures of international competitiveness: A critical survey. *J. Mark. Manag.* **1988**, *4*, 175–200. [CrossRef]

110. White, H.D. Co-cited author retrieval and relevance theory: Examples from the humanities. *Scientometrics* **2014**, *102*, 2275–2299. [CrossRef]

111. Strobl, A.; Bauer, F.; Matzler, K. The impact of industry-wide and target market environmental hostility on entrepreneurial leadership in mergers and acquisitions. *J. World Bus.* **2020**, *55*, 100931. [CrossRef]

112. Li, Y.; Xu, L.; Sun, T.; Ding, R. The impact of project environmental practices on environmental and organizational performance in the construction industry. *Int. J. Manag. Proj. Bus.* **2019**, *13*, 367–387. [CrossRef]

113. Strunz, S.; Marselle, M.; Schröter, M. Leaving the "sustainability or collapse" narrative behind. *Sustain. Sci.* **2019**, *14*, 1717–1728. [CrossRef]

114. Wright, P.M.; Nyberg, A.J. Academic research meets practice: Why controversial results are not controversial. *Manag. Res. J. Iberoam. Acad. Manag.* **2018**, *16*, 66–74. [CrossRef]

115. Vachon, S.; Klassen, R. Environmental management and manufacturing performance: The role of collaboration in the supply chain. *Int. J. Prod. Econ.* **2008**, *111*, 299–315. [CrossRef]

116. Dragos, C.M.; Dragos, S.L. Bibliometric approach of factors affecting scientific productivity in environmental sciences and ecology. *Sci. Total. Environ.* **2013**, *449*, 184–188. [CrossRef] [PubMed]

117. Zhuchenko, O. Control system of carbon production. *Sci. Notes Taurida Natl. V.I. Vernadsky Univ. Series Tech. Sci.* **2020**, *1*, 72–78. [CrossRef]

118. Feinstein, N.W.; Kirchgasler, K. Sustainability in Science Education? How the Next Generation Science Standards Approach Sustainability, and Why It Matters. *Sci. Educ.* **2014**, *99*, 121–144. [CrossRef]

119. Pauliuk, S. Making sustainability science a cumulative effort. *Nat. Sustain.* **2019**, *3*, 2–4. [CrossRef]

120. Robinson-Garcia, N.; Sugimoto, C.R.; Murray, D.; Yegros-Yegros, A.; Larivière, V.; Costas, R. The many faces of mobility: Using bibliometric data to measure the movement of scientists. *J. Inf.* **2019**, *13*, 50–63. [CrossRef]

121. Checchi, D.; Malgarini, M.; Sarlo, S. Do performance-based research funding systems affect research production and impact? *High. Educ. Q.* **2018**, *73*, 45–69. [CrossRef]

122. Freeth, R.; Caniglia, G. Learning to collaborate while collaborating: Advancing interdisciplinary sustainability research. *Sustain. Sci.* **2019**, *15*, 247–261. [CrossRef]

123. Van Eck, N.J.; Waltman, L. Citation-based clustering of publications using CitNetExplorer and VOSviewer. *Scientometrics* **2017**, *111*, 1053–1070. [CrossRef] [PubMed]

124. Campbell, M.C. Sustainable economic growth and environmental conservation. *Environ. Pr.* **2018**, *20*, 1–2. [CrossRef]

125. Paulsson, A. Making the sustainable more sustainable: Public transport and the collaborative spaces of policy translation. *J. Environ. Policy Plan.* **2018**, *20*, 419–433. [CrossRef]

126. Nakic, D. Environmental evaluation of concrete with sewage sludge ash based on LCA. *Sustain. Prod. Consum.* **2018**, *16*, 193–201. [CrossRef]

127. Arora, N.K. Biodiversity conservation for sustainable future. *Environ. Sustain.* **2018**, *1*, 109–111. [CrossRef]

128. Ellerbrock, M.J. Sustainable development requires environmental justice. *Int. J. Sustain. Dev. Plan.* **2018**, *13*, 208–214. [CrossRef]

129. Kakouris, A.; Sfakianaki, E. Motives for implementing ISO 9000 – does enterprise size matter? *Int. J. Prod. Perform. Manag.* **2019**, *68*, 447–463. [CrossRef]

130. Zeiger, B.; Gunton, T.; Rutherford, M. Toward sustainable development: A methodology for evaluating environmental planning systems. *Sustain. Dev.* **2018**, *27*, 13–24. [CrossRef]

131. Schlosberg, D. From postmaterialism to sustainable materialism: The environmental politics of practice-based movements. *Environ. Politi.* **2019**, 1–21. [CrossRef]

132. Berawi, M.A. The Role of Industry 4.0 in Achieving Sustainable Development Goals. *Int. J. Technol.* **2019**, *10*, 644. [CrossRef]

133. De Gooyert, V. Long term investments in critical infrastructure under environmental turbulence; Dilemmas of infrastructure responsiveness. *Sustain. Futur.* **2020**, *2*, 100028. [CrossRef]

134. Mersal, A. Sustainable Urban Futures: Environmental Planning for Sustainable Urban Development. *Procedia Environ. Sci.* **2016**, *34*, 49–61. [CrossRef]

135. Mathews, R. A Six-Step Framework for Ecologically Sustainable Water Management. *J. Contemp. Water Res. Educ.* **2009**, *131*, 60–65. [CrossRef]

136. Belmonte-Ureña, L.J.; Román-Sánchez, I.M. Appraisal of environmental regulations on sustainable consumption of water in the wine-producing spanish industry. *Environ. Eng. Manag. J.* **2013**, *12*, 1979–1987. [CrossRef]

137. Adam, I.H.D.; Jusoh, A.; Mardani, A.; Streimikiene, D.; Nor, K.M. Scoping research on sustainability performance from manufacturing industry sector. *Probl. Perspect. Manag.* **2019**, *17*, 134–146. [CrossRef]

138. Krdžalić, A.; Hodžić, L. Sustainable engineering challenges towards Industry 4.0: A comprehensive review. *Sustain. Eng. Innov. ISSN 2712-0562* **2019**, *1*, 1–23. [CrossRef]

139. Jena, M.C.; Mishra, S.K.; Moharana, H.S. Application of Industry 4.0 to enhance sustainable manufacturing. *Environ. Prog. Sustain. Energy* **2019**, *39*, 13360. [CrossRef]

140. Miranda, J.; Ponce, P.; Molina, A.; Wright, P. Sensing, smart and sustainable technologies for Agri-Food 4.0. *Computers in Industry* **2019**, *108*, 21–36. [CrossRef]

141. Fatma, M.; Khan, I.; Rahman, Z. Striving for legitimacy through CSR: An exploration of employees responses in controversial industry sector. *Soc. Responsib. J.* **2019**, *15*, 924–938. [CrossRef]

142. Song, B.; Wen, J.; Ferguson, M.A. Toward effective CSR communication in controversial industry sectors. *J. Mark. Commun.* **2018**, *26*, 243–267. [CrossRef]

143. Lovrinčević, Ž.; Mikulić, D.; Orlović, A. Economic Aspects of Sin Industry in Croatia. *Drustvena Istraz.* **2015**, *24*, 175–196. [CrossRef]

144. Malone, R.E. "Sustainable" tobacco industry? *Tob. Control.* **2019**, *28*, e85. [CrossRef] [PubMed]

145. Miller, T. Explanation in artificial intelligence: Insights from the social sciences. *Artif. Intell.* **2019**, *267*, 1–38. [CrossRef]

146. Sharma, G.D.; Yadav, A.; Chopra, R. Artificial intelligence and effective governance: A review, critique and research agenda. *Sustain. Futures* **2020**, *2*, 100004. [CrossRef]

Article

Does Strategic Corporate Social Responsibility Drive Better Organizational Performance through Integration with a Public Sector Scorecard? Empirical Evidence in a Developing Country

Pham Quang Huy [1,*] and **Vu Kien Phuc** [2]

1   School of Accounting, University of Economics Ho Chi Minh City, 59C Nguyen Dinh Chieu Street, District 3, Ho Chi Minh City 70000, Vietnam
2   School of Accounting, University of Economics Ho Chi Minh City, Vinh Long Campus, 1B Nguyen Trung Truc Street, Vinh Long 85000, Vietnam; phucvk@ueh.edu.vn
*   Correspondence: pquanghuy@ueh.edu.vn; Tel.: +84-908-231260

Received: 25 March 2020; Accepted: 12 May 2020; Published: 16 May 2020

Abstract: This paper sets its sights on propounding a structural model to delve into the interrelationship between the impact of the integration of corporate social responsibility activities into the public sector scorecard management framework on the corporate social responsibility disclosure and enhancement of the organizational performance among public sector organizations. The conceptual framework in company with hypothesis framing were established after examining the related literature. Data were gathered from a sample of 723 respondents in public sector organizations in South Vietnam via convenience sampling method. Structural equation modeling was employed to validate the goodness of model fit and examine the hypotheses. These findings revealed that integration of corporate social responsibility activities into the public sector scorecard management framework was significantly and positively related to the corporate social responsibility disclosure and organizational performance. Additionally, it also asserted that corporate social responsibility disclosure was considerably associated in a positive manner with organizational performance. Thus, some detailed implications in connection with each causal relationship and several orientations were underlined to ameliorate the capacity of managing and measuring the organizational corporate social responsibility practices in a strategic manner.

Keywords: corporate social responsibility; corporate social responsibility disclosure; organizational performance; public sector scorecard

1. Introduction

Corporate social responsibility (CSR) practices have been broadly adopted throughout the organizational community in light of the value creation characteristic of CSR [1]. CSR's implementation can help an organization easily achieve the approval of local communities, good staff attraction and retention [2], far-reaching risk mitigation [3] and so on.

CSR has been widely applied in both developed and developing economies [4]. Surprisingly, the numerous scholars have largely paid attention to the developed countries while placing less concern on the developing economies [5].

In the developing economies, CSR represents the formal and informal ways in which a business creates a significant contribution to ameliorating the governance, social, ethical, labor and environmental conditions of the countries [6]. In addition, these are the regions where globalization, economic growth, investment and business activities are likely to cause the most significant social and environmental impacts [7]. Furthermore, as stated by [8], the CSR report earned divergent attention from stakeholders in developed and developing countries. This is because the organizations in developing nations receive lower pressures in corporate social responsibility disclosure (CSRD). Asia has been the region most often covered in the literature in relation to CSR in developing nations [7]. Unfortunately, the research about CSR in this region within and among countries has still been sparse [9].

As a member of Asia, Vietnam first introduced CSR activities in 2000. The change in awareness of Vietnamese organizations was marked in 2010. Although a public sector organization (PSO) is more likely to perform socially responsible activities, these organizations are often found to concentrate on philanthropic activities rather than prioritizing environmental CSR initiatives to attain better organizational performance (OP). Admittedly, the biggest barriers and challenges to the implementation faced by PSO include the insufficient awareness of the concept of social responsibility; lack of financial and technical resources; and multiple sets of codes of conduct. Notably, the evaluation on the actual impact of CSR implementation was also another concern among many organizations [10]. However, PSOs have been lacking a clear, quantitative, consistent instrument for evaluating their goals. Moreover, according to government regulations, an alternation should be taken into consideration to meet the transparent and fair requirements when measuring the PSO's performance, especially in CSR's implementation.

In this regard, the management and measurement framework that has been proposed for this application is the public sector scorecard (PSS) due to the main advantages of outcome improvement economically; measurement development; public sector (PS) characteristics' conformity; emphasizing the expectations of users and stakeholders; re-designing phases and ameliorating service delivery; dealing with capability and organizational matters; and promoting a performance management culture based on improvement, innovation and learning [11]. Indeed, the PSS is considered as an effective framework for ensuring that strategy, processes and performance measures in relation to CSR practices are aligned with each other. In particular, these activities are also aligned with the demands and expectations of service users and stakeholders. Besides, PSS can create better conditions for the collaborative work between the leaders and staff and service users and other key stakeholders. It can also deal with risk management and organizational culture, and has the capability to assure that staff and processes are facilitated to attain the targets. In addition, PSS can give rise to the improvement of CSR practices and concentrate on creating desired outcomes, including value for money.

Building on these abovementioned analysis, this research generated a significant contribution to the literature in several aspects. Firstly, this paper placed an emphasis on the benefits and the likelihood of the integration between CSR and PSS (ICP) because the scrupulous CSR governance would lead to success via organizational structure enhancement, CSR performance accomplishment and sustainable development [12]. On the other hand, due to lacking independent verification, voluntary CSRD used to be criticized as being unreliable and reported under a selected manner [13]. Thus, the association between CSRD and ICP was handled to determine the effectiveness of ICP on the disclosure activities. Thirdly, the impact of ICP on PSOs' performance was also accentuated in terms of sustainable development. Although Asia has been broadly acknowledged as a region with numerous studies focusing on the CSR issues, the studies about CSR in the region within and among nations are still limited [9]. In addition, based on the assumption of [14], there have been limited theoretical and empirical investigations on CSR practices in Vietnam up to date. Thus, this research added new empirical evidence to ease the limitations of the amount of empirical research and the association between CSR and organizational outcomes [15]. These findings could help both practitioners and academicians attain a deep understanding on ICP through a general strategic management framework in connection with the CSR practices. In addition, these results also gain several useful insights into CSRD in PSO, as CSRD reporting has been shown to fail to attain homogeneous designs between countries [16]. Importantly, the proposed framework of the current research can be treated as an example for PSOs in other developing countries, especially the countries belonging to Asia, due to these following reasons. Firstly, as PSOs have been established with the primary role of serving the public, they are required to deliver the highest levels of compliance with the law during their operations. Secondly, there have been several similarities in the characteristics of CSR practices [7]: the geo-economic development conditions and the same size of geo-economic flow between the countries in Asia [17]. To that end, this paper endeavors to illustrate how to adopt the theoretical model through addressing these research questions below.

RQ1. Does the ICP have a significant effect on the CSRD in PSO? How far does it influence?

RQ2. Does the ICP have a significant effect on the OP in PSO? How far does it influence?

RQ3. Does the CSRD have a significant effect on the OP in PSO? How far does it influence?

The remaining parts of this research are structured as follows. The review of the prior research is sketched out in Section 2. The next section institutes the theoretical background and develops hypotheses for the research. Subsequently, the methodology employed in empirical research is carefully elucidated in Section 3. Our main section in which the findings of the study are included is Section 4. Eventually, theoretical and managerial implications, and useful directions for future research based on the inherent limitations, are foregrounded in Section 5.

2. Overview of Prior Research

2.1. The Association between Corporate Social Responsibility and Management

Several scholars have recently become interested in the integration between CSR components and organizational management due to the perception on the improvement in corporate CSR target controlling [18]. In particular, [19] propounded the combination of financial and strategic control procedures to apply for managing the environment. Additionally, the sustainability ingredients were integrated into organizational strategy in terms of a management control system [20]. Analogously, [21] added a new finding by means of putting the management control system into CSR strategy management.

2.2. The Linkage between Corporate Social Responsibility and Organizational Performance

A large amount of research has been concentrated on exploring the impacts of CSR on the various facets of an organization [22], which provoked a variety of homogeneous results. Particularly, [23] indicated the causal link between corporate reputation, CSR and OP, whereas [24] argued that good CSR could lead to a good corporate reputation and gained the performance of the organization. Additionally, more involvement in the satisfaction of stakeholders was also another concern to ensure the OP, as with financial performance (FP) was also closely related to the image of the entities [25]. On the other hand, CSR implementation was supposed to lead to the satisfaction of the consumers regarding the quality of service and retain the highly qualified workers [26]. Thus, ultimately higher profits were promised. On the contrary, CSR has been argued to cause a negative effect on FP (i.e., [27]). In particular, there was less likelihood that FP could take place simultaneously with abounding CSR activities [28], as supplemental resources and capacity during the process of CSR process would cause high expenditures and lower profit [29]. As such, an inverse association between organizational CSR activities and FP has been ascribed to the entities with better CSR performance but failing to attain the financial facet [30]. Besides, in the investigation on the relationship between CSR, public service motivation (PSM) and organizational citizenship behavior (OCB) in PS, [31] proved that employee perceptions of both internal and external CSR influenced the development of a desire to serve the public in a positive manner. In addition, these findings indicated that PSM not only partially mediated the interconnection between internal CSR perceptions and employee OCB, but also fully mediated the association between external CSR perceptions and OCB.

2.3. The Relationship between Corporate Social Responsibility Disclosure and Organizational Performance

Researchers have recently placed their sharp-witted concern on the disclosures of CSRs [32]. However, the outcome of this subject has come to a conflict [33] in which the variety of measures of

CSRD, research methodologies and FP measures were ruminated to be the main causes [34]. While [35] manifested the evidence of the positive association between levels of CSRD in the annual statements and OP in relation to FP and corporate reputation, the CSRD-performance relationship was experimentally found to be less significant for practical purposes in the work of [36]. Furthermore, the association between CSRD and FP was also scrutinized through numerous empirical studies and was proved to be tight [37]. On the other hand, the converse results on the relationship between CSRD and FP were also highlighted in several studies (i.e., [38]). The neutral association between CSRD and FP also occurred in the findings of numerous scholars [39]. In the meanwhile, the evidence between CSRD and FP has not yet even been detected in several research (i.e., [40]).

3. Theoretical Background and Hypothesis Development

3.1. Theoretical Background

3.1.1. Legitimacy Theory

Legitimacy theory. There has been a growing consensus among numerous scholars on putting legitimacy theory (LT) in place for explaining the driving force behind corporate social and environmental disclosures [41]. Therefore, this theory has been broadly applied in several studies related to the organizational environmental disclosures and has gradually assumed the central position in this type of research [42]. As stated by [43], the social contract is defined as multitudinous social expectations on the way which an entity should undertake its operations. LT is calculated based on the perception of a "social contract" existing between an entity and its operational surroundings [44]. According to [45], LT originated from the perception that an organization was supposed to consummate within the bounds and norms of with socially responsible behaviors. Simultaneously, organizational legitimacy leaned on the prolongation of reciprocal associations with its stakeholders [46], comprising implementing moral obligations to numerous stakeholders [47]. Building on the LT viewpoint, legitimacy and power have been consented to the organization by the society [48]. Hence, any particularly organizational behavior should be carefully investigated within its context and hunted for substitute driving forces [49], as these powers would have been lost if the organization had not utilized them in an appropriate manner. In doing so, the exploitable CSR activities could be effectively employed in almost all of organizations. On the other hand, as CSRD was considered to act as a critical mechanism to gain the impact of CSR on organizational reputation [50], it has been turned into a vital instrument for organizational management through integrating CSR activities into strategic risk management for the best result of CSR activities [51]. The more involvement in CSR activities, the more success in operation that this organization could reap [52].

The application of Legitimacy theory in this study. Within the association between organization and society, the CSR activities are persistently presented, investigated, identified and adjusted. The CSR reporting practices have become a primary instrument for organizational management [53] and maintaining legitimacy [43]. As such, if the chances of adverse shifts in community expectations become higher, the organization should make more effort regarding conducting the CSRD [43]. In other words, organizations attempt to legitimize their actions by engaging in CSR reporting to achieve acceptance from society. Integrating the CSR activities into the PSS framework should help by managing, measuring and improving the CSR activities. In doing so, the performance of CSR activities could be maximized. Accordingly, the PSO could determine which tactics and disclosure options would be available and suitable for managing legitimacy. Moreover, the PSO could take up numerous public disclosure strategies to gain the OP.

3.1.2. Resource-Based Theory

Resource-based theory. Building on the resource-based viewpoint, the choice and accumulation of resources was considered as a function of internal decision making and external strategic determinants [54]. Therefore, the resource-based theory could be used as an instrument for undertaking the analysis of social policy refinement [55]. As stated by [56], the organizational intangible resources consisted of technology, human capital and reputation. On the other hand, intangible resources could be comprised of the assets, capabilities, processes, attributes, information and knowledge managed by the organization [57]. Through combining intangible resources into strategic planning process, several researchers detached corporate social performance from OP and endeavored to bridge the former to organizational FP [58]. The CSR activities could give rise to a significant support for establishing and reinforcing a solid, sustainable, long-term reputation to enhance the competitive advantage [59]. Additionally, CSR practices could encourage the workforce, ameliorating productivity and facilitating improvement of performance [60].

The application of resource-based theory in this study. The considerable changes in the business environment have set a demand on organizational changes in terms of different resources and capabilities. An organization should have the capability to spread out its resources rather than only possessing unique resources. Thus, resource-side matters had to be addressed by the practicing strategists [61]. The resource-based view of the organization proposed that an organization should constitute its internal capabilities to match the conditions of the external environment. On the other hand, the theory inferred that the right combination of resources should be developed, progressively evaluated and managed for the specific OP intended. As such, PSOs have been advised to choose an appropriate framework to manage and measure the CSR activities based on organizational resources. In this case, PSS is considered as the most suitable framework for PSO [11]. Based on integrating the CSR practices into the PSS framework, CSR activities are aligned with each other. Additionally, these activities are also aligned with the expectations of service users and stakeholders. Besides, PSS can create better conditions for the collaborative work between the leaders and staff and service users and other key stakeholders. It can also deal with risk management and organization culture, and has the capability of assuring that staff and processes are enabled to attain certain targets. Furthermore, PSS can create improvements on CSR practices and concentrate on generating desired outcomes, including value for money. In doing so, it can make a significant contribution to the improvement in CSRD and enhancement in OP.

3.1.3. Corporate Social Responsibility and CSR Disclosure

Corporate social responsibility. CSR refers to a course of action in which the agreement to make a contribution for society and a cleaner environment was adopted in a voluntary manner by the organization [62]. It could be an approach for public image and reputational improvement through the activities that meet the needs of society [63]. The definition of CSR employed in the present research was insinuated by [64] which was mentioned as a consistent course of specific action and policies involved in stakeholders' satisfaction and the pivotal triple aspects of economic, social and environmental performance [64]. As propounded by [65], the two categories of CSR strategies for organizations to participate included the CSR governance in a serious and strict manner and CSR governance in a symbolic and opportunistic way. Particularly, a course of serious and rigorous operations was undertaken in terms of serious and strict CSR governance with the support of vital resources, which led to fruitful CSR outcomes [32]. Conversely, only corporate image or emergent matters were put in place for being addressed rather than dealing with essential resource allocation for a deep and strategic CSR program [66].

Corporate social responsibility disclosure. CSRD refers to the action of supplying financial and non-financial information presented in an annual statement or isolated social reports that was concerned with organizational interaction with its physical and social environment [67]. Besides, CSRD typically does not only consist of information on the physical environment, energy, human resources, products and community (presented in detail [68]), but also organizational operations, aspirations and public image in association with the environment, staff, consumer matters, energy utilization, corporate governance issues and so on [40].

3.1.4. Public Sector Scorecard

Being set up from the balanced scorecard adaption and extended for cultural and value establishment for public and voluntary sectors, PSS has been excogitated as a fructiferous framework with effective contributions to outcome improvement for service users and stakeholders without increasing overall expenditures [11]. PSS is administered through the three main junctures, including strategy mapping, service improvement and measurement and evaluation [11].

Strategy mapping is the process dealing with the association between outcome, process and capability components [69]. As such, a draft strategy mapping is built up after the series of interactive workshops on several matters regarding anticipated outcomes—strategic, service user, stakeholder, FP and capability outputs of the internal departments (i.e., senior managers, staff) and external components (i.e., service users and organizational stakeholders)—have been finished. A risk-management workshop transpiring through identification, lessening and eliminating of risks is subsequently used in the draft strategy map. Those efforts are finally complemented officially into the strategy map in terms of risk-management culture. Besides, disagreement on strategic drivers and diverge requirements and priorities should be balanced out [70] due to the diversity of targets and stakeholders [71].

The service improvement phase is set up with the aim of fostering the workshop participants to raise their voices and reveal the evidence or data for the generation of tools consisting of process maps, systems thinking and lean management to be complemented. Additionally, the next workshop confirms the capability outputs' attendance in the strategy with the purpose of bringing it to the attention of supporting staff, thereby creating a culture of improvement, innovation and learning rather than a quarrel culture.

In the measurement and evaluation step, the negotiation among workshop participants is held to find out the appropriate performance assessments for each constituents of the strategy map in which the potential evaluation approaches will be under investigation and selection. In addition, a comprehensive understanding on OP can be achieved through careful analysis and learning performance measures in the light of cause and effect definition opportunities and addressing issues of creation.

The cycle will end with utilizing the performance information for strategy map modification, the determination of further service renovations and the promotion of better performance evaluations. Nonetheless, in view of several changes and the association between performance measures and changing strategy appearances, the cycle will still go on [72].

3.1.5. Organizational Performance

As stated by [73], the success of an organization is reflected by the degree of its performance during operations. Owing to the multidimensional characteristics and multiple measurement approaches in existence, OP was argued to lead to numerous challenges in evaluation [74]. OP was established by the weighted combination of perceived and objective performance information which was proven to have a strong degree of convergence [75] and named such primary categories as the financial (i.e., objective) and nonfinancial (i.e., subjective) measures in a slightly different manner in the literature [76]. Notwithstanding the huge contribution to the examination on the relationship between CSR activities and OP based on the financial indicators [77], unfortunately, the financial indicators have no longer been sufficient for OP assessment, as the financial prosperity of an organization cannot be separated from social, environmental and governance activities [78]. Besides, the economic performance not been cared for properly, although the such vital components of the CSR concept as environmental, social and governance dimensions and the integration between CSR and environmental, social and governance implementation has been indicated to create the positive effect on the organizational value and performance [79]. Furthermore, there was an urgent call for environmental performance to be tacked on in OP evaluation in terms of sustainable development [80]. Apparently, a comprehensive framework, one which is multi-dimensional and qualitatively-based, has been in demand as a replacement [81].

3.2. Hypothesis Development

Owing to the same main target of serving the stakeholders, ICP in organizational CSR implementation management and measurement as follows.

Service user and stakeholder involvement. The effective public service design and provisions can have better outcomes if the participation of experience and knowledge provided by service users is highlighted [82]. Customers are considered to possess the power on account of the competition of discrepant dimensions among various entities [34]. As such, suitable methods for better customer treatment renovation are put in the CSRD [83]. Besides, owing to the vital role of stakeholders in organizational strategy, there has been growing attention on constructing and maintaining a close association, increasing the interaction between the organization and stakeholders to gather useful information [84] and creating a better chance for more innovative activity and achievement [85].

Working across organizational boundaries. Eliminating the boundaries between entities in operational processes is the most crucial affair to take into consideration, because the focus of service users is on the available service distribution through many organizations or departments [86] and the solution for Governmental achievement in terms of the main outcomes is the cooperation between numerous organizations [11]. Admittedly, the diversity in CSR practices in numerous developing regions based on their own perceptions about the CSR has led to a variety of CSRD designs. This, therefore, raises an urgent claim regarding PSS's application into a CSR program as it does not only allow the people coming from various departments or entities to put their attention toward common essential outcomes instead of narrower targets, but also helps people, regarding the measurement and assessment, to make precise evaluations of the outcomes, processes and capability components [11].

Improvement and capability process. The bulk of organizational resource consumption being from undertaking CSR activities has led to a barrier for organizations regarding investment in CSR programs [12]. In this regard, PSS is recommended to be adopted for this program due to the advantage of process improvement dealing with the different outcomes demanded, including financial outcomes in an overall performance management framework [11]. On the other hand, almost all CSR strategies are also covered with interior and exterior components [87]. In particular, the internal ingredients mentioned on the method were applied by the staff, while external components were related to the requirements and expectations of outside stakeholders [88]. Thus, human resource management (HRM) has been argued to make significant contributions to employee commitment boosting; the organizational commitment with CSR practices; and integrating creation between the CSR principles and HRM processes together with stakeholder alignment establishment [88]. Additionally, the leaders who possess the superior external expertise and knowhow would become the good observers for environmental protection regulations and place sufficient attention on corporate stakeholders [89]. As such, they may devote themselves to the social responsibility implementation and gaining information [90].

Integrating risk management. As stated by [86], the combination between identification and addressing the primary risks has been well-recognized among the effective performance entities. The comprehensive CSRD also leads to the accurate evaluation on the operation situation and organizational risk factors [89] and good reputation building and reputational risk avoidance [91]. On account of the effect of the customer on the organizational market risk, the treatment with organizational customers would be significantly improved and the risk of losing its share of market would be decreased considerably when the CSRD has been put in place [34]. Nevertheless, there are still numerous risks related to the operational process related to CSR practices which should be integrated into PSS management framework to attain a better performance of CSRD.

Improvement, innovation and learning. As asseverated by [92], CSR practices have positively related to an entity's performance in terms of the innovation process. As such, CSR activities have been supposed to assist with organizational innovation capacity which leads to the enhancement of distinction creation and competitive advantages [93] and the capacity for process and product innovation [94]. In doing so, socially responsible organizations have been asserted to present the quantity of CSRD in an in depth and high-quality manner [95].

Based on the aforementioned information, the study hypothesis was formed.

Hypothesis 1 (H1). *ICP has caused an impact on the CSRD in a significant and positive manner.*

CSR activities have been well-recognized to give rise to the enhancement in FP [96] and economic performance [97]. Indeed, CSR implementation was reported to provide a competitive advantage and help an organization to accomplish sustainable growth goals [98]. On the other hand, an organization which places more concern on CSR activities would increase its reputation, customer loyalty and employee satisfaction [99]. In terms of the association with OP in the stakeholder aspect, more involvement in CSR activities would positively impact OP due to the effective support in establishing a good relationship with stakeholders and better care for society [100]. In like manner, [101] also underlined that gaining better understanding among stakeholders of the measures implemented by the organization for their wellbeing would generate a rapidly increasing on OP. Numerous researchers have simultaneously placed an emphasis on the investment in stakeholder engagement and management due to its benefits of a positive image, high quality employee recruitment and employee retainment [102]. Thus, a research hypothesis was supposed as follows.

Hypothesis 2 (H2). *ICP has caused an impact on the OP in a significant and positive manner.*

Owing to the vital role of information provided by the entities, CSRD was proved to bring a great deal of benefit for the organizations. Regarding the financial facet, CSRD implementation would lead to the prosperity in the economy [103]. In terms of the management aspect, CSRD was considered to have a significant influence on the effectiveness of interior decision-making [104] and exterior relationship management [105]. Importantly, proficiency in stakeholder management also helped the organization receive much useful support from their stakeholders [106] to accomplish better FP [107]. Besides, involvement in the CSRD was revealed to yield a competitive advantage and enhance a company's value [108]. Thus, a research hypothesis was considered as follows.

Hypothesis 3 (H3). *CSRD has caused an impact on the OP in a significant and positive manner.*

The research model which was established on the LT and Resource-based theory to investigate the interrelationship between the ICP, CSRD and OP was depicted in Figure 1.

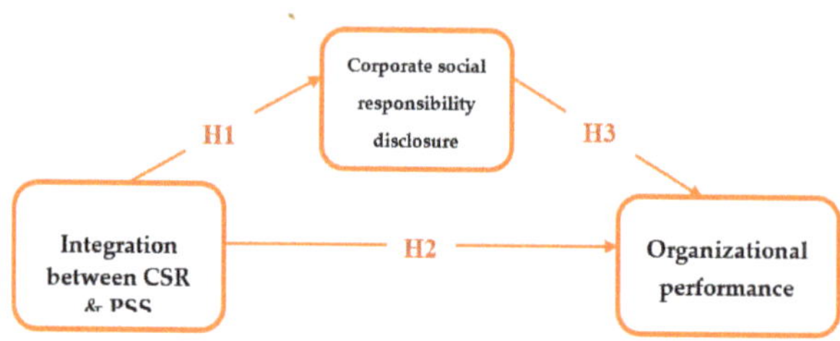

Figure 1. Proposed research model.

4. Methodology Design

4.1. Procedure and Item Generation

An empirical study was employed to verify the assumption model. Owing to its major role in the economic development of Vietnam, the southern region was selected to investigate the CSR practices among PSOs in the present study. Additionally, as PSOs have been established with the primary role of serving the public, they are required to deliver the highest compliance levels with the law during their operations. Hence, almost all of the PSOs in Vietnam could take the results of this study as a reference. Moreover, due to the advantages of favorable environmental and economic conditions, this region has faciliated foreign investment much more than any other regions in the country. It is not surprisingly that this has been the region with the greatest developments in advanced management and modern technology adoption. Taken together, the findings on the research conducted in this region could serve as a reference for several developing countries in light of the similiarities in economic conditions.

In light of the adaption from English literature, all the scales applied in this study were used after translation and back-translation by a variety of bilingual experts. The questionnaire utilized in this research was set up with seven-point Likert scale ranging from 1 "completely disagree" to 7 "completely agree." Building on the above-mentioned research questions, the current study went through several procedures as follows.

Semi-structured interviews were employed in this study as they were considered to be suitable for gathering qualitative data from professionals [109]. Through employing semi-structured interviews, researchers could draw up a previous framework of themes to be investigate. Nonetheless, this type of interview could also help to emerge new ideas during the interview. The semi-structured interviews were done with several experts to consult their advice. The interviewees included four leaders of PSOs and four faculty members. Based on their suggestions, adjustments were made for several items which could not describe the current state of CSR practices in PSOs or were very hard to understand. Then, revisions with the experts were conducted again to create a complete questionnaire.

To improve sentence structure and layout of the instrument in relation to the PSOs in South Vietnam, a pilot test was performed. The pilot survey was undertaken with 100 participants randomly picked up from the target population. The Cronbach's α value was used to check the degree of internal consistency of each construct [110]. The Cronbach's α value of the pilot test were found to be above 0.7 [111], substantiating that the variables and dimensions of this research enclosed with acceptable reliabilities.

To examine the newly developed scale empirically, a cross-sectional study was employed for primary data collection through a questionnaire survey. As stated by [18], the accountants played an important role in measuring, disclosing and assuring all the organizational information, especially information about CSR. Hence, accounting staff were considered to participate in the generation, assurance, publication and analysis report on CSR regardless of the lack of formal stipulated structure [18]. As propsed by [112], the ideal sample size to estimate parameters for covariance-based SEM (CB-SEM) was 10:1. In the meanwhile, [113] suggested the optimal sample size which had an items to participants' ratio ranging from 1:4 to 1:10. The theoretical model contained 52 parameters and 3 indicators; thus, the number of 520 respondents was considered to meet the demand. Through the convenience sampling technique, these questionnaires were distributed from September 2019 to February 2020 and personally collected by the researchers. Unfortunately, some erroneous and incomplete questionnaires were excluded from the analysis. Finally, 723 complete responses were obtained, corresponding to a response rate of 87.84 % of the respondents. Hence, the sample was representative of the general population in the target region.

4.2. Data Analysis

As stated by [114], SEM was an effective statistical instrument for analyzing the interconnection between multiple variables by the measurement and structural models [115]. In the current research, SPSS version 25.0 was employed for evaluating the item-total correlations and exploratory factor analysis (EFA). In the meanwhile, AMOS version 25.0 was utilized for SEM. Maximum likelihood estimation method was applied to evaluate both measurement and structural model [116].

4.3. Measures and the Questionnaire

With the aim of explorating the impact of ICP on the CSRD and the OP in PSO, the measurement scales applied in this study were taken from the previous works in private sector due to several reasons.

Firstly, a thorough review of literature illustrated that far more concerns regarding CSR practices have been placed in the private sector [117]; it was not surprisingly that there have been numerous works devoted to finding out the appropriate measurement scales for CSR practices in this sector.

Secondly, the CSR practices in the developing countries have been most commonly related to philanthropy and charity [7]. In particular, these nations tend to conduct the corporate social investment in education, health, sport development and service communication. As stated by [118], PSO was underlighted to conduct more social and environmental commitments in comparison to that of private sector. PSOs have also undertaken numerous activities related to CSR pratices; namely, adhering to strict regulations, helping the poor, operating in a manner in line with the philanthropic and charitable expectations of society, contributing toward bettering the local communities and so on.

Thirdly, under budgetary pressures and program effectiveness enhancement, public leaders typically seek a new approach to managingg the organization through adopting practices from the other sectors [119]. In addition, owing to the demand of adherence to strict regulations, PSOs have been made to experience the structural and procedural changes applied in the private sector [120].

Taken together, the measurement scales employed in this study were inherited from the previous works investigated in private sector. Nevertheless, the measurement scales used the revisions of experts and the pilot test to achieve appropriateness with the context of this study. In a nutshell, the measure scales employed in the current research were set up as follows.

4.3.1. The Integration between Corporate Social Responsibility and Public Sector Scorecard

There has been a growing concern in measuring the actual impact of CSR implementation [10]. The most critical demands of an effective performance management system included quality management, service redesign and performance measurement [69]. Through integrating with the PSS, the organizational strategies, processes and performance measures on CSR practices should accord with the outcomes related to service users and other key stakeholders.

As stated by [121], concentrating on the outcomes could guide the organization towards the true goals and enhance the accountability. Additionally, concentrating on the outcomes and being able to evaluate them lets the organization measure what organizational activities are actually being achieved [122]. As such, the integration of PSS into the organizational operation could bring several advantages. According to [11], the PSS project was begun with identifying the outcomes of PSO, its service users and other stakeholders and value for money which helped the PSO concentrate on these outcomes. Based on those things, the measurement scales for the outcomes of CSR activities when these activities were integrated with PSS were established as follows.

Key performance outcomes. The key performance outcomes were defined as the main performance outcomes demanded by the relevant organization [69]. Thus, the items for measuring the key performance outcomes in PSO in the research were developed from the works of [123,124].

Service user/stakeholder. The service user/stakeholder was referred to the outcomes relating to the service users and other key stakeholders. Therefore, the measurement scales of service user/stakeholder in this study were designed from those propounded by [123–125].

Financial. The financial was identified as the accomplishment of value for money or decreasing overall cost. The criteria applied to measure the financial were taken as reference from the contributions of [123,124].

It was distinguished from planned service and policies because it dealt with the actual experience of users and stakeholders [11]. The organization would try to determine how the processes could be improved to generate better performance.

Service delivery. The items for measuring the Service delivery in this study were modified from the works of [123,124].

The capability in PSS framework focused on what could be conducted to assure that redesigned processes performed smoothly. This might relate to extra resources to enable the organization to accomplish the required outcomes [69]. These extra resources typically comprised innovation and learning; effective leadership; and people, partnerships and resources. To that end, the measurement scales for these components were set up as follows.

Leadership. The significance of leadership towards success in CSR practices has been underlined ([126–129]). The criteria applied to measure the leadership in this study were adjusted from the contributions of [130].

People, partnerships and resources. Because CSR implementation also raise the demand on by generous amount (of organizational resources; [12]), the organization had to allocate resources in an appropriate manner [131]. Thus, in this study, the measurement scales of the People, partnerships and resources were modified from the works of [123,124].

Innovation and learning. As proposed by [132,133] CSR could serve as a source of innovation and competitive advantage, analyzing and learning from performance measures could offer deep understandings on how effectively the organizations performed [69]. Hence, the measurement scale of Innovation and learning were adjusted from the works of [26,125,134].

4.3.2. Corporate Social Responsibility Disclosure

Apart from the above-mentioned reasons for the selection of measurement scales in this research, our choice of modifying the measurement scales established by [135] was due to the correspondence between the two research contexts. To put it simply, the context of this study was the developing country and the research undertaken by [135] was also in the developing region. In doing so, CSRD was made up by the five primary components including Community Welfare, Contribution to Education, Environmental and Energy Importance, Services, Customers and Stakeholders and Workforce. Accordingly, Community Welfare included three sub-scale items was modified from the study of [40,135–137], and [33]. In the meanwhile, the rest of the measurement scales was adapted from the findings of [135].

4.3.3. Organizational Performance

The organizational financial prosperity should be fulfilled with the appearance of social, environmental, and governance activities [78]. Thus, the measurement scales for ORG employed in this paper were established by the four key elements namely Economic performance, Environment performance, Human performance, Governance performance.

Economic performance connected to the economic condition which focused on the economic indicators instead of financial indicators presented on the annual statement [138]. As such, Economic performance was adapted from the works of [139].

Human performance signified to the association between the organization and its labor force [140]. Thus, Human performance was aligned from the studies implemented by [141].

Environmental performance described the endeavor which organizations utilize to insulate nature [142]. Environment performance was adjusted from the works of [125,134,143,144].

Governance performance implied for the systems and processes related to sheltering the organizational orientation, control and accountability [145]. Additionally, board composition and board behavior [146] and satisfying stakeholders [147] were supposed to be the main targets of organizational governance. Hence, Governance performance was taken from the outcomes of [148].

5. Result Analysis and Discussion

5.1. Demographic Characteristics

The demographic profile of the 723 respondents was covered with their gender, age, qualification and working experience. In terms of the gender, females constituted 75.10 per cent of the respondents while only 24.9 per cent of males were devoted to the main sample. With regard to the age, 196 respondents (7.47 per cent) belonged to "above 45" group, 328 respondents (45.37 per cent) were the "35–45" group, 145 respondents (20.06 per cent) were "25–35" group and the remaining 54 respondents (7.47 per cent) were classified as "below 5." The work experience ranged from below 5 years (7.47 per cent) to 5–10 years (20.75 per cent), 10–15 years (49.41 per cent) and more than 15 years (23.62 per cent). Moreover, respondents having an undergraduate background accounted for 94.47 per cent, whereas respondents having a postgraduate degree took up a tiny minority (5.53 percent) of the target population.

5.2. Assessment of Convergent Validity

The reliability analysis of the scale was firstly carried out through evaluating the Cronbach's α. Hence, value of the Cronbach's α was recommended at 0.7 or more to demonstrate the trustworthiness of the scale [149]. Given that convergent validity illustrated the extent to which the scale correlated positively with other measures of the same constructs [150], factor loadings, composite reliability (CR) and average variance extracted (AVE) were employed in this study for convergent validity measurement [151]. Thus, standardized factor loadings were suggested to exceed the value of 0.6 [152]. Besides, CR was requested to be over the cutoff value of 0.82 [153]. The acceptable level of AVE was expected to be above 0.5 [154]. The results depicted in Table 1 indicated that the model obtained good convergent validity.

Table 1. Results summary for the measurement model.

Model Construct	Items	Factor Loadings Ranges	AVE	Cronbach's Alpha	Composite Reliability	Discriminant Validity	Source
Integration Between Csr and Pss							
Key performance outcome	4	0.748–0.854	0.614	0.861	0.864	Yes	
Financial	2	0.839–0.854	0.719	0.835	0.837	Yes	[123,124]
Service delivery	4	0.722–0.805	0.595	0.852	0.854	Yes	
People, partnerships and resources	3	0.823–0.858	0.698	0.871	0.874	Yes	
Service user/stakeholder	3	0.817–0.877	0.714	0.882	0.882	Yes	[123–125]
Leadership	3	0.806–0.889	0.710	0.878	0.880	Yes	[130]
Innovation and Learning	3	0.826–0.888	0.734	0.892	0.892	Yes	[26,125,134]
Corporate Social Responsibility Disclosures							
Community Welfare	3	0.795–0.888	0.696	0.872	0.873	Yes	[33,40,135–137]
Contribution to Education	2	0.832–0.910	0.759	0.862	0.863	Yes	
Environmental and Energy Importance	3	0.814–0.874	0.723	0.884	0.887	Yes	[135]
Services, Customers and Stakeholders	3	0.832–0.883	0.745	0.893	0.898	Yes	
Workforce	3	0.802–0.889	0.713	0.877	0.881	Yes	
Organizational Performance							
Economic performance	4	0.707–0.821	0.622	0.866	0.868	Yes	[139]
Environment performance	3	0.790–0.835	0.669	0.858	0.858	Yes	[125,134,143,144]
Human performance	5	0.705–0.804	0.562	0.864	0.865	Yes	[141]
Governance performance	4	0.709–0.861	0.607	0.859	0.860	Yes	[148]

5.3. Assessment of Discriminant Validity

The discriminant validity was considered as the extent to which measures of a given construct distinguished from measures of other constructs in the same model [112]. The AVE could be wielded to ascertain the discriminant validity [154]. Accordingly, when the AVE of each of the latent constructs was higher than the highest squared correlation compared with any other latent variable, discriminant validity of the construct level was set up [155]. The Table 2 displays that the square root of the AVE values were well above the correlation values; thus discriminant validity was achieved.

Table 2. Results of discriminant validity.

	HP	EP	KPO	GP	SD	SCS	IL	SUS	EEI	WORK	LEAD	CW	PPR	ENP	CE	FINA
HP	1															
EP	0.034	1														
KPO	0.243	0.090	1													
GP	0.128	0.171	0.093	1												
SD	0.061	0.065	0.064	0.186	1											
SCS	0.097	0.071	0.073	−0.005	0.012	1										
IL	0.146	−0.010	0.040	0.076	0.133	0.065	1									
SUS	0.014	−0.026	0.156	−0.011	0.037	0.078	0.025	1								
EEI	0.318	0.145	0.121	0.108	0.037	0.150	0.078	−0.001	1							
WORK	0.146	0.191	0.112	0.047	0.030	0.228	0.124	0.107	0.013	1						
LEAD	−0.023	−0.017	0.044	0.006	0.172	0.027	0.213	0.160	0.037	0.100	1					
CW	0.123	0.051	0.106	0.070	0.063	0.136	−0.024	0.035	0.209	0.110	0.143	1				
PPR	0.158	0.038	0.099	0.008	0.229	0.016	0.174	−0.160	0.056	0.123	0.168	0.074	1			
ENP	0.121	0.212	0.131	0.147	0.049	0.043	0.065	0.062	0.067	−0.043	0.004	0.089	0.003	1		
CE	0.062	0.019	0.119	0.071	0.027	0.095	0.072	0.066	0.159	0.126	0.054	0.233	−0.017	0.067	1	
FINA	0.231	0.035	0.195	0.074	−0.023	0.047	0.000	0.194	0.036	0.190	−0.071	0.077	0.008	0.024	0.017	1

HP = human performance; EP = economic performance; KPO = key performance outcome; GP = governance performance; SD = service delivery; SCS = services, customers and stakeholders; IL = innovation and learning; SUS = service user/stakeholder; WORK = workforce; EEI = environmental and energy Importance; LEAD = leadership; PPR = people, partnerships and resources; CW = community welfare; ENP = environment performance; CE = contribution to education; FINA = financial.

5.4. Assessment of Overall Model Fit

The generally lowest values suggested for GFI, TLI, AGFI and CFI were 0.90 and the ratio of $\chi 2/df$ was proposed to be below 3.0 [156]. On the other hand, the value of GFI was also reported to be under 0.95 in several research namely the GFI index ranging from 0.774 to 0.923 [157]. The results in the Table 3 exposed that the measurement model and structural model met the goodness of fit requirements in the present context.

Table 3. Results of measurement and structural model analysis.

The Goodness of Fit Measures	CMIN/DF	GFI	CFI	TLI	RMSEA
Recommended value	≤3	≥0.9	≥0.9	≥0.9	≤0.08
Measurement Model	1.810	0.903	0.953	0.964	0.033
Structural Model	1.946	0.887	0.940	0.937	0.036

5.5. Hypothesis Verification

5.5.1. Direct Effect

In order to test the research hypotheses, this study estimated the path coefficients of the research statistical structural model which revealed several noticeable results highlighted in Table 4 as follows.

Table 4. Structural coefficients (β) of the proposed model.

Hypothesis	Relationship			Estimate	S.E.	C.R.	P	Inference
Hypothesis 1 (H1)	CSRD	←	ICP	0.544	0.137	3.960	0.000	Supported
Hypothesis 2 (H2)	OP	←	ICP	0.291	0.117	2.482	0.013	Supported
Hypothesis 3 (H3)	OP	←	CSRD	0.272	0.112	2.427	0.015	Supported

The outputs illustrated that the positive effect of ICP (β = 0.544) was significant at the 95% confidence level, hence offering support for Hypothesis 1 (H1), which conjectured that ICP had a positive influence on the CSRD. This indicated that in the current study the effect of ICP on CSRD was significant. In other words, PSO could succeed in gaining the efficiency and effectiveness in disclosing CSR issues through putting the ICP into action.

The research results propped up Hypothesis 2 (H2) which was developed to investigate the effect of ICP on the ORG (β = 0.291). This hinted the positive effect of ICP on OP. In other words, the ICP would facilitate the PSO to enhance the overall performance. Accordingly, undertaking CSR practices under the PSS framework in a strategic manner could generate significant support for attaining higher performance in a variety of ways; namely, in economic, human, environmental and governance facets.

In order to examine Hypothesis 3 (H3), the impact of CSRD on OP was measured. The results showed that the effect of CSRD on OP was significant at the 95% confidence level (β = 0.272). Since gaining better understanding among stakeholders on the measures implemented by the organization for their wellbeing would generate a rapid increasing on OP [101]. Hence, the higher degree in the CSRD that could be achieved by the PSO, the better performance the PSO could reap.

5.5.2. Indirect Effect

The results showed the presence of the positive indirect effect of ICP on OP through CSRD was significant at the 95% confidence level. Therefore, if the level of CSRD gets higher, the mediating effect between ICP and OP is significant (β = 0.506), supporting the positive indirect effect of ICP on OP in PSO. By controlling the mediators, the direct effect of ICP on OP was significant but weaker (β = 0.257), indicating full mediation. To put it simply, PSOs could enhance their overall performance in such aspects as economic, human, environmental and governance facets when the disclosure practices were taken into consideration instead of concentrating only on performing the ICP.

6. Concluding Remarks

6.1. Discussion and Implication

From the academic standpoint, the current research has augmented on the studies related to the strategic CSR management toward the sustainable development. Although the combination between CSR and management has been not a new concept and substantiated to bring several certain effectiveness to the organizations, it has been widely applied in the private sector, which has many differences in characteristics compared to PSOs. In this regard, the research has deepened the insights on the likelihood and the advantages of integration between CSR practices into a management framework which is well-acknowledged to be best suited to the characteristics of a PSO. In light of the same target, stakeholder, PSS has been proved to be in appropriate to integrate the CSR practices in PSO. Importantly, this potential integration has empirically validated in this study to uplift the CSRD in the developing countries, which was regarded as the pressing concern in these regions as CSRD has not yet been compulsory regulation [158] and the demand on CSR programs due to the inherent social provisions and governance gaps [159]. On the other hand, the significant and positive impact that ICP effectuated on OP has invigorated the role of PSS adoption in relation to sustainable development. This framework is deliberated for workshop established approach in which the outcome definition has been implemented through raising spirits of the organizational internal and external components to participate to restructure and originate the new process to accomplish these outcomes and find out the effective solution for the capability and related matters to attain the outcome. As such, it can be taken as a reference for addressing the issue of lacking similarities in CSR practice and reporting as the boundary between regions and organization has eliminated. What is more, the outcomes mentioned in this model mainly come to grips with the target which is under the highest satisfaction from almost all of the users and stakeholders who played the most important role in the existence of the organization. Additionally, the financial outcomes are also involved to boost the financial situation of the organization, and simultaneously, generate a better solution to the prior expostulation based on the fact that the entities with better performance in CSR practices would suffer from the poor FP [30]. Importantly, service delivery chiefly revolve about actual experiences of users and stakeholders will lead to the higher opportunities to achieve loyalty of the users and the approval of the stakeholders. Besides, the capability in this framework also buttons down on the essential support for the employees (i.e., motivation and training) and processes (i.e., innovation and learning, leadership) in fulfilling the outcomes and outputs.

Additionally, the findings of this research have put accent on CSRD with the roles both on the independent and mediate variable. In terms of the role as an independent variable, the findings on the interconnection between CSRD and OP have elucidated and reinforced the works of several prior researchers (i.e., [35,160]). In the meanwhile, this study has given out a converse result to some prior works (i.e., [161]) in terms of the mediating role of CSRD.

Furthermore, the performance of PSO has been concentrated on the comprehensive aspects in terms of sustainable development, namely, the economic performance; environment performance, human resource performance and governance performance which highlighted the consistence of the ICP can be best suited for managing and measuring the CSR in a strategic manner toward the sustainable development in PSO in the developing regions.

In respect to the practical implication, the leaders are supposed to intensify the perception of CSR practices as a reasonable investment rather than an obligation to conduct. Besides, the result which accentuated the contribution of PSS into managing and measuring the CSR program in PSO in terms of the strategic manner toward the sustainable development also raises a demand on the consideration for adoption in PSO. As such, with the exception from the certain support of financial resources, the departmental communication also play an important role in facilitate for the introduction and application of PSS as the employees will be likely to adopt in effective manner with the comprehensive perception on the meaning and value of PSS and CSR practices. On the other hand, the findings on the mediating of CSRD in the ICP-OP linkage have raised an urgent demand for policymakers for a general regulation about CSRD towards sustainable development in PSO because environmental outcomes can only be achieved through policy instruments which based on the combination of laws, regulatory approaches and market signals to cause a positively significant impact on operations and behaviors [162] among PSOs.

6.2. Limitations and Further Research

Unfortunately, the current research still suffered from several limitations. The restriction in terms of small sample size and target population was considered as the primary barrier in the generalization of empirical outcomes. Nevertheless, this situation will probably be eliminated when such a manifold population has been targeted in further studies. Owing to the cross-sectional data collection, the findings of the present research were prevented from creating any strong causal requirements in relation to these effects [163]. Thus, future researchers should take the longitudinal designs into consideration to allow the research framework to be modeled over time. Importantly, the replication of this type of research in a variety of areas has been in demand due to desires to underline much more interesting findings on the relationships of the proposed model in detail; the introduction of PSS and the number of issues regarding CSR are still too much for PSOs to handle, especially in the developing economies. Besides, the secondary data related to CSR issues are also recommended for the future work to eradicate the inherent limitations in this study. Eventually, we request the involvement of several procedures in relation to the direct-indirect for the more accurate affirmation.

Author Contributions: P.Q.H. established the conception and design of the research. V.K.P. conducted the data acquisition, analysis and interpretation. V.K.P. prepared the drafted paper. P.Q.H. provided vital revision of the paper. P.Q.H. offered final approval of the version to publish. P.Q.H. was accountable for the accuracy or integrity of any part of the work. All authors have read and agreed to the published version of the manuscript.

Funding: This research was funded by University of Economics Ho Chi Minh city, Vietnam

Conflicts of Interest: The authors declare no conflict of interest.

References

1.   Saeed, M.M.; Arshad, F. Corporate social responsibility as a source of competitive advantage: The mediating role of social capital and reputational capital. *J. Database Mark. Cust. Strat. Manag.* **2012**, *19*, 219–232. [CrossRef]
2.   Mullerat, R. *International Corporate Social Responsibility: The Role of Corporations in the Economic Order of the 21st Century*; Wolters Kluwer: Alphen aan den Rijn, The Netherlands, 2010.
3.   Perry, P.; Towers, N. Conceptual framework development. *Int. J. Phys. Distrib. Logist. Manag.* **2013**, *43*, 478–501. [CrossRef]
4.   Khan, S.N. Making sense of the black box: An empirical analysis investigating strategic cognition of CSR strategists in a transitional market. *J. Clean. Prod.* **2018**, *196*, 916–926. [CrossRef]
5.   Wahba, H.; Elsayed, K. The mediating effect of financial performance on the relationship between social responsibility and ownership structure. *Futur. Bus. J.* **2015**, *1*, 1–12. [CrossRef]
6.   Visser, W.; Matten, D.; Polh, M.; Tolhurst, N. *The A to Z of corporate social responsibility: A Complete Reference Guide to Concepts, Codes and Organisations*; John Wiley & Sons Ltd: West Sussex, UK, 2007.
7.   Visser, W. Corporate Social Responsibility in Developing Countries. In *Corporate Social Responsibility in Developing Countries*; Oxford University Press (OUP): Oxford, UK, 2009; pp. 473–479.
8.   Ali, W.; Frynas, J.G.; Mahmood, Z. Determinants of Corporate Social Responsibility (CSR) Disclosure in Developed and Developing Countries: A Literature Review. *Corp. Soc. Responsib. Environ. Manag.* **2017**, *24*, 273–294. [CrossRef]
9.   De Oliveira, J.A.P. Introduction: Corporate Citizenship in Latin America. *J. Corp. Citizsh.* **2006**, *2006*, 17–20. [CrossRef]
10.   Moon, J. *Corporate Social Responsibility: A Very Short Introduction*; Oxford University Press (OUP): Oxford, UK, 2014.
11.   Moullin, M. Improving and evaluating performance with the Public Sector Scorecard. *Int. J. Prod. Perform. Manag.* **2017**, *66*, 442–458. [CrossRef]
12.   Wang, Z.; Sarkis, J. Corporate social responsibility governance, outcomes, and financial performance. *J. Clean. Prod.* **2017**, *162*, 1607–1616. [CrossRef]
13.   Tilt, C. The Influence of External Pressure Groups on Corporate Social Disclosure. *Account. Audit. Account. J.* **1994**, *7*, 47–72. [CrossRef]
14.   Van Thanh, P.; Podruzsik, S. CSR in Developing Countries: Case Study in Vietnam. *Management* **2018**, *13*, 287–300. [CrossRef]
15.   Cook, L.; LaVan, H.; Žilić, I. An exploratory analysis of corporate social responsibility reporting in US pharmaceutical companies. *J. Commun. Manag.* **2018**, *22*, 197–211. [CrossRef]
16.   Adnan, S.M.; Hay, D.; Van Staden, C. The influence of culture and corporate governance on corporate social responsibility disclosure: A cross country analysis. *J. Clean. Prod.* **2018**, *198*, 820–832. [CrossRef]
17.   Hu, W.; Ge, Y.; Dang, Q.; Huang, Y.; Hu, Y.; Ye, S.; Wang, S. Analysis of the Development Level of Geo-Economic Relations between China and Countries along the Belt and Road. *Sustainability* **2020**, *12*, 816. [CrossRef]
18.   Huang, X.; Watson, L. Corporate social responsibility research in accounting. *J. Account. Lit.* **2015**, *34*, 1–16. [CrossRef]
19.   Henri, J.-F.; Journeault, M. Eco-control: The influence of management control systems on environmental and economic performance. *Account. Organ. Soc.* **2010**, *35*, 63–80. [CrossRef]
20.   Gond, J.-P.; Grubnic, S.; Herzig, C.; Moon, J. Configuring management control systems: Theorizing the integration of strategy and sustainability. *Manag. Account. Res.* **2012**, *23*, 205–223. [CrossRef]
21.   Arjaliès, D.-L.; Mundy, J. The use of management control systems to manage CSR strategy: A levers of control perspective. *Manag. Account. Res.* **2013**, *24*, 284–300. [CrossRef]
22.   Sila, I.; Cek, K. The Impact of Environmental, Social and Governance Dimensions of Corporate Social Responsibility on Economic Performance: Australian Evidence. *Procedia Comput. Sci.* **2017**, *120*, 797–804. [CrossRef]

23. Carroll, A.B.; Buchholtz, A.K. *Business & Society: Ethics and Stakeholder Management*, 5th ed.; Thomson Learning: Mason, OH, USA, 2003.
24. Coldwell, D.A. Perceptions and expectations of corporate social responsibility: Theoretical issues and empirical findings. *S. Afr. J. Bus. Manag.* **2001**, *32*, 49–55. [CrossRef]
25. Freeman, R.E. *Strategic Management: A Stakeholder Approach*; Pitman: Boston, MA, USA, 1984.
26. Mishra, S.; Suar, D. Does Corporate Social Responsibility Influence Firm Performance of Indian Companies? *J. Bus. Ethics* **2010**, *95*, 571–601. [CrossRef]
27. Crisóstomo, V.L.; Freire, F.D.S.; De Vasconcellos, F.C. Corporate social responsibility, firm value and financial performance in Brazil. *Soc. Responsib. J.* **2011**, *7*, 295–309. [CrossRef]
28. Preston, L.E.; O'Bannon, D.P. The Corporate Social-Financial Performance Relationship. *Bus. Soc.* **1997**, *36*, 419–429. [CrossRef]
29. Brammer, S.; Millington, A. Does it pay to be different? An analysis of the relationship between corporate social and financial performance. *Strateg. Manag. J.* **2008**, *29*, 1325–1343. [CrossRef]
30. Gatsi, J.G.; Anipa, C.A.A.; Gadzo, S.G.; Ameyibor, J. Corporate social responsibility, risk factor and financial performance of listed firms in Ghana. *J. Appl. Financ. Bank.* **2016**, *6*, 21–38.
31. Abdelmotaleb, M.; Saha, S.K. Corporate Social Responsibility, Public Service Motivation and Organizational Citizenship Behavior in the Public Sector. *Int. J. Public Adm.* **2018**, *42*, 929–939. [CrossRef]
32. Clarkson, P.; Overell, M.B.; Chapple, L. Environmental Reporting and its Relation to Corporate Environmental Performance. *Abacus* **2011**, *47*, 27–60. [CrossRef]
33. Dabor, A.O.; Kaka, M.; Idogen, K. Corporate Social Responsibility and Financial Performance: A Two Least Regression Approach. *Int. J. Soc. Sci. Educ. Stud.* **2017**, *4*, 44–54. [CrossRef]
34. Ahmed, M.N.; Zakaree, S.; Kolawole, O.O. Corporate Social Responsibility Disclosure and Financial Performance of Listed Manufacturing Firms in Nigeria. *Res. J. Financ. Account.* **2016**, *7*, 47–58.
35. Bayoud, N.S.M.; Kavanagh, M.; Slaughter, G. An Empirical Study of the Relationship between Corporate Social Responsibility Disclosure and Organizational Performance: Evidence from Libya. *Int. J. Manag. Mark. Res.* **2012**, *5*, 69–82.
36. Gallardo-Vázquez, D.; Méndez, M.J.B.; Pajuelo-Moreno, M.-L.; Sánchez-Meca, J. Corporate Social Responsibility Disclosure and Performance: A Meta-Analytic Approach. *Sustainability* **2019**, *11*, 1115. [CrossRef]
37. Yahya, H.Y.; Ghodratollah, B. The effect of disclosure level of CSR on corporate financial performance in Tehran stock exchange. *Int. J. Account. Res.* **2014**, *1*, 43–51.
38. Mittal, R.; Sinha, N.; Singh, A. An analysis of linkage between economic value added and corporate social responsibility. *Manag. Decis.* **2008**, *46*, 1437–1443. [CrossRef]
39. Kimbro, M.B.; Melendy, S.R. Financial performance and voluntary environmental disclosures during the Asian Financial Crisis: The case of Hong Kong. *Int. J. Bus. Perform. Manag.* **2010**, *12*, 72. [CrossRef]
40. Gray, R.; Javad, M.; Power, D.; Sinclair, C.D. Social and Environmental Disclosure and Corporate Characteristics: A Research Note and Extension. *J. Bus. Financ. Account.* **2001**, *28*, 327–356. [CrossRef]
41. Castelló, I.; Lozano, J.M. Searching for New Forms of Legitimacy Through Corporate Responsibility Rhetoric. *J. Bus. Ethics* **2011**, *100*, 11–29. [CrossRef]
42. Lim, S.; Wilmshurst, T.; Shimeld, S. *Blowing in the Wind: Legitimacy Theory: An Environmental Incident and Disclosure*; Research paper; University of Tasmania: Hobart, Australia, 2009.
43. Deegan, C. Introduction. *Account. Audit. Account. J.* **2002**, *15*, 282–311. [CrossRef]
44. Deegan, C.; Unerman, J. *Finanical Accounting Theory*; McGraw-Hill: Sydney, Australia, 2011.
45. O'Donovan, G. Environmental disclosures in the annual report. *Account. Audit. Account. J.* **2002**, *15*, 344–371. [CrossRef]
46. Maignan, I.; Ralston, D.A. Corporate Social Responsibility in Europe and the U.S.: Insights from Businesses' Self-presentations. *J. Int. Bus. Stud.* **2002**, *33*, 497–514. [CrossRef]
47. Adams, C.; Hill, W.-Y.; Roberts, C.B. Corporate Social Reporting Practices in Western Europe: Legitimating Corporate Behaviour. *Br. Account. Rev.* **1998**, *30*, 1–21. [CrossRef]
48. Merkelsen, H. The double-edged sword of legitimacy in public relations. *J. Commun. Manag.* **2011**, *15*, 125–143. [CrossRef]
49. Moir, L. What do we mean by corporate social responsibility? *Corp. Gov.* **2001**, *1*, 16–22. [CrossRef]

50. Branco, M.C.; Rodrigues, L. Factors Influencing Social Responsibility Disclosure by Portuguese Companies. *J. Bus. Ethics* **2008**, *83*, 685–701. [CrossRef]
51. Omran, M.; Ramdhony, D. Theoretical Perspectives on Corporate Social Responsibility Disclosure: A Critical Review. *Int. J. Account. Financ. Rep.* **2015**, *5*, 38. [CrossRef]
52. Margolis, J.D.; Walsh, J.P. Misery Loves Companies: Rethinking Social Initiatives by Business. *Adm. Sci. Q.* **2003**, *48*, 268. [CrossRef]
53. Kytle, B.; Hamilton, B.A.; Ruggie, J.G. *Corporate Social Responsibility as Risk Management: A Model for Multinationals*; Social Responsibility Initiative Working Paper: Cambridge, MA, USA, 2005.
54. Oliver, C. Sustainable competitive advantage: Combining institutional and resource-based views. *Strateg. Manag. J.* **1997**, *18*, 697–713. [CrossRef]
55. Russo, M.V.; Fouts, P.A. A resource-based perspective on corporate environmental performance and profitability. *Acad. Manag. J.* **1997**, *40*, 534–559. [CrossRef]
56. Gomez-Mejia, L.R.; Balkin, D.B. *Management*; McGraw-Hill: New York, NY, USA, 2002.
57. Barney, J. Firm Resources and Sustained Competitive Advantage. *J. Manag.* **1991**, *17*, 99–120. [CrossRef]
58. Surroca, J.; Tribo, J.A.; Waddock, S. Corporate responsibility and financial performance: The role of intangible resources. *Strateg. Manag. J.* **2009**, *31*, 463–490. [CrossRef]
59. Orlitzky, M.; Siegel, D.S.; Waldman, D.A. Strategic Corporate Social Responsibility and Environmental Sustainability. *Bus. Soc.* **2011**, *50*, 6–27. [CrossRef]
60. Fombrun, C.; Shanley, M. What's in a name? Reputation building and corporate strategy. *Acad. Manag. J.* **1990**, *33*, 233–258. [CrossRef]
61. Priem, R.L.; Butler, J.E. Tautology in the Resource-Based View and the Implications of Externally Determined Resource Value: Further Comments. *Acad. Manag. Rev.* **2001**, *26*, 57–66. [CrossRef]
62. European Commission. *Promoting a European Framework for Corporate Social Responsibility*; Office for Official Publications of the European Communities: Luxembourg, 2011.
63. Lii, Y.-S.; Lee, M. Doing Right Leads to Doing Well: When the Type of CSR and Reputation Interact to Affect Consumer Evaluations of the Firm. *J. Bus. Ethics* **2011**, *105*, 69–81. [CrossRef]
64. Aguinis, H. Organizational responsibility: Doing good and doing well. In *Handbook of Industrial and Organizational Psychology 3*; Zedeck, S., Ed.; APA, American Psychological Association: Washington, DC, USA, 2011; pp. 855–879.
65. Kim, E.E.K.; Kang, J.; Mattila, A.S. The impact of prevention versus promotion hope on CSR activities. *Int. J. Hosp. Manag.* **2012**, *31*, 43–51. [CrossRef]
66. Wang, J.; Zhang, Y.; Goh, M. Moderating the Role of Firm Size in Sustainable Performance Improvement through Sustainable Supply Chain Management. *Sustainability* **2018**, *10*, 1654. [CrossRef]
67. Hackston, D.; Milne, M. Some determinants of social and environmental disclosures in New Zealand companies. *Account. Audit. Account. J.* **1996**, *9*, 77–108. [CrossRef]
68. Said, R.; Zainuddin, Y.H.; Haron, H. The relationship between corporate social responsibility disclosure and corporate governance characteristics in Malaysian public listed companies. *Soc. Responsib. J.* **2009**, *5*, 212–226. [CrossRef]
69. Moullin, M. Using the Public Sector Scorecard to measure and improve healthcare services. *Nurs. Manag.* **2009**, *16*, 26–31. [CrossRef]
70. Pidd, M. *Measuring the Performance of Public Services*; Cambridge University Press (CUP): Cambridge, UK, 2012.
71. McAdam, R.; Casey, C.; Hazlett, S.-A. Performance management in the UK public sector. *Int. J. Public Sect. Manag.* **2005**, *18*, 256–273. [CrossRef]
72. Johnston, R.; Pongatichat, P. Managing the tension between performance measurement and strategy: Coping strategies. *Int. J. Oper. Prod. Manag.* **2008**, *28*, 941–967. [CrossRef]
73. Al-Samman, E.; Al-Nashmi, M.M. Effect of corporate social responsibility on nonfinancial organizational performance: Evidence from Yemeni for-profit public and private enterprises. *Soc. Responsib. J.* **2016**, *12*, 247–262. [CrossRef]
74. Talbot, C. *Theories of Performance: Organizational and Service Improvement in the Public Domain*; Oxford University Press: New York, NY, USA, 2010.
75. Venkatraman, N.; Ramanujam, V. Measurement of Business Economic Performance: An Examination of Method Convergence. *J. Manag.* **1987**, *13*, 109–122. [CrossRef]

76. Hancott, D.E. The Relationship between Transformational Leadership and Organizational Performance in the Largest Public Companies in Canada. Ph.D. Thesis, ProQuest database, UMI No. 3159704. Morrisville, NC, USA, 2005.
77. Akanbi, P.A.; Ofoegbu, O.E. Impact of corporate social responsibility on bank performance in Nigeria. *J. US China Public Adm.* **2012**, *9*, 374–383.
78. E Porter, M.; Van Der Linde, C. Toward a New Conception of the Environment-Competitiveness Relationship. *J. Econ. Perspect.* **1995**, *9*, 97–118. [CrossRef]
79. Broadstock, D.C.; Matousek, R.; Meyer, M.; Tzeremes, N.G. Does corporate social responsibility impact firms' innovation capacity? The indirect link between environmental & social governance implementation and innovation performance. *J. Bus. Res.* **2019**. [CrossRef]
80. Zhu, Q.; Lowe, E.A.; Wei, Y.-A.; Barnes, D. Industrial Symbiosis in China: A Case Study of the Guitang Group. *J. Ind. Ecol.* **2008**, *11*, 31–42. [CrossRef]
81. Rojas, R.R. A Review of Models for Measuring Organizational Effectiveness Among For-Profit and Nonprofit Organizations. *Nonprofit Manag. Leadersh.* **2000**, *11*, 97–104. [CrossRef]
82. Osborne, S.P.; Radnor, Z.; Nasi, G. A New Theory for Public Service Management? Toward a (Public) Service-Dominant Approach. *Am. Rev. Public Adm.* **2012**, *43*, 135–158. [CrossRef]
83. Maignan, I.; Ferrell, O.; Ferrell, L. A stakeholder model for implementing social responsibility in marketing. *Eur. J. Mark.* **2005**, *39*, 956–977. [CrossRef]
84. Costa, C.; Lages, L.F.; Hortinha, P. The bright and dark side of CSR in export markets: Its impact on innovation and performance. *Int. Bus. Rev.* **2015**, *24*, 749–757. [CrossRef]
85. McWilliams, A.; Siegel, D. Corporate social responsibility and financial performance: Correlation or misspecification? *Strateg. Manag. J.* **2000**, *21*, 603–609. [CrossRef]
86. Moullin, M. The design of an alternative balanced scorecard framework for public and voluntary organizations. *Perspect. Perform.* **2006**, *5*, 10–12.
87. Goergen, M.; Chahine, S.; Wood, G.; Brewster, C. The relationship between public listing, context, multi-nationality and internal CSR. *J. Corp. Financ.* **2019**, *57*, 122–141. [CrossRef]
88. Stahl, G.K.; Brewster, C.; Collings, D.; Hajro, A. Enhancing the role of human resource management in corporate sustainability and social responsibility: A multi-stakeholder, multidimensional approach to HRM. *Hum. Resour. Manag. Rev.* **2019**, *100708*, 100708. [CrossRef]
89. Liu, X.; Zhang, C. Corporate governance, social responsibility information disclosure, and enterprise value in China. *J. Clean. Prod.* **2017**, *142*, 1075–1084. [CrossRef]
90. Forker, J.J. Corporate Governance and Disclosure Quality. *Account. Bus. Res.* **1992**, *22*, 111–124. [CrossRef]
91. Branco, M.C.; Rodrigues, L. Corporate Social Responsibility and Resource-Based Perspectives. *J. Bus. Ethics* **2006**, *69*, 111–132. [CrossRef]
92. Hull, C.E.; Rothenberg, S. Firm performance: The interactions of corporate social performance with innovation and industry differentiation. *Strateg. Manag. J.* **2008**, *29*, 781–789. [CrossRef]
93. McGahan, A.M.; Porter, M.E. HOW MUCH DOES INDUSTRY MATTER, REALLY? *Strateg. Manag. J.* **1997**, *18*, 15–30. [CrossRef]
94. Bocquet, R.; Le Bas, C.; Mothe, C.; Poussing, N. Are firms with different CSR profiles equally innovative? Empirical analysis with survey data. *Eur. Manag. J.* **2013**, *31*, 642–654. [CrossRef]
95. Aras, G.; Aybars, A.; Kutlu, O.; Aybars, A. Managing corporate performance. *Int. J. Prod. Perform. Manag.* **2010**, *59*, 229–254. [CrossRef]
96. Hillman, A.; Keim, G. Shareholder value, stakeholder management, and social issues: What's the bottom line? *Strateg. Manag. J.* **2001**, *22*, 125–139. [CrossRef]
97. Oeyono, J.; Samy, M.; Bampton, R. An examination of corporate social responsibility and financial performance. *J. Glob. Responsib.* **2011**, *2*, 100–112. [CrossRef]
98. Gorski, H. Leadership and Corporate Social Responsibility. In Proceedings of the International Conference Knowledge Based Organization, Warsaw, Poland, 12 June 2017; Volume 23, pp. 372–377. [CrossRef]
99. Asrar-Ul-Haq, M.; Kuchinke, K.P.; Iqbal, A. The relationship between corporate social responsibility, job satisfaction, and organizational commitment: Case of Pakistani higher education. *J. Clean. Prod.* **2017**, *142*, 2352–2363. [CrossRef]

100. Suganthi, L. Examining the relationship between corporate social responsibility, performance, employees' pro-environmental behavior at work with green practices as mediator. *J. Clean. Prod.* **2019**, *232*, 739–750. [CrossRef]

101. Lindgreen, A.; Swaen, V.; Johnston, W. The Supporting Function of Marketing in Corporate Social Responsibility. *Corp. Reput. Rev.* **2009**, *12*, 120–139. [CrossRef]

102. Humphrey, J.; Lee, D.D.; Shen, Y. The independent effects of environmental, social and governance initiatives on the performance of UK firms. *Aust. J. Manag.* **2012**, *37*, 135–151. [CrossRef]

103. Luo, X.; Bhattacharya, C.B. Corporate Social Responsibility, Customer Satisfaction, and Market Value. *J. Mark.* **2006**, *70*, 1–18. [CrossRef]

104. Vurro, C.; Perrini, F. Making the most of corporate social responsibility reporting: Disclosure structure and its impact on performance. *Corp. Gov. Int. J. Bus. Soc.* **2011**, *11*, 459–474. [CrossRef]

105. Waddock, S.; Bodwell, C. Managing Responsibility: What Can Be Learned from the Quality Movement? *Calif. Manag. Rev.* **2004**, *47*, 25–37. [CrossRef]

106. Yusoff, H.; Mohamad, S.S.; Darus, F. The Influence of CSR Disclosure Structure on Corporate Financial Performance: Evidence from Stakeholders' Perspectives. *Procedia Econ. Financ.* **2013**, *7*, 213–220. [CrossRef]

107. Choongo, P. A Longitudinal Study of the Impact of Corporate Social Responsibility on Firm Performance in SMEs in Zambia. *Sustainability* **2017**, *9*, 1300. [CrossRef]

108. Dhaliwal, D.S.; Li, O.Z.; Tsang, A.; Yang, Y.G. Voluntary Nonfinancial Disclosure and the Cost of Equity Capital: The Initiation of Corporate Social Responsibility Reporting. *Account. Rev.* **2011**, *86*, 59–100. [CrossRef]

109. Bernard, H.R. *Social Research Methods: Qualitative and Quantitative Approaches*; Sage: Thousand Oaks, CA, USA, 2000.

110. Dunn, T.J.; Baguley, T.; Brunsden, V.; Baguley, T. From alpha to omega: A practical solution to the pervasive problem of internal consistency estimation. *Br. J. Psychol.* **2013**, *105*, 399–412. [CrossRef] [PubMed]

111. Hair, J.F.; Ringle, C.M.; Sarstedt, M. PLS-SEM: Indeed a Silver Bullet. *J. Mark. Theory Pr.* **2011**, *19*, 139–152. [CrossRef]

112. Hair, J.F.; Anderson, R.E.; Tatham, R.L.; Black, W.C. *Multivariate Data Analysis with Readings*, 7th ed.; Prentice Hall: Englewood Cliffs, NJ, USA, 2010.

113. Hinkin, T.R. A review of scale development practices in the study of organizations. *J. Manag.* **1995**, *21*, 967–988. [CrossRef]

114. Hair, J.F.; Anderson, R.E.; Tatham, R.L.; Black, W.C. *Multivariate Data Analysis*; Prentice Hall: London, UK, 1998.

115. Vinodh, S.; Joy, D. Structural Equation Modelling of lean manufacturing practices. *Int. J. Prod. Res.* **2012**, *50*, 1598–1607. [CrossRef]

116. Moon, J.-W.; Kim, Y.-G. Extending the TAM for a World-Wide-Web context. *Inf. Manag.* **2001**, *38*, 217–230. [CrossRef]

117. Petkoski, D.; Nigel, T. (Eds.) *Public Policy for Corporate Social Responsibility. WBI Series on Corporate Responsibility, July 7–25 2003*; World Bank Institute: Washington, DC, USA, 2003.

118. Mansi, M.; Pandey, R.; Ghauri, E. CSR focus in the mission and vision statements of public sector enterprises: Evidence from India. *Manag. Audit. J.* **2017**, *32*, 356–377. [CrossRef]

119. Townsend, W. Innovation and the perception of risk in the public sector. *Int. J. Organ. Innov.* **2013**, *5*, 21–34.

120. Khan, A.R.; Khandaker, S. Public and Private Organizations: How Different or Similar are They. *J. Sib. Fed. Univ. Humanit. Soc. Sci.* **2016**, *9*, 2873–2885. [CrossRef]

121. Niven, P.R. *Balanced Scorecard Step-by-Step for Government and Non-Profit Agencies*; John Wiley and Sons: Hoboken, NJ, USA, 2003.

122. Ellis, J. *The Case for an Outcomes Focus*; Charities Evaluation Services: London, UK, 2009.

123. Carroll, A.B. The pyramid of corporate social responsibility: Toward the moral management of organizational stakeholders. *Bus. Horizons* **1991**, *34*, 39–48. [CrossRef]

124. Schwartz, M.S.; Carroll, A.B. Corporate Social Responsibility: A Three-Domain Approach. *Bus. Ethics Q.* **2003**, *13*, 503–530. [CrossRef]

125. Rettab, B.; Ben Brik, A.; Mellahi, K. A Study of Management Perceptions of the Impact of Corporate Social Responsibility on Organisational Performance in Emerging Economies: The Case of Dubai. *J. Bus. Ethics* **2008**, *89*, 371–390. [CrossRef]

126. Al-Salem, A.; Speece, M. Women in leadership in Kuwait: A research agenda. *Gend. Manag. Int. J.* **2017**, *32*, 141–162. [CrossRef]

127. Avina, J. The Evolution of Corporate Social Responsibility (CSR) in the Arab Spring. *Middle East J.* **2013**, *67*, 76–91. [CrossRef]

128. Hodges, J. Cracking the walls of leadership: Women in Saudi Arabia. *Gend. Manag. Int. J.* **2017**, *32*, 34–46. [CrossRef]

129. Sidani, Y. Ibn Khaldun of North Africa: An AD 1377 theory of leadership. *J. Manag. Hist.* **2008**, *14*, 73–86. [CrossRef]

130. Asemah, E.; Okpanachi, R.; Olumuji, E. Universities and Corporate Social Responsibility Performance: An Implosion of the Reality. *Afr. Res. Rev.* **2013**, *7*, 195. [CrossRef]

131. Engert, S.; Baumgartner, R.J. Corporate sustainability strategy – bridging the gap between formulation and implementation. *J. Clean. Prod.* **2016**, *113*, 822–834. [CrossRef]

132. Porter, M.E.; Kramer, M.R. Strategy and society: The link between competitive advantage and corporate social responsibility. *Harv. Bus. Rev.* **2006**, *84*, 78–92.

133. Porter, M.E.; Kramer, M.R. Creating Shared Value. *Manag. Sustain. Bus.* **2018**, *89*, 323–346. [CrossRef]

134. Heslin, P.A.; Vandewalle, D.; Latham, G.P. Keen to help? Managers' implicit person theories and their subsequent employee coaching. *Pers. Psychol.* **2006**, *59*, 871–902. [CrossRef]

135. Ehsan, S.; Nazir, M.S.; Nurunnabi, M.; Khan, Q.R.; Tahir, S.; Ahmed, I. A Multimethod Approach to Assess and Measure Corporate Social Responsibility Disclosure and Practices in a Developing Economy. *Sustainability* **2018**, *10*, 2955. [CrossRef]

136. Guthrie, J.; Parker, L.D. Corporate Social Reporting: A Rebuttal of Legitimacy Theory. *Account. Bus. Res.* **1989**, *19*, 343–352. [CrossRef]

137. Hamid, F.Z.; Atan, R. Corporate social responsibility by the Malaysian Telecommunication firms. *Int. J. Bus. Soc. Sci.* **2011**, *2*, 1–11.

138. Wiklund, J.; Shepherd, D. Entrepreneurial orientation and small business performance: A configurational approach. *J. Bus. Ventur.* **2005**, *20*, 71–91. [CrossRef]

139. Helm, S.T.; Andersson, F.O. Beyond taxonomy. *Nonprofit Manag. Leadersh.* **2010**, *20*, 259–276. [CrossRef]

140. Chang, H.-T.; Chi, N.-W. Human resource managers' role consistency and HR performance indicators: The moderating effect of interpersonal trust in Taiwan. *Int. J. Hum. Resour. Manag.* **2007**, *18*, 665–683. [CrossRef]

141. Calantone, R.J.; Cavusgil, S.T.; Zhao, Y. Learning orientation, firm innovation capability, and firm performance. *Ind. Mark. Manag.* **2002**, *31*, 515–524. [CrossRef]

142. Andersson, L.; Jackson, S.E.; Russell, S.V. Greening organizational behavior: An introduction to the special issue. *J. Organ. Behav.* **2013**, *34*, 151–155. [CrossRef]

143. Junior, F.H.; Gabriel, M.L.D.S.; Gallardo-Vázquez, D.A. Triple bottom line and sustainable performance measurement in industrial companies. *Revista de Gestão* **2018**, *25*, 413–429. [CrossRef]

144. De La Cuesta-González, M.; Rodríguez, D.M.; Fernández-Izquierdo, M. Ángeles Analysis of Social Performance in the Spanish Financial Industry Through Public Data. A Proposal. *J. Bus. Ethics* **2006**, *69*, 289–304. [CrossRef]

145. Cornforth, C. Nonprofit Governance Research: The Need for Innovative Perspectives and Approaches. In *Nonprofit Governance, Innovative Perspectives and Approaches*; Cornforth, C., Brown, W.A., Eds.; Routledge: Abingdon, UK, 2014; pp. 1–14.

146. Hambrick, D.C.; Werder, A.V.; Zajac, E.J. New Directions in Corporate Governance Research. *Organ. Sci.* **2008**, *19*, 381–385. [CrossRef]

147. Chan, M.C.; Watson, J.; Woodliff, D. Corporate Governance Quality and CSR Disclosures. *J. Bus. Ethics* **2013**, *125*, 59–73. [CrossRef]

148. Jackson, D.K.; Holland, T.P. Measuring the Effectiveness of Nonprofit Boards. *Nonprofit Volunt. Sect. Q.* **1998**, *27*, 159–182. [CrossRef]

149. Hundleby, J.D.; Nunnally, J. Psychometric Theory. *Am. Educ. Res. J.* **1968**, *5*, 431. [CrossRef]

150. Malhotra, N.K. *Marketing Research*, 4th ed.; Pearson Education: Singapore, 2002.

151. Anderson, J.C.; Gerbing, D.W. Structural Equation Modeling in Practice: A Review and Recommended Two-Step Approach. *Psychol. Bull.* **1988**, *103*, 411–423. [CrossRef]

152. Suh, B.; Han, I. Effect of trust on customer acceptance of Internet banking. *Electron. Commer. Res. Appl.* **2002**, *1*, 247–263. [CrossRef]

153. Raykov, T. Estimation of Composite Reliability for Congeneric Measures. *Appl. Psychol. Meas.* **1997**, *21*, 173–184. [CrossRef]

154. Fornell, C.; Larcker, D.F. Evaluating structural equation models with unobservable variables and measurement error. *J. Mark. Res.* **1981**, *18*, 39–50. [CrossRef]

155. Henseler, J.; Dijkstra, T.K.; Sarstedt, M.; Ringle, C.M.; Diamantopoulos, A.; Straub, D.W.; Ketchen, D.J.; Hair, J.F.; Hult, G.T.M.; Calantone, R.J. Common beliefs and reality about partialleast squares: Comments on Rönkkö & Evermann (2013). *Organ. Res. Methods* **2014**, *17*, 182–209.

156. Segars, A.H.; Grover, V. Re-Examining Perceived Ease of Use and Usefulness: A Confirmatory Factor Analysis. *MIS Q.* **1993**, *17*, 517. [CrossRef]

157. Motawa, I.; Oladokun, M. Structural equation modelling of energy consumption in buildings. *Int. J. Energy Sect. Manag.* **2015**, *9*, 435–450. [CrossRef]

158. Coffie, W.; Aboagye-Otchere, F.; Musah, A. Corporate social responsibility disclosures (CSRD), corporate governance and the degree of multinational activities. *J. Account. Emerg. Econ.* **2018**, *8*, 106–123. [CrossRef]

159. Fifka, M.S. Corporate Responsibility Reporting and its Determinants in Comparative Perspective—A Review of the Empirical Literature and a Meta-analysis. *Bus. Strat. Environ.* **2011**, *22*, 1–35. [CrossRef]

160. Bayoud, N.S.M. Corporate Social Responsibility Disclosure and Organisational Performance: The Case of Libya, a Mixed Methods Study. Ph.D. Thesis, University of Southern Queensland, Toowoomba, Australia, 2012.

161. Machdar, N.M. Corporate Social Responsibility Disclosure Mediates the Relationship between Corporate Governance and Corporate Financial Performance in Indonesia. *Acad. Account. Financ. Stud. J.* **2019**, *23*, 1–14.

162. Kellett, P. Securing High Levels of Business Compliance with Environmental Laws: What Works and What to Avoid. *J. Environ. Law* **2019**. [CrossRef]

163. Stone-Romero, E.F. *Implications of Research Design Options for the Validity of Inferences Derived from Organizational Research*; The Sage Handbook of Organizational Research Methods, Sage: Thousand Oaks, CA, USA, 2009; pp. 302–327.

MDPI
St. Alban-Anlage 66
4052 Basel
Switzerland
Tel. +41 61 683 77 34
Fax +41 61 302 89 18
www.mdpi.com

Processes Editorial Office
E-mail: processes@mdpi.com
www.mdpi.com/journal/processes